Advances in Remediation of Environmental Pollutants for Sustainable Development

Advances in Remediation of Environmental Pollutants for Sustainable Development

Editors

Mohd Rafatullah
Masoom Raza Siddiqui

MDPI • Basel • Beijing • Wuhan • Barcelona • Belgrade • Manchester • Tokyo • Cluj • Tianjin

Editors
Mohd Rafatullah
School of Industrial
Technology
Universiti Sains Malaysia
Penang
Malaysia

Masoom Raza Siddiqui
Chemistry Department
King Saud University
Riyadh
Saudi Arabia

Editorial Office
MDPI
St. Alban-Anlage 66
4052 Basel, Switzerland

This is a reprint of articles from the Special Issue published online in the open access journal *Sustainability* (ISSN 2071-1050) (available at: www.mdpi.com/journal/sustainability/special_issues/AREPSD).

For citation purposes, cite each article independently as indicated on the article page online and as indicated below:

LastName, A.A.; LastName, B.B.; LastName, C.C. Article Title. *Journal Name* **Year**, *Volume Number*, Page Range.

ISBN 978-3-0365-6603-0 (Hbk)
ISBN 978-3-0365-6602-3 (PDF)

© 2023 by the authors. Articles in this book are Open Access and distributed under the Creative Commons Attribution (CC BY) license, which allows users to download, copy and build upon published articles, as long as the author and publisher are properly credited, which ensures maximum dissemination and a wider impact of our publications.

The book as a whole is distributed by MDPI under the terms and conditions of the Creative Commons license CC BY-NC-ND.

Contents

About the Editors vii

Ahmad Zia Ul-Saufie, Nurul Haziqah Hamzan, Zulaika Zahari, Wan Nur Shaziayani, Norazian Mohamad Noor and Mohd Remy Rozainy Mohd Arif Zainol et al.
Improving Air Pollution Prediction Modelling Using Wrapper Feature Selection
Reprinted from: *Sustainability* **2022**, *14*, 11403, doi:10.3390/su141811403 1

Artur Szwalec, Paweł Mundała and Renata Kedzior
Suitability of Selected Plant Species for Phytoremediation: A Case Study of a Coal Combustion Ash Landfill
Reprinted from: *Sustainability* **2022**, *14*, 7083, doi:10.3390/su14127083 17

Mohd Fazly Yusof, Mohd Remy Rozainy Mohd Arif Zainol, Andrei Victor Sandu, Ali Riahi, Nor Azazi Zakaria and Syafiq Shaharuddin et al.
Clean Water Production Enhancement through the Integration of Small-Scale Solar Stills with Solar Dish Concentrators (SDCs)—A Review
Reprinted from: *Sustainability* **2022**, *14*, 5442, doi:10.3390/su14095442 33

Jia Hui Ang, Yusri Yusup, Sheikh Ahmad Zaki, Ali Salehabadi and Mardiana Idayu Ahmad
Comprehensive Energy Consumption of Elevator Systems Based on Hybrid Approach of Measurement and Calculation in Low- and High-Rise Buildings of Tropical Climate towards Energy Efficiency
Reprinted from: *Sustainability* **2022**, *14*, 4779, doi:10.3390/su14084779 61

Riti Thapar Kapoor, Mohd Rafatullah, Masoom Raza Siddiqui, Moonis Ali Khan and Mika Sillanpää
Removal of Reactive Black 5 *Dye* by Banana Peel Biochar and Evaluation of Its Phytotoxicity on Tomato
Reprinted from: *Sustainability* **2022**, *14*, 4176, doi:10.3390/su14074176 83

Mohammad Nishat Akhtar, Mohd Talha Anees, Emaad Ansari, Jazmina Binti Ja'afar, Mohammed Danish and Elmi Abu Bakar
Baseline Assessment of Heavy Metal Pollution during COVID-19 near River Mouth of Kerian River, Malaysia
Reprinted from: *Sustainability* **2022**, *14*, 3976, doi:10.3390/su14073976 101

Hamza S. AL-Shehri, Hamdah S. Alanazi, Areej Mohammed Shaykhayn, Lina Saad ALharbi, Wedyan Saud Alnafaei and Ali Q. Alorabi et al.
Adsorption of Methylene Blue by Biosorption on Alkali-Treated *Solanum incanum*: Isotherms, Equilibrium and Mechanism
Reprinted from: *Sustainability* **2022**, *14*, 2644, doi:10.3390/su14052644 117

Lucas Rafael Santana Pinheiro, Diana Gomes Gradíssimo, Luciana Pereira Xavier and Agenor Valadares Santos
[-15]Degradation of Azo Dyes: Bacterial Potential for Bioremediation
Reprinted from: *Sustainability* **2022**, *14*, 1510, doi:10.3390/su14031510 133

Muthia Elma, Amalia Enggar Pratiwi, Aulia Rahma, Erdina Lulu Atika Rampun, Mahmud Mahmud and Chairul Abdi et al.
Combination of Coagulation, Adsorption, and Ultrafiltration Processes for Organic Matter Removal from Peat Water
Reprinted from: *Sustainability* **2021**, *14*, 370, doi:10.3390/su14010370 157

Kah Aik Tan, Japareng Lalung, Norhashimah Morad, Norli Ismail, Wan Maznah Wan Omar and Moonis Ali Khan et al.
Post-Treatment of Palm Oil Mill Effluent Using Immobilised Green Microalgae *Chlorococcum oleofaciens*
Reprinted from: *Sustainability* **2021**, *13*, 11562, doi:10.3390/su132111562 . . . . **169**

Abdassalam A. Azamzam, Mohd Rafatullah, Esam Bashir Yahya, Mardiana Idayu Ahmad, Japareng Lalung and Sarah Alharthi et al.
Insights into Solar Disinfection Enhancements for Drinking Water Treatment Applications
Reprinted from: *Sustainability* **2021**, *13*, 10570, doi:10.3390/su131910570 . . . . **187**

Mashur Mashur, Muhammad Roil Bilad, Hunaepi Hunaepi, Nurul Huda and Jumardi Roslan
Formulation of Organic Wastes as Growth Media for Cultivation of Earthworm Nutrient-Rich *Eisenia foetida*
Reprinted from: *Sustainability* **2021**, *13*, 10322, doi:10.3390/su131810322 . . . . **209**

Emaad Ansari, Mohammad Nishat Akhtar, Mohamad Nazir Abdullah, Wan Amir Fuad Wajdi Othman, Elmi Abu Bakar and Ahmad Faizul Hawary et al.
Image Processing of UAV Imagery for River Feature Recognition of Kerian River, Malaysia
Reprinted from: *Sustainability* **2021**, *13*, 9568, doi:10.3390/su13179568 . . . . **223**

Sharjeel Waqas, Muhammad Roil Bilad, Nurul Huda, Noorfidza Yub Harun, Nik Abdul Hadi Md Nordin and Norazanita Shamsuddin et al.
Membrane Filtration as Post-Treatment of Rotating Biological Contactor for Wastewater Treatment
Reprinted from: *Sustainability* **2021**, *13*, 7287, doi:10.3390/su13137287 . . . . **237**

Yuhei Kobayashi, Fumihiko Ogata, Chalermpong Saenjum, Takehiro Nakamura and Naohito Kawasaki
Adsorption/Desorption Capability of Potassium-Type Zeolite Prepared from Coal Fly Ash for Removing of Hg^{2+}
Reprinted from: *Sustainability* **2021**, *13*, 4269, doi:10.3390/su13084269 . . . . **253**

About the Editors

Mohd Rafatullah

Dr. Rafatullah presently works as an Associate Professor of Environmental Technology in the School of Industrial Technology, Universiti Sains Malaysia (USM), Malaysia. He joined this school in the year 2008 as a Post Doctoral Fellow. He has completed his education; Ph. D. in Environmental Chemistry, Master of Science in Analytical Chemistry and Bachelor of Science in Chemistry from Aligarh Muslim University (AMU), India. His research interest is in the areas of environmental water pollutants and their safe removal; preparation of various nanomaterials to protect the environment; water and wastewater treatment; adsorption and ion exchange; microbial fuel cells; advance oxidation process; activated carbons and their electrochemical properties. His contribution was recognized by Guest Editors and Members of the Editorial Boards of various scientific journals. He is listed in the World's top 2% Scientists by Stanford University, listed among the Top 1% peer reviewers, in Chemistry, Environmental Science and cross-field on Publons global reviewer, Web of Science. He is a life time fellow member of the International Society of Sustainable Developments and member of various professional international societies. He has published several reviews articles and regular research papers in journals of international repute and presented his research work in various national and international conferences. He has also attended many workshops and seminars of environmental chemistry, based on his performance and contribution in research (total citations: >8600 and h index: 40 @ Scopus).

Masoom Raza Siddiqui

Prof. Siddiqui received his B.Sc, M.Sc and Ph.D in chemistry from Aligarh Muslim University, Aligarh India. After 2.5 years'employment in Pharmaceutical R&D, Dr. Siddiqui joined King Saud University as an assistant professor in 2011 and was subsequently promoted to full professor in May 2022. His research interest includes water treatment, removal of environmental pollutants, synthesis of cost adsorbents and its application in solid phase extraction, development of analytical methods for determination of pharmaceutical, environmental and food additives. Prof. Siddiqui has more than 100 research papers to his credit.

Article

Improving Air Pollution Prediction Modelling Using Wrapper Feature Selection

Ahmad Zia Ul-Saufie [1,*], Nurul Haziqah Hamzan [1], Zulaika Zahari [1], Wan Nur Shaziayani [1], Norazian Mohamad Noor [2], Mohd Remy Rozainy Mohd Arif Zainol [3], Andrei Victor Sandu [4,5,6,*], Gyorgy Deak [6] and Petrica Vizureanu [4,7]

1 Faculty of Computer and Mathematical Sciences, Universiti Teknologi MARA, Shah Alam 40450, Selangor, Malaysia
2 Faculty of Civil Engineering Technology, Universiti Malaysia Perlis, Kompleks Pengajian Jejawi 3, Arau 02600, Perlis, Malaysia
3 School of Civil Engineering, Engineering Campus, Universiti Sains Malaysia, Nibong Tebal 14300, Pulau Pinang, Malaysia
4 Faculty of Material Science and Engineering, Gheorghe Asachi Technical University of Iasi, 61 D. Mangeron Blvd., 700050 Iasi, Romania
5 Romanian Inventors Forum, St. P. Movila 3, 700089 Iasi, Romania
6 National Institute for Research and Development in Environmental Protection INCDPM, Splaiul Independentei 294, 060031 Bucharest, Romania
7 Technical Sciences Academy of Romania, Dacia Blvd 26, 030167 Bucharest, Romania
* Correspondence: ahmadzia101@uitm.edu.my (A.Z.U.-S.); sav@tuiasi.ro (A.V.S.)

Abstract: Feature selection is considered as one of the essential steps in data pre-processing. However, all of the previous studies on predicting PM_{10} concentration in Malaysia have been limited to statistical method feature selection, and none of these studies used machine-learning approaches. Therefore, the objective of this research is to investigate the influence variables of the PM_{10} prediction model by using wrapper feature selection to compare the prediction model performance of different wrapper feature selection and to predict the concentration of PM_{10} for the next day. This research uses 10 years of daily data on pollutant concentrations from two stations (Klang and Shah Alam) obtained from the Department of Environment Malaysia (DOE) from 2009 until 2018. Six wrapper methods (forward selection, backward elimination, stepwise, brute-force, weight-guided and genetic algorithm evolution and the predictive analytics multiple linear regression (MLR) and artificial neural network (ANN)) were implemented in this study. This study found that brute-force is the dominant wrapper method in most of the best models in selecting important features for MLR. Moreover, compared to MLR, ANN provides more advantages regarding model accuracy and permits feature selection in predicting PM_{10}. The overall results revealed that the RMSE value for next day prediction in Klang is 20.728, while the AE value is 15.69. Furthermore, the RMSE value for next day prediction in Shah Alam is 10.004, while the AE value is 7.982. Finally, all of the predicted models in Klang and Shah Alam can be used to predict the PM_{10} concentrations. This proposed model can be used as a tool for an early warning system in giving air quality information to local authorities in order to formulate air-quality-improvement strategies.

Keywords: hybrid models; air pollution modelling; feature selection; wrapper method; artificial neural network

Citation: Ul-Saufie, A.Z.; Hamzan, N.H.; Zahari, Z.; Shaziayani, W.N.; Noor, N.M.; Zainol, M.R.R.M.A.; Sandu, A.V.; Deak, G.; Vizureanu, P. Improving Air Pollution Prediction Modelling Using Wrapper Feature Selection. *Sustainability* **2022**, *14*, 11403. https://doi.org/10.3390/su141811403

Academic Editors: Tin-Chih Toly Chen and Amir Mosavi

Received: 7 June 2022
Accepted: 7 September 2022
Published: 11 September 2022

Publisher's Note: MDPI stays neutral with regard to jurisdictional claims in published maps and institutional affiliations.

Copyright: © 2022 by the authors. Licensee MDPI, Basel, Switzerland. This article is an open access article distributed under the terms and conditions of the Creative Commons Attribution (CC BY) license (https://creativecommons.org/licenses/by/4.0/).

1. Introduction

Malaysia is an increasingly developed country. In line with this progress, there are plenty of advances in technology that indirectly contribute to air pollution. Moreover, open burning, power plants, motor vehicle emissions and industrial process emissions are the major sources of particulate matter less than or equal to 10 micrometers (PM_{10}) in Malaysia [1].

In observing air quality, Malaysia has been following the Malaysian Ambient Air Quality Standard for allowable air pollutant levels. According to the Malaysian Ambient Air Quality Standard, the acceptable threshold levels of PM_{10} are 50 µg/m^3 per year and 100 µg/m^3 per 24 h, which are considered to be safe [2]. These particulate matters can become dissolved and absorbed into the bloodstream, which can later trigger serious biological effects. In addition, it is also one of the factors that cause lung cancer and cardiopulmonary deaths. Thus, in facing this hazardous situation, building optimized forecasting models of PM_{10} is the best solution in controlling these particle concentrations, and this also helps to prepare for the worst circumstances.

Feature selection is considered as one of the data pre-processing essential steps and is important in solving problems of high dimensionality dataset. This method is significant in discovering correlated features and in removing uncorrelated or redundant features from the original data set. By implementing the feature selection method, the performance of the model will be improved as this method will reduce the error by removing irrelevant and redundant features. However, all of the previous studies in Malaysia were only limited to statistical methods, such as backward, forward and stepwise analysis, and none of these studies uses machine-learning approaches, such as brute-force, weight-guided and GA evolution. Therefore, this study will investigate which approaches are better at selecting features in predicting the PM_{10} concentration.

Various methods have been used by previous researchers in predicting PM_{10} concentrations in Malaysia. For instance, a study by [3] determined the best loss function in boosted regression trees (BRT) for the prediction of the PM_{10} concentration in Alor Setar, Klang and Kota Bharu, Malaysia. A study conducted by [4] suggested that the prediction of PM_{10} concentrations can be made by considering the conditions of the previous day event. In China, [5,6] applied deep-learning-network models to predict air pollution. Most studies do not focus on optimizing the number of inputs in predicting the PM_{10} concentration. Therefore, this study investigates the optimal number of inputs and identifies the influence factors for which predictive analytics are suitable for predicting PM_{10} to compare the performances.

In summary, this study investigates which variables influence the PM_{10} concentration and which approaches are better in selecting features for predicting the PM_{10} concentration. Next, this study will investigate which predictive analytics for statistical and machine-learning methods are commonly used in predicting the PM_{10} and compare which method is best in predicting PM_{10}.

According to [7], the goal of feature selection is to discover features that can precisely and concisely describe the original dataset and later generate new features based on the original dataset. Feature selection is a method using an algorithm or procedure to retain the most vital features and their application domain. Feature selection is beneficial in performance accuracy and complexity reduction as this method removes irrelevant features from the model. It also reduces the integration time and produces a simpler model, which is much easier to debug [8–10].

In machine learning, feature-selection techniques are mainly divided into supervised techniques and unsupervised techniques. The difference between these two techniques is whether to select features based on the target variable. The supervised techniques use the target variable in choosing its features. On the other hand, the unsupervised techniques ignore the target variable in selecting its features [11]. Filter methods, wrapper methods and embedded methods are among the feature-selection techniques.

A study conducted by Ibrahim et al. [12] aimed to compare the wrapper and filter methods to maximize the classifier accuracy. Correlation-based and information gain are the filter methods used in this study. The wrapper methods are sequential forward and sequential backward elimination. The study [12] applied the selected feature selection methods obtained from the UCI Machine Learning Repository to measure its performance and the datasets are Pima Indians Diabetes, Breast Cancer Wisconsin and Spam base.

As a result, all of the datasets showed that the wrapper method had higher significant features compared with the filter method. The results also indicated that the logistic regression performed the best with the highest accuracy, specificity and sensitivity using the wrapper methods features. Thus, based on the evidence provided by the previous study, the wrapper method performs better in selecting the features compared to the filter method.

Lastly, a gap in the study was regarding predicting PM_{10} concentrations using different types of features. The common method is still limited to statistical model approaches in feature selection, such as forward, backward and stepwise selection. When compared to overseas, none of the studies in Malaysia used machine-learning approaches, such as brute-force, weight-guided, or GA evolution for feature selection in predicting the air pollutant concentration. On the other hand, MLR is the most commonly used statistical method for predicting PM_{10} concentrations, whereas ANN is the most commonly used machine-learning method. Therefore, this study implements both common statistical and machine-learning approaches in selecting its features and uses MLR and ANN in predicting PM_{10} concentrations.

2. Materials and Methods

As shown in Figure 1, data acquisition, exploration, cleaning, transform and partitioning the data set are part of the data preparation. As for the feature selection, the partition data will be implemented in six wrapper methods, which are forward selection, backward elimination, stepwise, brute-force, weight-guided and genetic algorithm evolution. The significant variables obtained according to each method later will be used to develop predictive models, and MLR and ANN and will be evaluated using performance indicators. The performance of each model in MLR and ANN later will be compared and ranked according to its performance. The best model for each day of MLR and ANN will be compared again between the predictive analytics model. The best model obtained will be used to predict the concentration of PM_{10} for the next day.

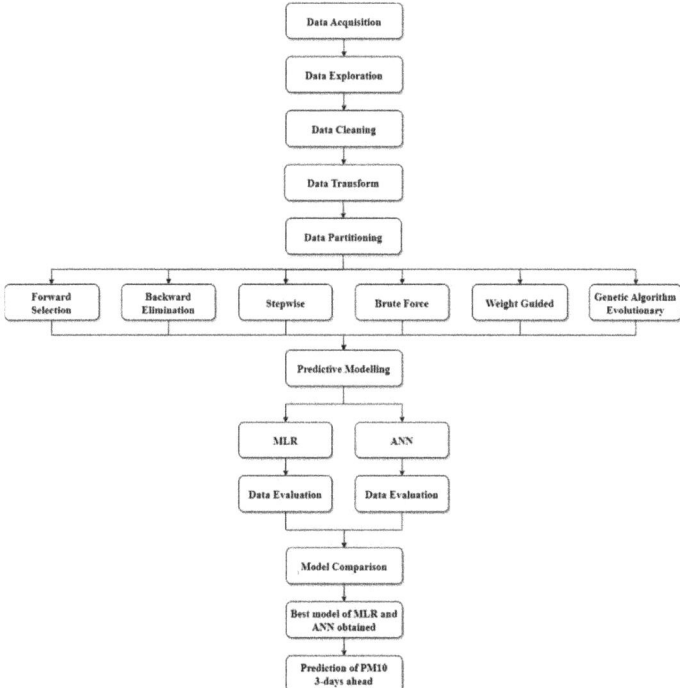

Figure 1. Research work flow.

This section consists of data acquisition, data exploration, data cleaning, data transform and partitioning the dataset. The data acquisition will explain the information of data and parameters included in this study. Second, this study will conduct descriptive analysis in data exploration. Third, data cleaning will explain the technique involved in imputing the data. Next, data transform will explain the transformation on the data before being analyzed. Lastly, data partitioning will explain the partition of the dataset.

As for the data acquisition, this research used ten years of daily data on pollutant concentrations from two stations obtained from the Department of Environment Malaysia (DOE) from 2009 until 2018. The stations included in this study are Klang station located at Sekolah Menengah Perempuan Raja Zarina, Klang and Shah Alam station located at Sekolah Kebangsaan TTDI Jaya, Shah Alam. These two stations were selected because they are surrounded by major roads that experience heavy traffic, particularly during the morning rush hour.

Based on the Exploratory Data Analysis (EDA), this analysis measured the central tendency, dispersion, skewness and graphical representation. This study measured the central tendency of the data by estimating the mean, mode and median. This study also computed the variance, standard deviation and range in measuring the dispersion. Moreover, this research evaluated the skewness to check for the probability distribution of the data.

In this study, linear interpolation and the series mean are used to impute the missing values in the data as suggested by others [13]. Linear interpolation is an interpolation method for single-dimensional data. This method estimates the data point value needed to be interpolated based on the two data points adjacent to that point in the single-dimensional data sequence. Equation (1) shows the formula and graph of linear interpolation. While the series means method was used to impute all missing values with the mean value of the data. Therefore, the data was imputed using linear interpolation first, and the rest was imputed using the series mean.

$$y = y_i + \frac{(x - x_i)(y_2 - y_1)}{(x_2 - x_1)} \quad (1)$$

Based on the data retrieved, the readings of each parameter were recorded hourly. As this study predicts the PM_{10} concentration by day, the data is transformed into a daily format. This study used the average PM_{10} concentration of hourly data as the daily data. Next, the wind direction parameters in this study were split into two variables following [14]. The variables were the sinusoidal (sinWD) and the cosinusoidal (cosWD).

Before developing the model, the original dataset was divided into three datasets for training, validation and verification so that there would be new data to assess the model. The dataset collected for this study was divided chronologically into 80% for training data (2009–2016) and 20% for validation data (2016–2017). The training data was for estimating the predictive method parameters, while the validation data is for analyzing the accuracy. The proposed model was verified using the new dataset for 2018.

2.1. Feature Selection

Feature selection is the process of minimizing the number of input variables when building a predictive model [15]. In this research, the wrapper method was used to develop air pollution prediction modelling consisting forward selection, backward selection, stepwise selection, brute-force feature selection, Genetic Algorithm (GA) evolutionary and weight-guided.

Forward selection is a type of stepwise regression that begins with a null model. The approach initiates with no variables in the model and step by step adds variables to the model until no variable not included in the model can make a significant contribution to the model's conclusion. The variable with the highest test statistic that is more than the cut-off value or the lowest *p*-value with less than the cut-off value is chosen and added to the model.

Backward elimination is the most basic approach to variable selection. This technique begins with a complete model that includes all of the variables in the model. Variables are subsequently removed from the whole model one by one until all remaining variables are sure to have a meaningful impact on the result. The variable with the lowest test statistic or the highest p-value more than the cut-off value is removed from the model. This procedure is repeated until every remaining variable is statistically significant at the cut-off value.

The stepwise selection method is the mixture of forward selection and backward elimination procedures that allow one to go in both directions while adding and eliminating variables at various stages. Forward selection and backward elimination can be applied to begin the process. If stepwise selection begins with forward selection, variables are added to the model one at a time according to the statistical significance. After each step, the model is analyzed. Any variable that is not significant will be removed from the model. The process repeats until every variable in the model is statistically significant.

If stepwise selection begins with backward elimination, the variables are removed from the full model based on statistical significance and then re-added if they show statistical significance afterward. Brute-force is a straightforward approach to solving a problem. It involves iterating through all possible features until the best feature selection is found. Brute-force feature selection tries every potential combination of the variables and provides the highest performing subset. The best subset is chosen by maximizing a defined performance metric in the presence of an arbitrary regressor or classifier. The algorithm will choose each combination and compute its score before selecting the optimal combination based on its score.

For this study, the number of possible subsets is calculated using the best subset regression formula, 2 p, where p equals 11 (the number of predictors). As a result, this method will generate 2048 subsets. Essentially, this method starts with the generation of possible subsets, beginning with one variable, two variables, three variables and so on until eleven predictors are generated. Each subset will have its own regression equation, which will be evaluated using the adjusted R square ($\overline{R}^2$). The reason for using $\overline{R}^2$ rather than R^2 to compare the performance of subsets is that $\overline{R}^2$ values are often artificially inflated as more variables are chosen. The formula for $\overline{R}^2$ is stated below:

$$\overline{R}^2 = 1 - \left(1 - R^2\right) \frac{n-1}{n-p-1} \qquad (2)$$

where p is the number of predictors and n is the number of samples. The best subset, which contains the most significant factors to predict PM_{10} concentration, will be the subset with the highest R^2 value.

Next, GA evolution is a type of optimization technique that mimics the concepts of natural evolution. There are three basic concepts in this process, which are selection, crossover and mutation. The first step of an evolutionary algorithm is an initialization phase where it creates a population of air pollution models, each with their unique set of chromosomes. The chromosomes are binary strings; 1 means the feature is included, and 0 means the feature is excluded. The models for the starting population are randomly generated.

A good rule of thumb is to use between 5% and 30% of the total number of features as the population size [16]. For the second step, each model in the population will have their fitness calculated. Models with better fitness have a higher chance of being chosen for recombination. After calculating the fitness value, the third step is that the models will be selected randomly using the roulette wheel method and selected according to their fitness level. The number of selected models is half of the population size.

After selecting the models, the fourth step is the crossover, in which the selected models are recombined to create a new population. In this step, two models will be chosen at random and their features will be combined to produce offspring for the new population until the new population is the same size as the old one. In the crossover, offspring that are genetically identical to their parents may be produced, resulting in a low-diversity new generation. Therefore, in step five, the mutation is done by changing the value of some features in the offspring at random. Lastly, the process is looped to step two until a stopping criterion is met and the best feature selection is obtained.

Finally, the wrapper method used weight-guided via correlation. Equation (3) is the formula of the correlation. A weight is given to the variables, and the highest weight of the variable is considered.

$$r = \frac{n(\sum xy - (\sum x)(\sum y))}{\sqrt{[n\sum x^2 - (\sum x)^2][n\sum x^2 - (\sum x)^2]}} \quad (3)$$

2.2. Model Development and Model Evaluation

In this model development, the process of developing MLR and ANN models is explained to conduct predictive modelling. Figure 2 shows the process of the wrapper method in developing a predictive model.

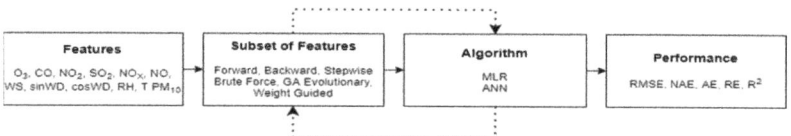

Figure 2. Process of the wrapper method in model development.

There are four steps in developing a multiple linear regression (MLR) model. First, the development of the MLR model will be based on 80% of the data. Second, the assumption of the MLR models is checked using certain methods and tests, such as histograms and scatter plots. Next, the model is validated based on the performance indicator value using 20% of the data. Finally, the best model of MLR is obtained. The expected MLR models are as shown in Equation (4). The past PM_{10} daily average concentration was used to predict the next day's PM_{10} concentration.

$$PM_{10,D+1} = \beta_0 + \beta_1 PM_{10,D} + \beta_2 CO_D + \beta_3 NO_{2,D} + \beta_4 NO_{x,D} + \beta_5 NO_D + \beta_6 SO_{2,D} + \beta_7 RH_D + \beta_8 T_D + \beta_9 WS_D + \beta_{10} \cos WD_D + \beta_{11} \sin WD_D \quad (4)$$

where

$PM_{10,D+1}$ = Next day prediction of PM_{10} concentration.
$PM_{10,D}$ = Particulate matter (µg/m³).
CO_D = Carbon monoxides (ppm).
$NO_{2,D}$ = Nitrogen dioxide (ppm).
$NO_{x,D}$ = Nitrogen oxide (ppm).
NO_D = Nitric oxide (ppm).
$SO_{2,D}$ = Sulfur dioxide (ppm).
RH_D = Relative humidity (%).
T_D = Temperature (°C).
WS_D = Wind speed (km/h).
$\cos WD_D$ = Cosine Wind direction (units).
$\sin WD_D$ = Sine Wind direction (units).
β_0 = regression constant.
$\beta_1, \ldots, \beta_{11}$ = regression coefficient for each predictors used.

A feed forward backpropagation neural network (FFBP) was used in this study. The structure of FFBP was composed of three layers of neurons called the input, hidden and output layers. The first layer of neurons consisted of an input layer, representing independent variables. The input layer contained 12 independent variables—namely, O_3, CO, NO_2, SO_2, NO, sinWD, cosWD, NOx, PM_{10}, T, WS and RH. The second layer was the hidden layer, which is responsible for processing the input weight from the input layer and transferring it to the output layer. The third layer was the output layer, which represents the $PM_{10,D+1}$ concentrations.

The maximum error used as a criterion for stopping was set at 0.05. In this study, the training process was set to 10,000 iterations or until the maximum error was reached, as suggested by [17]. As a network training function, Levenberg–Marquardt optimization was used to update weight and bias values. As sigmoid units are easier to train than other activation functions, [18] proposed using them. In this case, the layer size was 2 number of attributes + number of classes)/2 + 1 = 8 hidden nodes, as recommended by [19]. This study fitted models with varying learning rate lr (0.01) values, which [20] proposed for the study of air pollution datasets.

Furthermore, [21] stated that changing the momentum rate and learning rate from 0.05 to 1 had no effect on the training and prediction networks' errors. Performance indicators in this research are used to identify the best method to predict the concentration of $PM_{10,D+1}$. The Root Mean Square Error (RMSE), Normalized Absolute Error (NAE), Absolute Error (AE) and Relative Error (RE) are the error measures used to determine the error of the model, while the Coefficient of Determination (R^2) is the accuracy measure used to determine the accuracy of the model outcome.

Regarding the model deployment, the dataset from 2009 until 2017 was used to produce prediction models. The prediction models were later deployed on the 2018 dataset. However, there were extreme outliers in ozone concentration for the 2018 dataset for the Klang station. Based on the 2009 until 2017 dataset, the maximum ozone concentration value was 0.056 ppb. Some of the ozone concentration in the 2018 dataset exceeded 0.7 ppb. This situation may happen due to technical errors. Therefore, a total of 29 data were removed before deployment.

3. Results and Discussion

3.1. Descriptive Analysis

Table 1 is the descriptive statistics of each parameter in Klang and Shah Alam, Selangor. Based on the table, the maximum concentration of PM_{10} was 551.542 µg/m^3 (Klang) and 1332.814 µg/m^3 (Shah Alam). The high concentration was taken on 25 of June 2013 for Klang, while Shah Alam happened during April 2017. Next, the skewness value of PM_{10} is 4.62 (Klang) and 17.11 (Shah Alam). Since the value is more than 1, the data of PM_{10} are highly skewed to the right. This may be due to the presence of extreme outliers in this data.

Table 1. Descriptive statistics for Klang and Shah Alam.

		Mean	Median	Standard Deviation	Skewness	Minimum	Maximum
	WS (km/h)	5.38	5.80	2.03	4.89	0.47	52.03
	WD (°)	176.78	168.83	46.05	0.579	42.17	318.17
	T (°C)	28.62	28.60	1.52	−0.12	22.56	33.12
K	RH (%)	69.74	69.62	7.07	0.01	46.46	94.11
L	NO_x (ppm)	0.037	0.035	0.029	25.60	<0.01	1.14
A	NO (ppm)	0.016	0.014	0.028	31.6	<0.01	1.15
N	SO_2 (ppm)	0.004	0.004	0.002	2.25	<0.01	0.03
G	NO_2 (ppm)	0.02106	0.021	0.007	0.28	0.001	0.05
	O_3 (ppm)	0.019	0.018	0.007	0.78	<0.01	0.06
	CO (ppm)	1.029	0.967	0.412	2.38	0.15	6.21
	PM_{10} (µg/m^3)	62.75	56.16	32.52	4.62	17.18	551.54
	WS (km/h)	4.828	4.775	1.908	0.198	0.314	16.529
S	WD (°)	157.15	153.71	52.76	0.50	9.75	337.17
H	T (°C)	28.10	28.09	1.32	0.02	23.31	33.29
A	RH (%)	77.35	77.46	6.11	−0.05	54	94.63
H	NO_x (ppm)	0.04	0.04	0.02	0.48	0.004	0.12
	NO (ppm)	0.02	0.02	0.01	1.11	<0.001	0.09
A	SO_2 (ppm)	0.003	0.003	0.002	1.74	<0.001	0.02
L	NO_2 (ppm)	0.02	0.02	0.01	0.45	0.002	0.06
A	O_3 (ppm)	0.02	0.02	0.01	0.86	<0.001	0.08
M	CO (ppm)	0.76	0.73	0.30	1.18	0.09	3.69
	PM_{10} (µg/m^3)	50.91	45.33	34.86	17.11	11.92	1332.81

Figures 3 and 4 below are heatmaps of the average PM_{10} concentrations according to the month and year for Klang and Shah Alam. The greenish parts have the lowest average concentration, and the reddish parts have the highest average concentration. Based on both Figures, it is indicated that October 2015 had the highest average concentration compared to the others. Moreover, the concentration of PM_{10} in September 2015 was the second-highest concentration. Referring to haze incidents in Malaysia, this supports the outline of the heatmap as there was a hazing incident in 2015 in August and September due to massive land and forest fires in Indonesia [22]. The high PM_{10} concentration in October 2015 may be due to the backlash of this incident.

	2009	2010	2011	2012	2013	2014	2015	2016	2017
JAN	64.390	60.087	71.897	51.698	60.182	59.363	60.933	48.892	54.042
FEB	71.085	68.076	78.813	54.500	58.160	91.354	65.842	50.911	56.266
MAR	48.852	62.414	56.777	53.776	70.563	137.918	68.915	65.789	59.719
APR	48.192	59.497	58.879	47.407	57.508	50.692	55.901	83.738	46.278
MAY	62.422	66.247	55.668	51.024	57.441	46.890	60.310	53.347	39.651
JUN	91.883	57.697	73.524	96.972	121.650	71.472	63.389	58.678	43.233
JUL	97.388	55.325	82.142	72.665	80.358	103.941	68.226	61.663	41.526
AUG	77.843	54.574	65.145	83.531	70.488	58.591	73.797	72.378	33.349
SEP	65.806	56.822	68.272	64.568	60.336	59.328	139.286	50.794	33.915
OCT	57.814	59.397	53.549	52.950	44.946	72.480	161.804	52.421	45.269
NOV	48.056	52.028	42.474	46.571	45.608	53.508	49.411	53.025	37.268
DEC	55.582	56.695	49.313	54.255	50.207	55.809	50.742	47.837	43.630

Figure 3. Average PM_{10} concentration by month and year for Klang.

	2009	2010	2011	2012	2013	2014	2015	2016	2017
JAN	52.71371	50.04288	45.81868	46.87345	43.73466	45.53154	43.07919	44.84227	38.01365
FEB	56.66667	51.39274	55.38003	48.82537	38.90179	59.36161	45.23958	41.75231	41.54456
MAR	40.31492	48.41317	46.95245	47.46291	48.93539	95.06989	52.21137	68.14919	43.77309
APR	43.25481	46.11206	52.20889	43.55972	41.04113	45.01111	40.83452	76.07639	70.28736
MAY	51.07661	51.89093	54.24194	47.58737	36.79971	41.6129	43.85857	44.93939	30.24706
JUN	72.88194	47.6625	65.35833	77.59583	83.30139	64.33889	56.40246	48.59916	36.20518
JUL	70.60618	45.96582	73.11694	49.92418	57.81586	76.92546	67.49791	55.30528	36.54162
AUG	57.42608	50.25412	59.44086	56.18011	50.58602	41.2836	65.95902	66.44355	30.85402
SEP	49.46111	45.57778	66.23194	52.34028	45.26045	53.55556	135.098	42.00139	29.35253
OCT	45.7493	55.88844	45.97606	39.97984	36.99059	60.16553	149.6801	35.67769	37.64508
NOV	38.22807	41.95028	39.80139	33.82917	35.93611	37.18333	42.77042	35.46037	29.40257
DEC	49.15319	41.06198	45.59945	35.70161	36.37366	39.2288	45.83994	40.34294	29.7766

Figure 4. Average PM_{10} concentration by month and year for Shah Alam.

In March 2014, there was a high concentration of PM_{10} in both locations where Klang was slightly higher than Shah Alam. It is also proved by the chronology of haze incidents in Malaysia as haze incidents happened between February and March 2014. This incident occurred due to forest and peatland fires. High PM_{10} concentration incidents were also detected in June 2013 from both locations. However, Klang had a higher concentration value compared to Shah Alam. This incident may be due to haze incidents that happened from 15 to 27 June 2013 [22]. In addition, both Figures show a low average of PM_{10} concentration starting from May until December 2017. This situation may be due to zero cases of haze incidents happening in 2017.

In conclusion, the heatmaps of both locations align with the haze chronology in Malaysia. The heatmaps also makes it easier to observe the condition of air pollution in Malaysia.

3.2. Correlation of PM_{10} Concentration with Other Parameters

Figures 5 and 6 show the heatmap of the Spearman's rank correlation coefficient (r) of the PM_{10} concentration with other parameters in Klang and Shah Alam, respectively. Figure 5 shows that all of the parameters have positive correlation with PM_{10} except for WD, RH and NO in Klang. It also indicates all of the parameters have moderate-to-very-weak correlations with the PM_{10} for Klang. CO has the highest correlation PM_{10} concentration with a positive moderate correlation (r = 0.498), while WD has the lowest correlation with a negative very weak correlation (r = -0.075).

	WS	WD	T	RH	NOX	NO	SO_2	NO_2	O_3	CO	PM_{10}
WS	1.000	-0.035	0.269	-0.421	-0.393	-0.359	0.208	-0.179	0.260	-0.142	0.220
WD	-0.035	1.000	-0.006	0.037	0.096	0.155	0.105	-0.021	-0.114	0.045	-0.075
T	0.269	-0.006	1.000	-0.566	-0.266	-0.283	0.125	-0.145	0.257	-0.058	0.197
RH	-0.421	0.037	-0.566	1.000	0.328	0.350	-0.079	0.159	-0.318	0.163	-0.306
NOX	-0.393	0.096	-0.266	0.328	1.000	0.864	0.174	0.823	-0.178	0.561	0.136
NO	-0.359	0.155	-0.283	0.350	0.864	1.000	0.164	0.460	-0.389	0.439	-0.084
SO_2	0.208	0.105	0.125	-0.079	0.174	0.164	1.000	0.196	0.062	0.183	0.205
NO_2	-0.179	-0.021	-0.145	0.159	0.823	0.460	0.196	1.000	0.137	0.542	0.374
O_3	0.260	-0.114	0.257	-0.318	-0.178	-0.389	0.062	0.137	1.000	0.080	0.373
CO	-0.142	0.045	-0.058	0.163	0.561	0.439	0.183	0.542	0.080	1.000	0.498
PM_{10}	0.220	-0.075	0.197	-0.306	0.136	-0.084	0.205	0.374	0.373	0.498	1.000

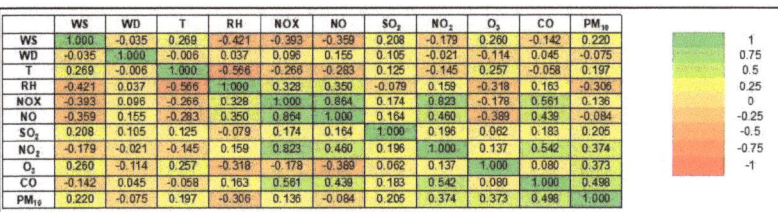

Figure 5. Correlation between he pParameters and PM_{10} for Klang.

Figure 6 shows that all of the parameters have a positive correlation with PM_{10} except for WD and RH in Shah Alam. All of the parameters have moderate, weak and very weak correlations with the PM_{10} concentration. It also indicates that NO_2 has the highest correlation with PM_{10} concentration with a positive moderate correlation (r = 0.437), while WD has the lowest correlation with PM_{10} concentration with a positive very weak correlation (r = 0.151).

	WS	WD	T	RH	NOX	NO	SO2	NO_2	O_3	CO	PM_{10}
WS	1	0.056	0.365	-0.365	-0.332	-0.295	0.269	-0.245	0.263	-0.365	0.159
WD	0.056	1	-0.00808	-0.074	-0.306	-0.292	-0.126	-0.293	0.010061	-0.146	-0.214
T	0.365	-0.00808	1	-0.684	-0.043	-0.087	0.198	0.020659	0.441	0.012683	0.317
RH	-0.365	-0.074	-0.684	1	0.116	0.127	-0.087	0.075	-0.332	0.033644	-0.166
NOX	-0.332	-0.306	-0.043	0.116	1	0.9	0.148	0.809	-0.196	0.689	0.306
NO	-0.295	-0.292	-0.087	0.127	0.9	1	0.111	0.507	-0.329	0.605	0.151
SO_2	0.269	-0.126	0.198	-0.087	0.148	0.111	1	0.181	0.097	0.083	0.358
NO_2	-0.245	-0.293	0.020659	0.075	0.809	0.507	0.181	1	0.031983	0.567	0.437
O_3	0.263	0.010061	0.441	-0.332	-0.196	-0.329	0.097	0.031983	1	-0.044	0.277
CO	-0.365	-0.146	0.012683	0.033644	0.689	0.605	0.083	0.567	-0.044	1	0.366
PM_{10}	0.159	-0.214	0.317	-0.166	0.306	0.151	0.358	0.437	0.277	0.366	1

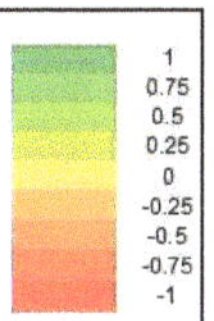

Figure 6. Correlation between the parameters and PM_{10} for Shah Alam.

3.3. *Performance Model and Feature Selection*

Performance measures for this section used the validation dataset (20%). Table 2 shows that the performance of all MLR model was compared to find the best model for next day prediction. Based on Table 3, for Klang, brute-force has the lowest value of RMSE, AE, RE and NAE. The backward method has the highest R^2 value compared to others.

Therefore, it is shown that brute-force is the best model for Klang as it had the lowest value of error and the lowest total rank with WS, RH, SO_2, O_3 and PM_{10} as the parameters selected to predict the next day's PM_{10} concentration. Furthermore, the performance for all models in predicting PM_{10} in Shah Alam also shows that brute-force had the lowest error measures for RMSE, AE, RE and NAE and the highest accuracy for R^2. Therefore, brute-force is the best model with T, RH, SO_2, NO_2, sinWD, NO and PM_{10} as the parameters selected to predict the $PM_{10,D+1}$.

Table 2. Performance models of MLR for Klang and Shah Alam for the validation.

Location	Method	RMSE	AE	RE	NAE	R^2
Klang	Forward	14.224	10.292	17.97%	0.596	0.709
	Backward	14.186	10.447	18.17%	0.605	0.713
	Stepwise	14.224	10.292	17.97%	0.596	0.709
	Brute-Force	**13.694**	**10.005**	**17.52%**	**0.596**	**0.707**
	Weight-Guided	16.27	11.089	19.21%	0.658	0.591
	Evolution	14.367	10.701	18.84%	0.633	0.665
Shah Alam	Forward	15.176	10.932	23.99%	0.709	0.598
	Backward	15.23	11.346	24.77%	0.736	0.502
	Stepwise	15.176	10.932	23.99%	0.709	0.598
	Brute-Force	**12.338**	**9.559**	**21.77%**	**0.662**	**0.632**
	Weight-Guided	13.417	10.371	25.92%	0.722	0.596
	Evolution	14.530	11.411	26.84%	0.782	0.531

Table 3. Ranking performance models of MLR for Klang and Shah Alam.

Location	Method	RMSE	AE	RE	NAE	R^2	Total Rank
Klang	Forward	3	2	3	1	2	11
	Backward	2	3	4	2	1	12
	Stepwise	3	2	2	1	2	10
	Brute-Force	**1**	**1**	**1**	**1**	**3**	**7**
	Weight-Guided	5	5	6	4	5	25
	Evolution	4	4	5	3	4	20
Shah Alam	Forward	4	3	2	2	2	13
	Backward	5	4	3	4	5	21
	Stepwise	4	3	2	2	2	13
	Brute-Force	**1**	**1**	**1**	**1**	**1**	**5**
	Weight-Guided	2	2	4	3	3	14
	Evolution	3	5	5	5	4	22

Referring to Table 4, the ticked table means that the parameter is selected in that model. For the best model $PM_{10,D+1}$ for Klang (MLR-Brute-Force), RH, SO_2, NO_2, O_3 and PM_{10} were analyzed to predict the next day for Klang. For the best model $PM_{10,D+1}$ for Shah Alam (MLR-Brute -Force), T, RH, SO_2, NO_2, sinWD, NO and PM_{10} are the parameters selected to predict the next day for Shah Alam.

Table 4. Selected features of MLR for Klang and Shah Alam.

Location	Method	WS	T	RH	SO_2	NO_2	O_3	CO	Sin WD	Cos WD	NO_x	NO	PM_{10}
Klang	Forward		/	/									/
	Backward	/		/	/	/	/	/		/	/		/
	Stepwise			/									/
	Brute-Force			/	/		/						/
	Weight-Guided												/
	Evolution	/		/				/		/			/
Shah Alam	Forward	/		/				/		/			/
	Backward	/			/								/
	Stepwise		/		/			/	/	/	/	/	/
	Brute-Force	/		/	/				/			/	/
	Weight-Guided		/	/	/	/			/			/	/
	Evolution												/

Based on Table 5, the performances of all ANN models for next-day prediction in Klang and Shah Alam were compared to determine the best model. For Klang, brute-force had the lowest value of RMSE and RE, and backward had the lowest value AE and highest R^2 value compared to the others. Evolution had the lowest value of NAE. However, backward is the best model for Klang station as it had the lowest total rank compared to the others with T, RH, SO_2, NO_2, O_3, sinWD, cosWD, NO_x, NO and PM_{10} as the parameters selected to predict the $PM_{10,D+1}$.

Table 5. Performance models of ANN for Klang and Shah Alam for validation.

Location	Method	RMSE	AE	RE	NAE	R^2
Klang	Forward	15.517	11.067	19.57%	0.588	0.701
	Backward	**15.085**	**10.272**	**17.06%**	**0.574**	**0.742**
	Stepwise	15.517	11.067	19.57%	0.588	0.701
	Brute-Force	14.139	10.282	16.90%	0.605	0.701
	Weight-Guided	17,288	11.779	18.64%	0.699	0.581
	Evolution	14.228	10.335	17.92%	0.563	0.732
Shah Alam	Forward	13.448	9.317	18.93%	0.594	0.715
	Backward	14.071	9.595	20.19%	0.647	0.596
	Stepwise	12.966	9.505	22.30%	0.65	0.691
	Brute-Force	12.252	9.138	20.75%	0.614	0.687
	Weight-Guided	12.463	8.840	18.72%	0.615	0.615
	Evolution	14.003	10.005	21.83%	0.645	0.657

Next, the performance for all ANN models in predicting $PM_{10,D+1}$ in Shah Alam shows that brute-force had the lowest value for RMSE, weight-guided had the lowest value for AE and RE, and forward had the lowest value for NAE and the highest value for R^2. However, forward is the best model in predicting $PM_{10,D+1}$ since it hd the lowest total rank with WS, NO_X and PM_{10} as the parameters selected to predict the $PM_{10,D+1}$ as shown in Table 6.

Table 6. Ranking the performance models of ANN for Klang and Shah Alam.

Location	Method	RMSE	AE	RE	NAE	R^2	Total Rank
Klang	Forward	4	4	4	3	3	18
	Backward	3	1	2	2	1	9
	Stepwise	4	4	4	3	3	18
	Brute-Force	1	2	1	4	3	11
	Weight-Guided	5	5	5	5	4	24
	Evolution	2	3	3	1	2	11
Shah Alam	**Forward**	4	3	2	1	1	11
	Backward	6	5	3	5	6	25
	Stepwise	3	4	6	6	2	21
	Brute-Force	1	2	4	2	3	12
	Weight-Guided	2	1	1	3	5	12
	Evolution	5	6	5	4	4	24

Referring to Table 7, the ticked (/) table means that the parameter is selected in that model. For the best model $PM_{10,D+1}$ for Klang (ANN-Backward), RH, SO_2, NO_2, O_3, PM_{10}, sinWD, cosWD, NO_X and NO were analyzed to predict the next day. For the best model $PM_{10,D+1}$ for Shah Alam (ANN-Forward), WS, NO_X and PM_{10} are the parameters selected to predict the next day for Shah Alam using ANN.

Table 7. Selected features of ANN for Klang and Shah Alam.

Location	Method	WS	T	RH	SO_2	NO_2	O_3	CO	Sin WD	Cos WD	NO_X	NO	PM_{10}
Shah Alam	Forward	/									/		/
	Backward				/	/	/		/	/		/	/
	Stepwise	/											
	Brute-Force	/	/	/		/	/				/		/
	Weight-Guided												/
Klang	Forward								/		/		/
	Backward			/	/	/	/	/	/	/	/	/	/
	Stepwise								/		/		/
	Brute-Force	/			/	/					/	/	/
	Weight-Guided												/
	Evolution				/	/	/		/	/			/

As a conclusion, brute-force is the best feature selection to predict the next day's PM_{10} concentration in Klang and Shah Alam by using MLR, and the models fulfil the assumptions of MLR. The backward for Klang and forward for Shah Alam are the best feature selections for predicting the next day's PM_{10} concentration using the ANN model.

3.4. The Best Model

The best model to predict the $PM_{10,D+1}$ for each station was obtained by comparing the performance of models between MLR and ANN. For the overall performance, each predicted day shows that the ANN model had the best performance compared to the MLR model for both Klang and Shah Alam station. This result is supported with Table 8 as the ANN model for each predicted day for both stations shows the lowest total score. In Klang, ANN with backward elimination is the best model selected, while for Shah Alam, ANN with forward selection is the best model.

Furthermore, Table 9 summarizes the comparison results with other research. This indicates that the results in this study are similar to those in other studies. Regression is involved with linear dependencies, whereas neural networks are involved with nonlinearities. As a result, if the data contains nonlinear dependencies, neural networks should outperform regression.

According to studies [23–25], the ANN method predicts the dependent variable more accurately than MLR. Although ANN is regarded as a powerful technique for non-linear models [26], some researchers have used and reported on this linear model better than the regression model [27–29]. This showed that our ANN model can be used to predict PM_{10} concentrations since it improved the performance of the model.

Table 8. Performance ranking of the best model.

Station	Model	Method	RMSE	AE	RE	NAE	R^2	Total
Klang	MLR	Brute-Force	1	1	2	2	2	8
	ANN	Backward	2	2	1	1	1	7
Shah Alam	MLR	Brute-Force	1	2	2	2	2	9
	ANN	Forward	2	1	1	1	1	6

Table 9. Performance indicators results gained from other research.

Authors	Method	Result
[30]	MLR	R^2 = 0.347–0.614
[31]	MLR	RMSE = 126.728–164.978 NAE = 0.325–0.429 PA = 0.359–0.668
[32]	MLR	R^2 = 0.3239
[33]	MLR	R^2 = 0.586–0.715
This Study	ANN-Forward ANN-Backward MLR-Brute-Force	R^2 = 0.63–0.74 RMSE = 12.33–15.08

3.5. Model Verification

For the model verification, the dataset from 2009 until 2017 was used to develop prediction models. The proposed prediction models were used to predict the PM_{10} concentration using the 2018 dataset [30–35]. Figures 7 and 8 below show line charts of the observed and predicted values of the PM_{10} concentration in Shah Alam and Klang.

This predictive model used ANN with backward elimination using RH, SO_2, NO_2, O_3, PM_{10}, sinWD, cosWD, NO_X and NO as a parameters in Klang. For the best model $PM_{10,D+1}$ for Shah Alam (ANN-Forward), WS, NO_X and PM_{10} are the parameters selected to predict the next day for Shah Alam using ANN.

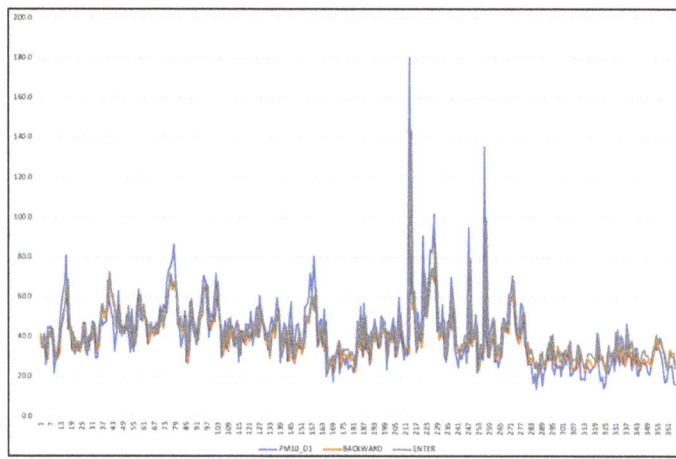

Figure 7. Klang line chart of the observed and predicted $PM_{10,D+1}$ for model ANN-forward selection and predicted ANN using all variables.

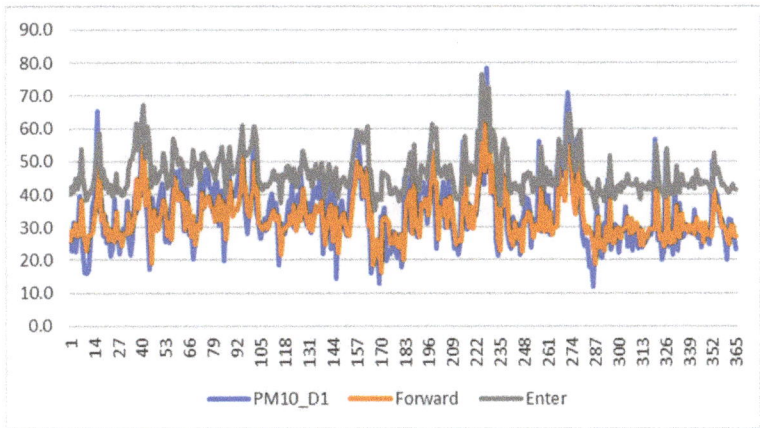

Figure 8. Shah Alam line chart of the observed and predicted $PM_{10,D+1}$ for model ANN-forward selection and predicted ANN using all variables.

Figure 7 shows the comparison of the line chart between the observed and predicted value for $PM_{10,D+1}$ for model ANN-forward selection and predicted ANN using all variables. Referring to the line chart, it shows that, on average, the observed and predicted values have a slight gap. Most of the prediction values exceed the observed value; however, in some cases, the observed value exceeds the prediction value. Furthermore, the enter method has a large gap since, in 2018, there was a slight increase in the value of ozone, causing the prediction using all parameters to be higher than the observed value. Therefore, it shows that the predicted values of the PM_{10} concentration were not notably affected by the ozone concentration.

Overall, the values of RMSE and AE of this model are 20.728 and 15.69, respectively. Hence, this model can be used for unseen data since there is no huge difference between the observed and predicted values. Figure 8 shows the comparison of the line chart between the observed and predicted values for $PM_{10\,D+1}$ for the model ANN-forward selection and predicted ANN using all variables. Referring to the line chart, it shows that, on average, the observed and predicted values have a minimum gap between each other with the value of RMSE at 10.004 and value of AE at 7.982. Most of the prediction values exceed the observed value, and in only a few cases does the observed value exceed the prediction value. Hence, this model can be used for unseen data since there is no great difference between the observed and predicted values.s

The prediction error in Klang is higher than in Shah Alam because the industrial area of Klang suffers from severe haze, while Shah Alam is only a residential area. Furthermore, if all variables based on previous studies are selected to predict PM_{10} concentrations for the next day, it will take more time to determine the best model and reduce the maintenance data cost for the future.

4. Conclusions

In this study, the wrapper methods of six different feature selections were analyzed and compared to determine the best feature selection method. The methods included were forward, backward, stepwise, brute-force, weight-guided and GA evolution. These methods were analyzed together with the predictive analytics methods MLR and ANN. The performance of the models determined the best model to predict the next day. This study found that the best feature selections were backward elimination, forward selection and brute-force in predicting the PM_{10} concentration in Malaysia.

Based on the results, the best feature selection method to predict the $PM_{10,D+1}$ in Klang was the backward method with the parameters T, RH, SO_2, NO_2, O_3, PM_{10}, sinWD, cosWD,

NOX and NO. For Shah Alam, the best feature selection method to predict $PM_{10,D+1}$ was the forward method with the parameters WS, NO_X and PM_{10}.

The prediction of the ANN model for $PM_{10,D+1}$ was deployed in the 2018 dataset. Based on the line chart in Figures 7 and 8, the gaps between the observed and predicted lines show a minimum difference. The RMSE value in Klang for $PM_{10,D+1}$ was 20.728, while the AE value was 15.69. In addition, the line chart of observed and predicted of each predicted day in Shah Alam also shows a minimum gap between each line with the RMSE value for $PM_{10,D+1}$ of 10.004, while the AE value was 7.982. In conclusion, all of the predicted models in Klang and Shah Alam can be used for unseen data.

There are a few recommendations for improving the performance of air pollution modelling that can be suggested to other researchers. This study used the cross-sectional method, and for future research, we suggest using time series, since the time-series forecasting method is better at predicting extreme events compared to the cross-sectional method. Apart from the MLR and ANN models, a new approach can be implemented to the predicted modelling to forecast the concentration of PM_{10} using machine-learning methods, such as long short-term memory (LSTM), gated recurrent units (GRU) and deep learning [36].

Other methods, aside from wrapper methods, can be applied to conduct feature selection for air pollution modelling, such as the filter method, embedded method and hybrid method. Hence, various approaches to predicted modelling and feature selection methods for air pollution modelling will be beneficial as they will produce better results. In addition, predictions for other particulate matter, such as $PM_{2.5}$, should be made since the DOE began to include $PM_{2.5}$ in calculating the API from 2018. In addition, $PM_{2.5}$ is more dangerous since the size of the particles is smaller compared to PM_{10}. Therefore, predicting $PM_{2.5}$ may help to improve the performance of air pollution modelling. Lastly, this output can be used by the authorities as it will be helpful to reduce the impact of air pollutants.

For example, the DOE's prediction of air pollutants can be used for early alertness to help in performing the relevant procedures. Hopefully, this recommendation will help improve air pollution modelling and help the authorities to pay early attention to the air pollutants in Malaysia. The limitation of this research is that the model can only be used when the sources and conditions of the characteristics of PM_{10} remain the same. Therefore, it may not be suitable for the other locations. For instance, if there is a sudden forest fire or storm in a selected area, this would affect the PM_{10} concentration.

Author Contributions: A.Z.U.-S. and W.N.S. designed the study concept and secured funding. A.Z.U.-S. is the project administrator. N.H.H., Z.Z. and A.Z.U.-S. performed the data analysis. N.H.H., Z.Z. and W.N.S. wrote the manuscript. A.Z.U.-S., W.N.S., N.M.N. and M.R.R.M.A.Z. reviewed and edited the manuscript, A.V.S., G.D. and P.V. data curation and validation of research. All authors have read and agreed to the published version of the manuscript.

Funding: The research was funded by Ministry of Science, Technology & Innovation (MOSTI) under Technology Development Fund 1 (TDF04211363). Thank you to Faculty of Computer and Mathematical Sciences, Universiti Teknologi MARA for their support and also thanks to the Department of Environment Malaysia for providing air quality monitoring data. This publication was also supported by TUIASI from the University Scientific Research Fund (FCSU).

Institutional Review Board Statement: Not applicable.

Informed Consent Statement: Not applicable.

Data Availability Statement: The data for this project are confidential but may be obtained with Data Use Agreements with the Department of Environment (DOE), Ministry of Environment and Water of Malaysia.

Acknowledgments: The authors thank Faculty of Computer and Mathematical Sciences, Universiti Teknologi MARA for their support and also the Department of Environment Malaysia for providing air quality monitoring data.

Conflicts of Interest: The authors declare no conflict of interest. The funders had no role in the design of the study; in the collection, analyses, or interpretation of data; in the writing of the manuscript; or in the decision to publish the results.

References

1. Department of Environment, Malaysia (DOE); Info Umum Kualiti Udara Kronologi Episod Jerebu di Malaysia. *Malaysia Environmental Quality Report*; Department of Environment Ministry of Energy, Science, Technology, Environment & Climate Change, Malaysia: Kuala Lumpur, Malaysia, 2014.
2. Department of Environment, Malaysia (DOE); Info Umum Kualiti Udara Kronologi Episod Jerebu di Malaysia. *Malaysia Environmental Quality Report*; Department of Environment Ministry of Energy, Science, Technology, Environment & Climate Change, Malaysia: Kuala Lumpur, Malaysia, 2018.
3. Shaziayani, W.N.; Ul-Saufie, A.Z.; Ahmat, H.; Al-Jumeily, D. Coupling of quantile regression into boosted regression trees (BRT) technique in forecasting emission model of PM10 concentration. *Air Qual. Atmos. Health* **2021**, *14*, 1647–1663. [CrossRef]
4. Mohamad, N.S.; Deni, S.M.; Ul-Saufie, A.Z. Application of the First Order of Markov Chain Model in Describing the PM10 Occurrences in Shah Alam and Jerantut, Malaysia. *Pertanika J. Sci. Technol.* **2018**, *26*, 367–378.
5. Du, S.; Li, T.; Yang, Y.; Horng, S.J. Deep Air Quality Forecasting Using Hybrid Deep Learning Framework. *IEEE Trans. Knowl. Data Eng.* **2021**, *33*, 2412–2424. [CrossRef]
6. Yan, R.; Liao, J.; Yang, J.; Sun, W.; Nong, M.; Li, F. Multi-hour and multi-site air quality index forecasting in Beijing using CNN, LSTM, CNN-LSTM, and spatiotemporal clustering. *Expert Syst. Appl.* **2020**, *169*, 114513. [CrossRef]
7. Zhou, H.; Han, S.; Liu, Y. A novel feature selection approach based on document frequency of segmented term frequency. *IEEE Access* **2018**, *6*, 53811–53821. [CrossRef]
8. Towards Data Science. An Introduction to Feature Selection. 2020. Available online: https://towardsdatascience.com/an-introduction-to-feature-selection-dd72535ecf2b (accessed on 2 February 2022).
9. Sukatis, F.F.; Noor, N.M.; Zakaria, N.A.; Ul-Saufie, A.Z.; Suwardi, A. Estimation of missing values In Air Pollution Dataset by Using Various Imputation Methods. *Int. J. Conserv. Sci.* **2019**, *10*, 791–804.
10. Shaziayani, W.N.; Harun, F.D.; Ul-Saufie, A.Z.; Samsudin, N.; Noor, N.M. Three-Days Ahead Prediction of Daily Maximum Concentrations of PM10 Using Decision Tree Approach. *Int. J. Conserv. Sci.* **2021**, *12*, 217–224.
11. Zhou, Z.; Liu, H. Spectral feature selection for supervised and unsupervised learning. In Proceedings of the 24th International Conference on Machine Learning, Corvallis, OR, USA, 20–24 June 2007; pp. 1151–1157. [CrossRef]
12. Ibrahim, N.; Hamid, H.A.; Rahman, S.; Fong, S. Feature selection methods: Case of filter and wrapper approaches for maximising classification accuracy. *Pertanika J. Sci. Technol.* **2018**, *26*, 329–340.
13. Libasin, Z.; Suhailah, W.; Fauzi, W.M.; Ul-Saufie, A.Z.; Idris, N.A.; Mazeni, N.A. Evaluation of Single Missing Value Imputation Techniques for Incomplete Air Particulates Matter (PM10) Data in Malaysia. *Pertanika J. Sci. Technol.* **2021**, *29*, 3099–3112. [CrossRef]
14. Kukkonen, J.; Partanen, L.; Karppinen, A.; Ruuskanen, J.; Junninen, H.; Kolehmainen, M.; Niska, H.; Dorling, S.; Chatterton, T.; Foxall, R.; et al. Extensive evaluation of neural network models for the prediction of NO_2 and PM10 concentrations, compared with a deterministic modelling system and measurements in central Helsinki. *Atmos. Environ.* **2003**, *37*, 4539–4550. [CrossRef]
15. Brownlee, J. How to Choose a Feature Selection Method For Machine Learning. Machine Learning Mastery. 2020. Available online: https://machinelearningmastery.com/feature-selection-with-real-and-categorical-data/ (accessed on 20 August 2020).
16. Jain, S. Genetic Algorithm | Application of Genetic Algorithm. Analytics Vidhya. 2017. Available online: https://www.analyticsvidhya.com/blog/2017/07/introduction-to-genetic-algorithm/ (accessed on 15 June 2022).
17. Shafie, A.S.; Masrom, S.; Ahmad, N. *Improved Neural Network Backpropagation with Genetic Algorithm Based Parameter Tuning for Classification Problem*; Research Report; Universiti Teknologi Mara: Alam, Malaysia, 2010.
18. Kamruzzaman, J.; Aziz, S.M. A Note on Activation Function in Multilayer Feedforward Learning. In Proceedings of the 2002 International Joint Conference on Neural Networks, Honolulu, HI, USA, 12–17 May 2002; pp. 519–523.
19. RapidMiner. RapidMiner Documetation. 2022. Available online: https://docs.rapidminer.com/latest/studio/operators/modeling/predictive/neural_nets/neural_net.html (accessed on 5 March 2022).
20. Guo, C.; Liu, G.; Chen, C.H. Air Pollution Concentration Forecast Method Based on the Deep Ensemble Neural Network. *Wirel. Commun. Mob. Comput.* **2020**, *2020*, 8854649. [CrossRef]
21. Al-Rashed, A.; Al-Mutairi, N.; Boureslli, A. Prediction of air pollution in al-hmadi city using artificial neural network (Ann). *J. Environ. Treat. Tech.* **2020**, *8*, 1390–1399. [CrossRef]
22. Department of Environment, Malaysia. Malaysia Environmental Quality Report. 2015. Available online: https://www.doe.gov.my/ (accessed on 1 January 2022).
23. Adielsson, S. Statistical and Neural Networks Analysis of Pesticide Losses to Surface Water in Small Agricultural Catchments in Sweden. Master's Thesis, Sweden University, Uppsala, Sweden, 2005.
24. Miao, Y.; Mulla, D.J.; Robert, P.C. Identifying important factors influencing corn yield and grain quality variability using artificial neural networks. *Precis. Agric.* **2006**, *7*, 117–135. [CrossRef]
25. Pastor, O. Unbased sensitivity analysis and pruning techniques in ANN for surface ozone modeling. *Ecol. Model.* **2005**, *182*, 149–158. [CrossRef]

26. Starett, S.K.; Najjar, Y.; Adams, S.G.; Hill, J. Modeling pesticide leaching from golf courses using artificial neural networks. *Commun. Soil Sci. Plant Anal.* **1998**, *29*, 3093–3106. [CrossRef]
27. Lek, S.; Delacoste, M.; Baran, P.; Dimopoulos, I.; Lauga, J.; Aulagnier, A. Application of neural networks to modeling nonlinear relationships in ecology. *Ecol. Model.* **1996**, *90*, 39–52. [CrossRef]
28. Lek, S.; Delacoste, M.; Baran, P.; Dimopoulos, I.; Lauga, J.; Aulagnier, S. Comparing discriminant analysis, neural networks and logistic regression for predicting species distributions: A case study with a Himalayan river bird. *Ecol. Model.* **1999**, *120*, 337–347.
29. Ozesmi, L.S.; Ozesmi, U. An artificial neural network appr oach to spatial habitat modeling with interspecific interaction. *Ecol. Model.* **1999**, *116*, 15–31. [CrossRef]
30. Mansor, A.A.; Abdullah, S.; Dom, N.C.; Napi, N.N.L.M.; Ahmed, A.N.; Ismail, M.; Zulkifli, M.F.R. Three-Hour-Ahead of Multiple Linear Regression (MLR) Models for Particulate Matter (PM10) Forecasting. *Int. J. Des. Nat. Ecodyn.* **2021**, *16*, 53–59. Available online: http://iieta.org/journals/ijdne (accessed on 5 May 2022). [CrossRef]
31. Abdullah, S.; Ismail, M.; Ahmed, A.N.; Abdullah, A.M. Forecasting particulate matter concentration using linear and non-linear approaches for air quality decision support. *Atmosphere* **2019**, *10*, 667. [CrossRef]
32. Ceylan, Z.; Bulkan, S.E.R.O.L. Forecasting PM10 levels using ANN and MLR: A case study for Sakarya City. *Glob. Nest J.* **2018**, *20*, 281–290.
33. Fong, S.Y.; Abdullah, S.; Ismail, M. Forecasting of Particulate Matter (PM10) concentration based on gaseous pollutants and meteorological factors for different monsoons of urban coastal area in Terengganu. *J. Sustain. Sci. Manag.* **2018**, *13*, 3–17.
34. Comite, V.; Pozo-Antonio, J.S.; Cardell, C.; Rivas, T.; Randazzo, L.; La Russa, M.F.; Fermo, P. Environmental Impact Assessment on the Monza Cathedral (Italy): A Multi-Analytical Approach. *Int. J. Conserv. Sci.* **2020**, *11 (SI1)*, 291–304.
35. Cazacu, M.M.; Pelin, V.; Radinschi, I.; Sandu, I.; Ciocan, V.; Sandu, I.G.; Gurlui, S. Effects of Meteorological Factors on the Hydrophobization of Specific Calcareous Geomaterials From Repedea—Iasi Area, Under the Urban Ambiental Air Exposure. *Int. J. Conserv. Sci.* **2020**, *11*, 1019–1030.
36. Wu, B.; Wang, L.; Zeng, Y.R. Interpretable wind speed prediction with multivariate time series and temporal fusion transformers. *Energy* **2022**, *252*, 123990. [CrossRef]

Article

Suitability of Selected Plant Species for Phytoremediation: A Case Study of a Coal Combustion Ash Landfill

Artur Szwalec, Paweł Mundała * and Renata Kędzior

Department of Ecology, Climatology and Air Protection, University of Agriculture in Krakow al. Mickiewicz 24/28, 30-059 Krakow, Poland; artur.szwalec@urk.edu.pl (A.S.); renata.kedzior@urk.edu.pl (R.K.)
* Correspondence: rmmundal@cyf-kr.edu.pl; Tel.: +48-607-762-035

Abstract: Coal bottom and fly ash waste continue to be generated as a result of energy production from coal in the amount of about 750 million tonnes a year globally. Coal is the main source of energy in Poland, and about 338 million tonnes of combustion waste has already been landfilled. The aim of the research was to identify factors determining the Cd, Pb, Zn and Cu phytostabilisation by vegetation growing on a coal combustion waste landfill. Soil and shoots of the following plants were analysed: wood small-reed, European goldenrod, common reed; silver birch, black locust, European aspen and common oak. The influence of the location where the plants grew and the influence of the interaction between the two factors (species and location) were significant. The tree species were more effective at accumulating heavy metals than the herbaceous plants. European aspen had the highest Bioaccumulation Factor (BCF) for cadmium and zinc. A high capacity to accumulate these elements was also demonstrated by silver birch, and in the case of cadmium, by common oak. Accumulation of both lead and copper was low in all plants. The Translocation Factors (TF) indicated that the heavy metals were accumulated mainly in the roots. European aspen, silver birch and European goldenrod were shown to be most suitable for stabilization of the metals analysed in the research.

Keywords: heavy metal phytostabilization; bioaccumulation and translocation factors; vegetation; coal fly and bottom ash landfill; interaction of plant species and location

Citation: Szwalec, A.; Mundała, P.; Kędzior, R. Suitability of Selected Plant Species for Phytoremediation: A Case Study of a Coal Combustion Ash Landfill. *Sustainability* 2022, 14, 7083. https://doi.org/10.3390/ su14127083

Academic Editors: Mohd Rafatullah and Masoom Raza Siddiqui

Received: 30 April 2022
Accepted: 2 June 2022
Published: 9 June 2022

Publisher's Note: MDPI stays neutral with regard to jurisdictional claims in published maps and institutional affiliations.

Copyright: © 2022 by the authors. Licensee MDPI, Basel, Switzerland. This article is an open access article distributed under the terms and conditions of the Creative Commons Attribution (CC BY) license (https:// creativecommons.org/licenses/by/ 4.0/).

1. Introduction

Despite the systematic reduction in the use of solid fuels, they still account for about 28.1% of world primary energy production and as much as 39.3% of electricity generation [1]. The use of coal for energy generates a relatively large amount of waste, mainly in the form of bottom ash. Samanli et al. [2] have reported that an estimated 750 million tonnes of this type of waste is produced globally per year. Although recent decades have seen an increase in the economic utilization of power plant waste, especially fly ash [3–5], a significant amount is still deposited in landfills. In Poland, over 338 million tonnes of combustion waste, mainly bottom ash, had been deposited in landfills by the end of 2020 [6]. Combustion waste landfills have been a potential source of uncontrolled secondary emission of toxins into the nearby environment, including topsoil, forests, surface water and groundwater, posing a threat to all living organisms [7–9]. Conditions for the development and life of plants at sites of bottom and fly ash deposition are generally unfavourable, mainly due to an unsuitable air-to-water ratio, a strongly alkaline reaction, and poor accessibility of basic nutrients, including a nearly complete lack of nitrogen, which is essential for the life of plants [10–12]. There have been many works devoted to experimental research on the development and chemistry of plant species on landfills under conditions of varied fertilization and variation in the species sown [13–16]. Fewer studies concern plants growing on combustion waste landfills under conditions of spontaneous succession [17,18]. Irrespective of the type of succession, developing vegetation plays an important role in limiting secondary emission of pollutants from these sites and thus mitigating their negative impact on the environment [18–20]. The most hazardous environmental toxins accumulated in combustion waste include heavy metals [21,22]. In the last century, power plant waste landfills in Poland generally appeared in the same way: close to the power plant, in an oxbow, in sand, gravel or clay excavations, without regard to any environmental rights. The landfill selected for this study is one of many similar sites in Eastern Europe.

In the case of a coal combustion waste landfill, phytoremediation as phytoextraction of heavy metals from an entire contaminated area is generally not feasible. The limitations mainly concern the slopes. The slope angle precludes the use of agrotechnical procedures needed for cultivation in practice, including harvesting of the crop. Furthermore, the removal of biomass would significantly disturb the circulation of matter, exacerbating the sterility of the nutrient-poor substrate. It should be noted that the fruits and seeds of plants could provide food for wildlife, whose existence increases the fertility of the substrate. For these reasons, our study focused on phytostabilization of heavy metals.

The aims of the study were as follows:
- to identify the species of plants with the highest phytostabilization capacity (defined by bioaccumulation and translocation factors).
- to determine the influence of the location of plant growth on the landfill on phytostabilization.

The following scientific hypotheses were put forth:
- Phytostabilization is influenced by the species (i).
- Phytostabilization is influenced by the location on the slope (ii).
- Phytostabilization is influenced by the interaction of species × location (iii).

2. Materials and Methods

2.1. Study Site Description

The Skawina Power Plant waste landfill (Figure 1), located in excavations remaining after the exploitation of natural aggregates, began operating in 1975. In the geological structure of the terrain, the superficial layer consists of quaternary formations represented by river sediments (coarse sands and gravel covered with alluvial soils in places, interbedded with loam, silt loam and mud sludge). Together with the accompanying infrastructure, the landfill occupies an area of 68 ha. Combustion waste in the form of coal bottom and fly ash was pumped into the landfill by hydraulic transport. Deposition of waste on the landfill was terminated in 2015. There is no specific documentation of biological reclamation of the landfill. However, differences in the botanical composition of vegetation growing on the embankment suggest that such attempts have been made in the past, as there are woody plants growing downwind of the westerly winds usually occurring in this region. These were probably planted in order to protect residents' homes and fields, situated on the north and east side of the landfill, from airborne dust [20,23]. The height of the landfill is from 22 to 26 m above the surrounding areas. The length of the embankment is 52 ± 5 m, and the slope angle is $28 \pm 3°$.

2.2. Plant and Soil Sampling

Soil and plants from the slopes of the landfill were analysed. The northern and eastern slopes were sampled. The southern and western slopes were omitted because they were not overgrown with vegetation. Three sampling areas were established for each of the embankments: the base, the middle and the top. Each sampling area was divided into three sampling plots of about 152 m^2, i.e., a rectangle 4 m × 38 m. The sampling plots were subdivided into seven sampling points of 22 m^2 (Figure 2.). The study complies with local and national guidelines. Permission was obtained for collection of plants.

Figure 1. Location of the Skawina Power Plant combustion waste landfill, Kopanka Village, Skawina Commune, South of Poland, Europe.

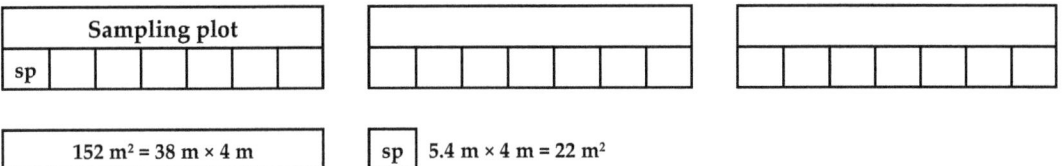

Figure 2. Relatioship between the sampling area, sampling plots and sampling points (sp).

Plant species were selected for the research on the basis of their percentage share in the coverage of the landfill slope, which was determined by the authors in their previous research [24]. The tree species black locust (b.l.; abbreviation used in figures and tables) (*Robinia pseudoacacia* L.), silver birch (s.b.) (*Betula pendula* Roth.), European aspen (e.a.) (*Populus tremula* L.) and common oak (c.a.) (*Quercus robur* L.), and the herbaceous species wood small-reed (w.s.r.) (*Calamagrostis epigejos* L.), European goldenrod (e.g.) (*Solidago virgaurea* L.) and common reed (c.r.) (*Phragmites australis* ((Cav.)Trin. ex Steud) were chosen. Shoots were sampled from each species. Primary samples of shoots were taken from five different trees of the same species at each sampling point. A primary herbaceous sample consisted of twenty plant shoots. Roots were taken from herbaceous plants, but tree roots were not sampled, as there was no way to collect a representative root sample from a tree species without seriously damaging the landfill slope. A garden fork was used to pull up the herbaceous plants with maximum volume of the root system. The roots and shoots were cut off with stainless steel secateurs. The roots were carefully washed in the laboratory in warm tap water to remove coal combustion ash from the roots. The shoots were washed as well. Following homogenization, the samples constituted an averaged sample of about 600 g fresh weight. The samples were collected in late summer. Soil samples were collected using a soil sampler, from a sampling depth of 0–0.1 m. Five primary samples were collected. After homogenization, the averaged soil samples were of about 500 g wet weight. The soil samples were taken from the sampling plots.

2.3. Chemical Analyses

The soil material was dried, ground and sieved, followed by wet mineralization using a mixture of concentrated HNO_3 and $HClO_4$ [25,26]. Cd, Pb, Zn and Cu concentrations were determined by FAAS. In addition, the pH of the samples was determined in a distilled water solution and 1N KCl by the potentiometric method [25]. The plant samples were dried at 60 °C and ground in a mill, followed by dry mineralization in a muffle furnace at 460 °C. Organic parts that remained were oxidized using HNO_3, and silica was precipitated with HCl [25]. The Cd, Pb, Zn and Cu concentrations were determined by FAAS. The spectrophotometer was calibrated with Merck standards (Merk Group, Darmstadt, Germany). In order to check the quality of calibration of atomic absorption, spectrophotometer references materials were used. They were trace metals–clay and trace metals–sandy loam 7 provided by Merck. Recovery was 94% for Cd, 95% for Cu, 96% for Pb and 99% for Zn.

Blind samples were used in a series of approximately 30 samples. Each soil and plant sample was mineralised and analysed independently twice. If Relative Standard Deviation (RSD) from two results were higher than 5%, the sample was mineralised and analysed again.

2.4. Assessment Methods

The assessment was based on analysis of the BCF [27–29] and the TF [30–32]. The Shapiro–Wilk test was used to estimate the data distribution. The Wald–Wolfowitz test (a nonparametric test) was used to test the hypotheses. As a post hoc test, the Kruskal–Wallis test was used to compare means. All statistical tests were performed at a significance level of $\alpha = 0.05$. All statistical analyses were performed in Statistica 12.0 software (StatSoft inc., Tulsa, OK, USA).

3. Results and Discussion

3.1. Results and General Assessment

The data in Section 3.1 are presented and discussed all together, i.e., are not subdivided into: the base, the middle and the top. The unfavourable ecological conditions prevailing in combustion waste landfills have a fundamental impact on the vegetation occurring at these sites as a result of either spontaneous succession or reclamation efforts. The soil samples had an alkaline reaction (mean pH_{KCl} =7.4). The relative standard deviation had medium values for lead (30%) and copper (28%) contents, low values for cadmium (16%) and zinc (21%), and an extremely low value of 3% for pH (Table 1). Jambhulkar and Juwarkar [16] reported a similarly low (RSD) of only 4% for a fly ash dump of the Khaperkheda thermal power plant in Maharashtra State, India.

Table 1. Average soil Cd, Pb, Zn and Cu content and pH_{KCl}, range and relative standard deviation (RSD) ($n = 54$).

	Cd	Pb	Zn	Cu	pH
	Range, Mean mg·kg^{-1} d.m. and RSD %				
Range	0.19–0.37	17.23–36.98	79.65–147.25	28.01–63.86	7.1–7.9
Mean	0.292	24.664	117.019	48.348	7.4
RSD	16	30	21	28	3

Cadmium content in the shoots ranged from 0.028 to 1.058 mg·kg^{-1} d.m. (Table 2), with the lowest content found in wood small-reed and common reed. The minimum and maximum cadmium contents in the roots of herbaceous plants were 0.038 and 1.24 mg·kg^{-1} d.m., which was a wider range than in the case of cadmium content in the shoots (Table 3). Variation in the concentration of this element in the aboveground parts of plants was very high in the case of European goldenrod (RSD = 103%) (Table 2) and high in the case of silver birch and European aspen (RSD 60% and 57%, respectively). Cadmium content was least varied in common reed (RSD = 26%). High variability of this metal was noted in the roots of the herbaceous plants: for European goldenrod RSD = 62%, for common reed RSD = 45% and for wood small-reed RSD = 67%. The lead content in the plants in the study area ranged from 0.207 to 1.4410 mg·kg^{-1} d.m.; the maximum for the roots was 4.450 mg·kg^{-1} d.m. (Table 3) Lead concentrations were much less variable than in the case of cadmium. The highest variations were noted for European goldenrod (28% and 34% for shoots and roots, respectively) and the lowest for common oak shoots (RSD = 4%). The zinc concentration in the plants ranged from 8.210 mg·kg^{-1} (common reed) to 222.294 mg·kg^{-1} (European aspen) (Table 2). The highest variation in the aerial parts of plants was found for the zinc concentrations in European goldenrod (RSD = 30%), followed by common reed (27%) and black locust (25%) (Table 2). The range of zinc concentrations for the roots was similar to the range in the shoots. Variation in zinc content in the roots was similar to the variation in the shoots for common reed and wood small-reed (Table 3).

Table 2. Cd, Pb, Zn and Cu average content, range, and RSD in shoots (n = 252).

Metal/Plant Species		Black Locust	Silver Birch	European Aspen	Common Oak	Common Reed	European Goldenrod	Wood Small Reed
		\multicolumn{7}{c}{Range, Mean mg·kg^{-1} d.m., RSD %}							
Cd	Range	0.069–0.326	0.149–0.650	0.262–1.058	0.253–0.578	0.028–0.052	0.047–0.590	0.028–0.120
	Mean	0.199	0.327	0.565	0.384	0.037	0.230	0.074
	RSD	46	60	57	31	26	103	44
Pb	Range	0.970–1.416	1.448–2.410	1.404–1.906	1.778–2.044	0.380–0.470	0.550–1.250	0.207–0.340
	Mean	1.169	1.897	1.686	1.893	0.422	0.921	0.265
	RSD	11	16	10	4	6	28	16
Zn	Range	28.924–53.618	139.156–169.340	148.852–222.294	83.764–105.050	8.210–19.580	28.470–62.780	11.050–15.220
	Mean	37.532	156.525	185.913	93.774	14.949	40.374	12.722
	RSD	25	5	12	6	27	30	8
Cu	Range	4.106–7.362	4.350–6.776	6.534–7.920	5.130–6.506	0.880–2.870	3.110–6.700	0.810–1.120
	Mean	5.290	5.900	7.022	5.851	1.841	4.667	0.931
	RSD	24	10	5	7	37	27	8

Table 3. Cd, Pb Zn and Cu average content, range and RSD in roots (n = 108).

Metal	Species	Common Reed	European Goldenrod	Wood Small-Reed
	Data	\multicolumn{3}{c}{Mean, Range mg·kg^{-1} d.m., RSD %}		
Cd	Range	0.038–0.150	0.195–1.240	0.046–0.900
	Mean	0.084	0.665	0.525
	RSD	45	62	67
Pb	Range	0.670–1.450	0.770–1.980	2.800–4.450
	Mean	1.071	1.241	3.373
	RSD	25	34	17
Zn	Range	14.500–32.780	30.680–50.850	71.550–110.170
	Mean	22.096	38.904	91.321
	RSD	28	13	12
Cu	Range	2.300–8.630	10.36–17.23	5.630–7.750
	Men	6.171	12.945	6.562
	RSD	42	19	9

Copper content in the shoots ranged from 0.810 mg·kg^{-1} (wood small-reed) (Table 2) to 7.920 mg·kg^{-1} (European aspen), whereas the maximum concentration in the roots was 17.23 mg·kg^{-1} (European goldenrod). The maximum soil copper content was 65.86 mg·kg^{-1}, with a mean of 48.348 mg·kg^{-1} ± 28% RSD (Table 1). The highest variation in the content of this element in the aerial parts of plants was found in common reed (37%), followed by European goldenrod (27%) and black locust (24%). A high RSD was found in the roots of common reed (42%). According to Alloway [33], the critical copper concentration in plants is 2 mg·kg^{-1} d.m. Wood small-reed copper content was much lower, in a range of 0.810–1.120 mg·kg^{-1} d.m. Another species near the threshold was common reed.

According to Szwalec et al. [34] the eluates from the waste were not found to pose a threat to surface water quality in terms of concentrations of macronutrients Na, K, Ca and

Mg, metals considered to be particularly harmful, i.e., Zn, Cu and Cr, or priority metals Cd and Pb.

3.2. Detailed Assessment

Biostabilization is an important part of bioremediation. Phytostabilization reduces the airborne spread of heavy metals and groundwater contamination by leaching [35]. It also reduces metal bioavailability, thereby improving conditions for less tolerant species. Phytostabilization has been described as a plant's ability to take up and accumulate metals from soil. The BCF and TF are useful tools here. Both factors can be used to assess a plant's potential for phytostabilisation [18,31]. Szöcs and Schäfer [36], recommend the use of factors in statistical analysis, particularly in the case of non-parametric data distribution. Both BCF and TF are good examples, as they make use of data pertaining to the content of metals in soil and plants (BCF) and in plant organs (TF), combining two sets of data in a representative manner. The Shapiro–Wilk test showed that only single subsets of data had a normal distribution. These were the data for five factors: BCFCu (silver birch and wood small-reed) and TFCu (wood small-reed, common reed and European goldenrod). The remaining data did not have a normal distribution. This type of distribution has been found in studies on phytoremediation [34,37–39]. The Wald–Wolfowitz runs test was used to verify the hypotheses on the influence of the species and location of the landfill on the values of BCF and TF. The following hypotheses were tested: BCF and TF are influenced by the species; by the location of plant growth on the landfill and by the interaction of the two factors. The analysis revealed that these three effects (hypothesis i, ii, iii) were statistically significant for each of the metals tested. However, for the location (i.e., hypothesis ii), the low Ws values of the Wald–Wolfowitz test calculated from our own data were similar to the critical values read from the theoretical distribution tables for this test. This indicates a weak effect, so we focused the discussion on the interactions (hypothesis iii).

3.2.1. Influence of Species (Regardless of Location) on Metal Phytostabilization

The effect of species on phytostabilization of heavy metals is the most important question in phytoremediation research [40,41]. It is particularly significant in areas contaminated with these elements, including post-industrial areas such as combustion waste landfills. In the landfill that was the subject of this study, a high degree of cadmium accumulation was noted in the shoots of aspen, common oak, and silver birch, while the other plants accumulated this element to a moderate degree. The statistical analysis using the Kruskal–Wallis test showed that the phytoaccumulative capacities of European aspen differed statistically considerably from those of the other two species (common oak and silver birch), with a high rate of accumulation of this metal (Table 4A). Similar relationships were noted for accumulation of zinc in the plants. Accumulation of this metal was high in European aspen and silver birch but moderate in the remaining plants, as in the case of cadmium. It should be noted that common oak was statistically considerably distinguished in the group of species with moderate accumulation of this metal (Table 4B). Accumulation of both lead and copper was low in all plants tested. Statistically considerably differences from the other species were found for aspen in the case of phytoaccumulation of copper and for common oak in the case of lead (Table 4C,D). It should be noted, however, that as in the case of cadmium and zinc, for these metals as well the bioaccumulation factors were higher in the shoots of trees than in the shoots of herbaceous plants. A comparable BCF range to those noted for the herbaceous plants (0.05–0.45) was reported [42] for these metals in *Saccharum munja* occurring in natural succession on a fly ash lagoon. Pandey [43] also describes moderate (<1) Cd, Pb, Zn and Cu bioaccumulation factors for the aerial parts of castor bean (*Ricinus communis*) growing on this type of landfill. The author, like many others [44–47], also draws attention to the role of the biomass growth of individual plant species and their ability to accumulate metals in relation to phytoextraction of these elements from contaminated areas. In this regard it is worth drawing attention to the tree species with high BCFs for cadmium and zinc, particularly European aspen and silver birch.

The hyperaccumulation capacity of silver birch, particularly in the case of zinc, has been confirmed by numerous studies [34,38,48]. Similar capacities of trees of the genus *Populus* to accumulate heavy metals have been pointed out by various authors [49–51].

Table 4. A–H. Statistical significance of differences in means for BCF (A–D) and TF (E–H) for species (Kruskal–Wallis test, α = 0.05). Letters (a, b, c, d, e, f) designate groups of plants for which there were no statistically considerable differences. The data is presented and calculated all together, i.e., are not subdivided into: the base, the middle and the top.

A/sp.	BCF_{Cd}	dif.		B/sp.	BCF_{Zn}	dif.		E/sp.	TF_{Cd}	dif.
c.r.	0.134	a		w.s.r.	0.131	a		w.s.r.	0.102	a
w.s.r.	0.267	a		c.r.	0.149	a		c.r.	0.534	b
b.l.	0.711	b		b.l.	0.394	a		e.g.	0.618	b
e.g.	0.766	b		e.g.	0.420	a				
s.b.	1.087	c		c.o.	0.950	b		F/sp.	TF_{Zn}	dif.
c.o.	1.312	c		s.b.	1.587	c		w.s.r.	0.141	a
e.a.	1.938	d		e.a.	1.935	c		c.r.	0.683	b
								e.g.	1.022	c
C/sp.	BCF_{Pb}	dif.		D/sp.	BCF_{Cu}	dif.				
w.s.r.	0.012	a		w.s.r.	0.017	a		G/sp.	TF_{Pb}	dif.
c.r.	0.019	b		c.r.	0.033	a		w.s.r.	0.079	a
e.g.	0.038	c		e.g.	0.084	b		c.r.	0.425	b
b.l.	0.050	d		b.l.	0.095	b,c		e.g.	0.785	c
e.a.	0.075	e		s.b.	0.105	c				
c.o.	0.083	f		c.o.	0.105	c		H/sp.	TF_{Cu}	dif.
s.b.	0.085	f		e.a.	0.126	d		w.s.r.	0.143	a
								c.r.	0.315	b
								e.g.	0.356	b

c.r. common reed; w.s.r. wood small-reed; b.l. black lotus; e.g. European goldenrod; s.b. silver birch; c.o. common oak; e.a. European aspen

In the case of the group of herbaceous plants, for which the content of metals was also analysed in the roots, it was in these organs that the elements primarily accumulated (Table 4E–H). This is confirmed by both the statistical analyses (Table 5) and the translocation factors (TF), which for most of the elements were within the range of $0.079 \leq TF \leq 0.785$. The only exception was zinc content in European goldenrod (TF = 1.264). In this case, the concentration of the element in the plant growing at the base of the landfill had a statistically considerable effect (Table 5F). At the other locations (the middle and top of the landfill), the value of the Zn translocation factor for this species was much lower ($TF_{Zn} \leq 0.853$). It should be noted, however, that this species had the highest TFs of all metals tested (TF_{Cd} = 0.47, TF_{Pb} = 1.107, TF_{Cu} = 0.391), differing statistically considerably from the other plants (Table 5E,F). Accumulation of heavy metals mainly in the roots of herbaceous plants (*Typha latifolia, Saccharum spontanum, Amaranthus deflexus* and *Fimbristylis dichotoma*) growing on a fly ash dump was confirmed by Maiti and Jasawal [17]. The authors also draw attention to differences in TF depending on the metal and the plant species. They report TF < 1 for Cu and Zn. The exception was TF calculated for Pb in *Amaranthus deflexus*, amounting to 1.970, whereas for the other species it ranged from 0.66 to 0.85. Similar values (TF < 1) are reported by Pandey [18], who analysed the suitability of *Ipomoea carnea* for Cd, Pb, Cu, Cr, Mn and Ni phytoremediation of ash dumps. According to the author, TF values were > 1 only for Cd and Cr (TF_{Cd} = 1.050, TF_{Cr} = 1.180), and the limited migration of the other metals in the root system may have been due to sequestration of these elements in the vacuoles of the root cells.

Table 5. A–F. Statistical significance of differences in means for BCF (A–D) and TF (E–H) for selected species × location interactions (Kruskal–Wallis test, α = 0.05). Letters (a, b, c, d) designate groups of plants for which there were no statistically considerable differences. The data is subdivided into the base, the middle and the top.

A/sp. × l.	BCFCd	dif.
e.g. × t.	0.156	a
e.g. × m.	0.340	a
s.b. × b.	0.741	b
e.a. × t.	0.932	b
c.o. × b.	1.107	b
e.a. × m.	1.629	c
c.o. × t.	1.691	c
e.g. × b.	1.800	c
s.b. × t.	1.862	c
e.a. × b.	3.252	d

B/sp. × l.	BCFPb	dif.
e.a. × t.	0.043	a
s.b. × t.	0.044	a
c.o. × t.	0.056	b
e.a. × b.	0.094	c
c.o. × b.	0.105	d
s.b. × b.	0.106	d

C/sp. × l.	BCFZn	dif.
c.o. × t.	0.676	a
c.o. × b.	1.018	a,b
e.a. × t.	1.119	b
s.b. × t.	1.122	b
s.b. × b.	1.891	c
e.a. × b.	2.452	d

D/sp. × l.	BCFCu	dif.
c.r. × b.	0.017	a
c.r. × m.	0.043	b
e.g. × m.	0.064	c
b.l. × t.	0.074	c
e.g. × b.	0.118	d
b.l. × b.	0.130	d

E/sp. × l.	TFCd	dif.
c.r. × b.	0.232	a
e.g. × b.	0.471	a
c.r. × t.	0.744	b
e.g. × t.	1.002	c

F/sp. × l.	TFZn	dif.
c.r. × b.	0.615	a
c.r. × t.	0.830	a,b
e.g. × m.	0.853	b
e.g. × b.	1.264	c

G/sp. × l.	TFPb	dif.
w.s.r. × b.	0.075	a
w.s.r. × t.	0.076	a
w.s.r. × m.	0.086	a
c.r. × b.	0.314	b
e.g. × m.	0.537	c
c.r. × t.	0.600	c
e.g. × b.	0.710	c
e.g. × t.	1.107	d

H/sp. × l.	TFCu	dif.
w.s.r. × b.	0.129	a
w.s.r. × t.	0.167	a
e.g. × m.	0.310	b
e.g. × t.	0.368	c
e.g. × b.	0.391	c

l. location; b base; m middle; t top; sp. species; × interaction.
c.r. common reed; w.s.r. wood small-reed; b.l. black loctus; e.g. European goldenrod; s.b. silver birch; c.o. common oak; e.a. European aspen.

3.2.2. Influence of the Interaction of Species and Location on Phytostabilization

All data presented, calculated and discussed in Section 3.2.2. is subdivided into location, i.e., the base, the middle and the top. Phytoremediation studies on combustion waste landfills have previously not taken into account the relationship between the stabilizing capacities of individual plant species and the site of their growth on the landfills. The statistical analysis made it possible to distinguish the following patterns of the combined effect of the two factors, i.e., the species and the location of growth on the landfill on the stabilizing potential of each plant species. The first was characteristic of wood small-reed and common reed, for which there were practically no statistically considerable differences between BCF and TF values depending on their location on the landfill (base, middle, or top). Both species also had low levels of phytostabilization of the metals tested (Figure 3A,C,E, Table 5G,H). This suggests that the bioaccumulation and translocation of trace elements was not a minimum factor for the development of these plants at the site, and therefore, despite their low phytostabilizing properties, the suitability of these species can be considered in revegetation of energy waste landfills. In the second-most common pattern of interaction, BCF and TF values were increased by the species' location at the base of the landfill and reduced by their situation at the top. This pattern was observed for European aspen and goldenrod for phytostabilization of cadmium. In the case of lead accumulation, these relationships were observed in silver birch, common oak and European aspen. Similar relationships for phytoaccumulation of zinc were noted for aspen, silver birch, European goldenrod and common reed. In the case of copper, these dependencies occurred in black locust, common oak, silver birch and European goldenrod (Figure 3B–J, Table 5B,D). This pattern was not observed for TF, except for phytoaccumulation of zinc in European goldenrod (Table 5F). The third pattern was such that the values of the indices describing phytostabilization of the metal in the plant were reduced by location of the species at the base of the landfall but increased by its location at the top. This was observed in the case of oak, birch and common reed for cadmium concentration, in common reed for copper content and in European goldenrod for lead content (Figure 3A,B,G,I,K, Table 5A,D,E,G,H). There weren't any patterns in the case of TFCu (Figure 3L). This might be attributable to the very low content of this element in tested plants [33]. Variation in the content of metals in ash deposited in landfills is relatively well described in the literature. Reported sources of variation include the geological source of the coal [52], combustion technology [53] and the stage of waste generation (combustion, transport and deposition) [54]. The content of trace elements in deposited waste directly affects their concentrations in the vegetation growing on landfills. Due to the negative effects of landfills on the environment, this problem continues to be a subject of research, particularly in terms of the eco-restoration potential of individual plant species [55,56]. To date, however, no studies have been conducted on the relationships between the phytostabilization capacity of individual plant species and their location on the landfill. Previous research conducted on the landfill that is the subject of this paper indicated that despite nearly 50 years of existence of the landfill, the ecosystem developing here is still in a relatively early phase of formation [24]. The differences between the phytoaccumulation properties of individual species depending on their location on the landfill may be due to biogeochemical processes taking place in the developing ecosystem. The research areas we selected were covered by the same vegetation, and the physicochemical properties of the soil were comparable (Table 1). However, their formation was closely linked to successive stages of formation of the landfill, and in this sense the area can be said to be heterogeneous. The base of the landfill was formed in the 1970s from the local soil. The middle and top were formed from power plant waste in the 1980s and 1990s.

The ecosystems in the vicinity of the landfill were the source of the plant species that recolonized the area. These areas were probably also a source of microbes and organic matter essential to the survival of heterotrophic microbes. According to numerous sources [11,57–59], microbial communities have a major influence on a forming ecosystem. From year to year, deposited ash was colonized by microbes—first the base, partially already built of colonized material, then the middle and finally the top of the landfill, the last and furthest situated from the natural source of microbes. Therefore, the time and starting conditions for development of microbial populations on the landfill were varied.

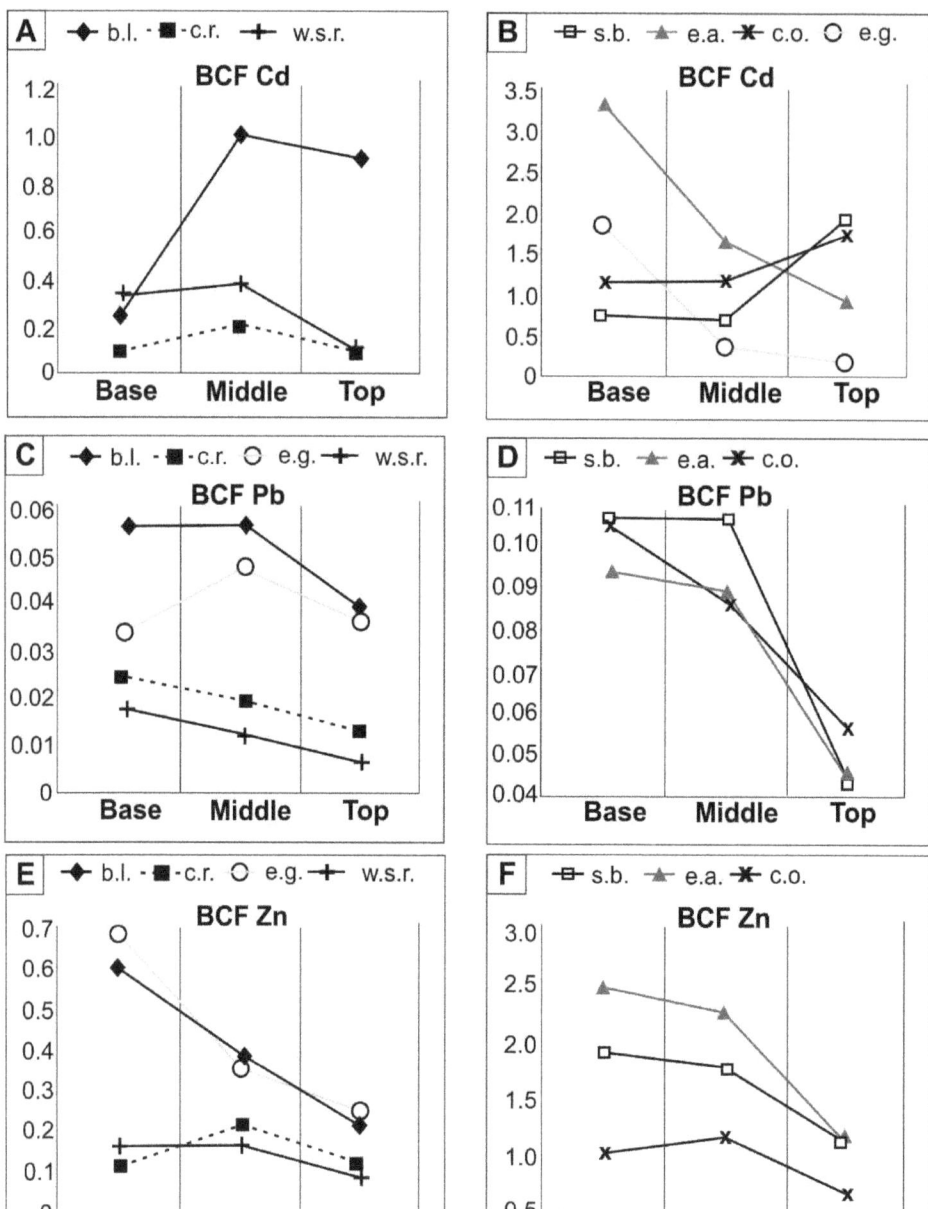

Figure 3. Cont.

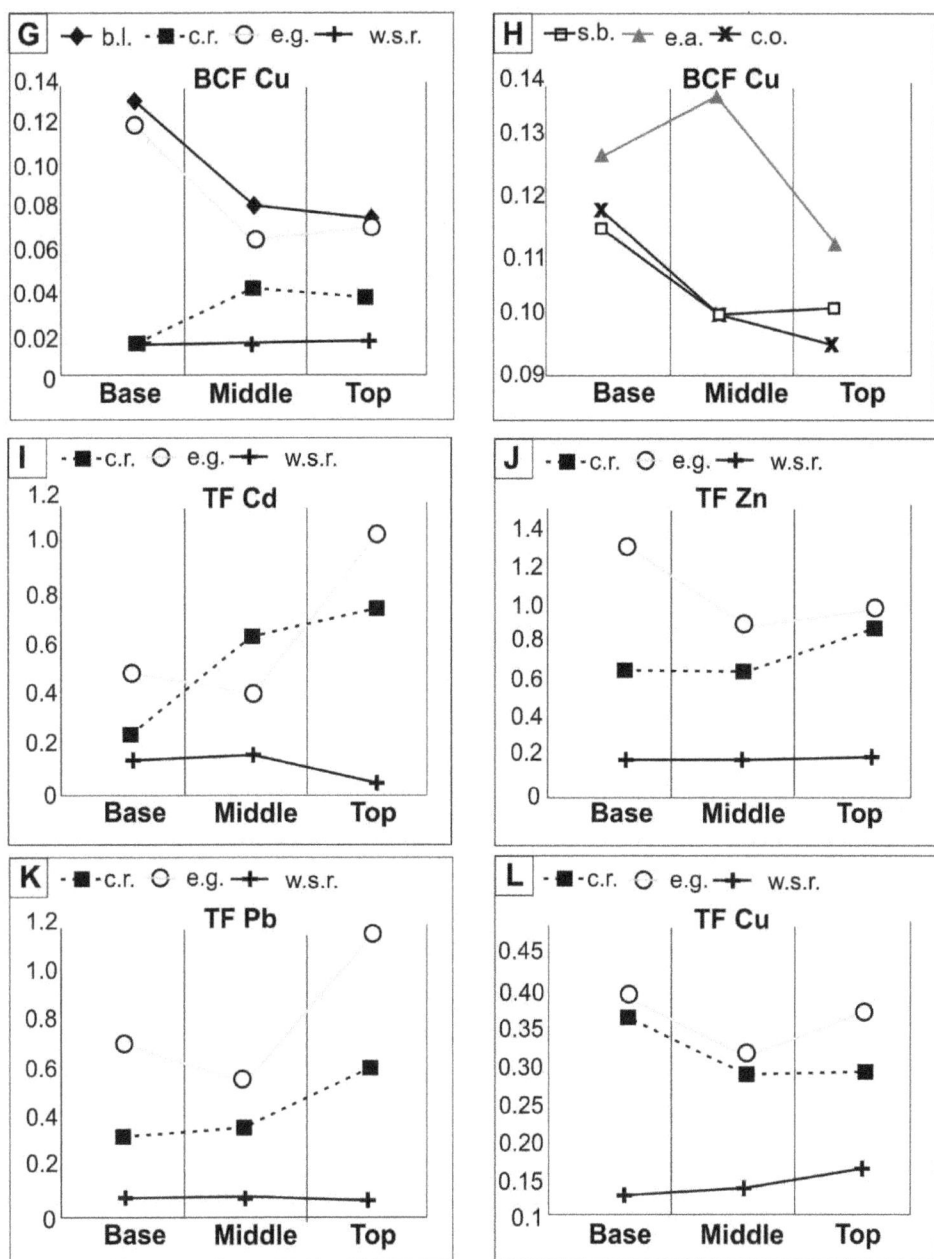

Figure 3. (**A–F**): Graphical comparison of cadmium, lead and zinc BCF for species and location interactions, divided into two groups: below median (**A,C,E**) and above median (**B,D,F**). (**G–L**): Continuum for copper ((**G**) below median, (**H**) above median) and graphical comparison of cadmium lead, zinc and copper TF for species and location interactions.

4. Conclusions

Differences in the phytostabilisation of metals (defined by bioaccumulation and translocation factors) between the analysed plants are statistically considerably different for all tested species and elements. The study made it possible to identify two distinct groups of vegetation: the first with high BCFs, i.e., European aspen, silver birch and common oak, and the second with low BCFs, i.e., common reed and wood small-reed. In the case of TF, the only species in the group with a low value was wood small-reed, whereas all other herbaceous species, including common reed, had a high TF. The translocation factors also indicated that for this group of plants all elements tested were accumulated mainly in the roots. However, the most important conclusion of our study is a fact of interaction (i.e., combined and simultaneous) influence of species and location on phytostabilisation. The interactions were statistically significant for all plants and metals. Location was understood as the set of undefined conditions. Location is divided in three groups: the base of the slope, the middle of the slope and the top of the slope. In general, at the base of the landfill, all species had the highest BCF values for lead and zinc, most species for copper, and only two species (aspen and oak) for cadmium. In contrast to BCF, the TF values for lead and cadmium increased upwards along the slope (except for wood small-reed). In the case of the physiological elements zinc and copper, the TF remained practically unchanged. Our results showed that European aspen (*Populus tremula* L.), silver birch (*Betula pendula* Roth) and European goldenrod (*Solidago virgaurea* L.) can be useful in eco-restoration of combustion waste landfills.

Author Contributions: Conceptualization, A.S.: selecting the landfill, the aim, scientific hypothesis discussion, researched plant species and slopes (as the localisation) selection, the use of BCF and TF factors, manuscript framework. P.M.: scientific hypothesis, framework of discussion chapter, use factors (BCF and TF) in statistical analysis. R.K.: idea of case study, scientific hypothesis discussion, statistical analysis. Methodology, A.S.: the experiment model, the set of the experiment, research areas settlement, samples collection, sample preparation, supervision of chemical analyses, literature studies. P.M.: the experiment model, the set of the experiment, research areas settlement, data collection, data interchange between co-Authors, corresponding author, literature studies. R.K.: application of statistical methods, data presentation in graphical forms, corresponding author. Writing and data investigation A.S.: the frame, first draft, discussion of the data investigation, hypothesis and literature, conducting a research and investigation process. Reading and correcting the text. P.M.: the frame change, the second, improved draft, discussion of the data, conducting a research and investigation process, hypothesis and literature, discussion of the statistical results (analysis). Reading and correcting the text. Prepared figures and tables. R.K.: reading and correcting the text, explanation and discussion of the statistical results (analysis). Correction and edition of figures. All authors have read and agreed to the published version of the manuscript.

Funding: This research received no external funding.

Institutional Review Board Statement: Not applicable.

Informed Consent Statement: Not applicable.

Data Availability Statement: All materials and data are available from the corresponding author rmmudal@cyf-kr.edu.pl.

Acknowledgments: The study was conducted with a subsidy of the Ministry of Science and Higher Education for the University of Agriculture in Kraków in 2022. The statistical analyses used in the paper were supported by the project 'Integrated Programme of the University of Agriculture in Krakow' with funding from the European Union under the European Social Fund.

Conflicts of Interest: The authors declare no conflict of interest.

References

1. Eurostat. EU Energy in Figures Statistical Pocketbook. 2018. Available online: https://ec.europa.eu/energy/sites/ener/files/documents/pocketbook_energy_2017_web.pdf (accessed on 29 April 2022).
2. Samanli, S.; Celik, H.; Oney, O.; Can, Y. A review on improvement of coal-fired power plants and environmental benefits of ash utilization. *Int. J. Global Warm.* **2017**, *11*, 67–86. [CrossRef]
3. Yao, Z.T.; Ji, X.S.; Sarker, P.K.; Tang, J.H.; Ge, L.Q.; Xia, M.S.; Xi, Y.Q. A comprehensive review on the applications of coal fly ash. *Earth-Sci. Rev.* **2015**, *141*, 105–121. [CrossRef]
4. Hirajima, T.; Petrus, H.T.B.M.; Oosako, Y.; Nonaka, M.; Sasaki, K.; Ando, T. Recovery of cenospheres from coal fly ash using a dry separation process: Separation estimation and potential application. *Int. J. Min. Process.* **2010**, *95*, 18–24. [CrossRef]
5. Blissett, R.S.; Rowson, N.A. A review of the multi-component utilization of coal fly ash. *Fuel* **2012**, *97*, 1–23. [CrossRef]
6. PCSO. Polish Central Statistical Office. Statistics Poland. Environment. 2021. Available online: https://stat.gov.pl/obszary-tematyczne/srodowisko-energia/srodowisko/ochrona-srodowiska-2020,1,21.html (accessed on 29 April 2022).
7. Pandey, V.C.; Singh, J.S.; Singh, R.P.; Singh, N.; Yunus, M. Arsenic hazards in coal fly ash and its fate in Indian scenario. *Resour. Conserv. Recycl.* **2011**, *55*, 819–835. [CrossRef]
8. Gruchot, A.; Szwalec, A.; Mundała, P. Chemical and geotechnical properties of ash-slag mixture from 'Czajka' landfill near Tarnow. *Environ. Prot. Nat. Res.* **2013**, *24*, 63–67.
9. Bhattacharyya, S.; Donahoe, R.J.; Patel, D. Experimental study of chemical treatment of coal fly ash to reduce the mobility of priority trace elements. *Fuel* **2009**, *88*, 1173–1184. [CrossRef]
10. Juwarkar, A.A.; Jambhulkar, H.P. Restoration of fly ash dump through biological interventions. *Environ. Monit. Assess.* **2008**, *139*, 355–365. [CrossRef] [PubMed]
11. Haynes, R.J. Reclamation and revegetation of fly ash disposal sites–challenges and research needs. *J. Environ. Manag.* **2009**, *90*, 43–53. [CrossRef]
12. Pandey, V.C.; Singh, N. Impact of fly ash incorporation in soil systems. *Agric. Ecosyst. Environ.* **2010**, *136*, 16–27. [CrossRef]
13. Dyguś, K.H. The role of plants in experimental biological reclamation in a bed of furnace waste from coal-based energy. *Ecol. Eng.* **2015**, *16*, 8–22. [CrossRef]
14. Técher, D.; Laval-Gilly, P.; Bennasroune, A.; Henry, S.; Martinez-Chois, C.; D'Innocenzo, M.; Falla, J. An appraisal of *Miscanthus × giganteus* cultivation for fly ash revegetation and soil restoration. *Ind. Crop. Prod.* **2012**, *36*, 427–433. [CrossRef]
15. Hrynkiewicz, K.; Baum, C.; Niedojadło, J.; Dahm, H. Promotion of mycorrhiza formation and growth of willows by the bacterial strain *Sphingomonas* sp. 23 L on fly ash. *Biol. Fert. Soils* **2009**, *45*, 385–394. [CrossRef]
16. Jambhulkar, H.P.; Juwarkar, A.A. Assessment of bioaccumulation of heavy metals by different plant species grow non fly ash dump. *Ecotoxicol. Environ. Saf.* **2009**, *72*, 1122–1128. [CrossRef] [PubMed]
17. Maiti, S.K.; Jaiswal, S. Bioaccumulation and translocation of metals in the natural vegetation growing on fly ash lagoons: A field study from Santaldih thermal power plant, West Bengal, India. *Environ. Monit. Assess.* **2008**, *136*, 355–370. [CrossRef] [PubMed]
18. Pandey, V.C. Invasive species based efficient green technology for phytoremediation of fly ash deposits. *J. Geoch. Explor.* **2012**, *123*, 13–18. [CrossRef]
19. Woch, M.W.; Radwańska, M.; Stanek, M.; Łopata, B.; Stefanowicz, A.M. Relationships between waste physicochemical properties, microbial activity and vegetation at coal ash and sludge disposal sites. *Sci. Total Environ.* **2018**, *642*, 264–275. [CrossRef] [PubMed]
20. Szwalec, A.; Mundała, P.; Kędzior, R. Cadmium, lead, zinc and copper content in herbaceous plants overgrowing furnace waste landfill. *Environ. Prot. Nat. Res.* **2013**, *24*, 33–37.
21. Chakraborty, R.; Mukherjee, A. Mutagenicity and genotoxicity of coal fly ash water leachate. *Ecotoxicol. Environ. Saf.* **2008**, *72*, 838–842. [CrossRef] [PubMed]
22. Gajić, G.; Djurdjević, L.; Kostić, O.; Jari, S.; Mitrović, M.; Pavlović, P. Ecological potential of plants for phytoremediation and ecorestoration of fly ash deposits and mine wastes. *Front. Environ. Sci.* **2018**, *6*, 124. [CrossRef]
23. DoEP. *Department of Environmental Protection of Skawina Power Plant, Characteristics of the Combustion Waste Landfill*; DoEP: Warsaw, Poland, 2012; (Typescript In Polish, Not Published).
24. Kędzior, R.; Szwalec, A.; Mundała, P.; Skalski, T. Ground beetle (*Coleoptera, Carabidae*) life history traits as indicators of habitat recovering process in postindustrial areas. *Ecol. Eng.* **2020**, *142*, 105615. [CrossRef]
25. Ostrowska, A.; Gawliński, S.; Zczubiałka, Z. *Methods of Analysis and Evaluation of Soil and Plant Properties. A Catalogue*; Institute of Environmental Protection: Warsaw, Poland, 1991. (In Polish)
26. Carter, M.R.; Gregorich, E.G. (Eds.) *Soil Sampling and Methods of Analysis*, 2nd ed.; Canadian Society of Soil Science: Pinawa, MB, Canada, 2021; Available online: https://www.niordc.ir/uploads%5C86_106_Binder1.pdf (accessed on 29 April 2022).
27. Li, M.S.; Luo, Y.P.; Su, Z.Y. Heavy metal concentrations in soils and plant accumulation in a restored manganese mine land in Guangxi. South China. *Environ. Pollut.* **2007**, *147*, 168–175. [CrossRef]
28. Marchiol, L.; Fellet, G.; Boscutti, F.; Montella, C.; Mozzi, R.; Guarino, C. Gentle remediation at the former 'Pertusola Sud' Zinc Smelter: Evaluation of native species for phytoremediation purposes. *Ecol. Eng.* **2013**, *53*, 343–353. [CrossRef]
29. Nannoni, F.; Rossi, S.; Protano, G. Potentially toxic element contamination in soil and accumulation in maize plants in a smelter area in Kosovo. *Environ. Sci. Pollut. Res.* **2016**, *23*, 11937–11946. [CrossRef] [PubMed]
30. Stanislawska-Glubiak, E.; Korzeniowska, J.; Kocn, A. Effect of peat on the accumulation and translocation of heavy metals by maize grown in contaminated soils. *Environ. Sci. Pollut. Res.* **2015**, *22*, 4706–4714. [CrossRef] [PubMed]

31. Galal, T.M.; Shehata, H.S. Bioaccumulation and translocation of heavy metals by *Plantago major* L. grown in contaminated soils under the effect of traffic pollution. *Ecol. Indic.* **2015**, *48*, 244–251. [CrossRef]
32. Boechat, C.L.; Pistóia, V.C.; Gianelo, C.; de Oliveira-Camargo, F.E. Accumulation and translocation of heavy metal by spontaneous plants growing on multi-metal-contaminated site in the Southeast of Rio Grande do Sul state. Brazil. *Environ. Sci. Pollut. Res.* **2016**, *23*, 2371–2380. [CrossRef] [PubMed]
33. Alloway, B.J. *Heavy Metals in Soil*; Blackie and Son Ltd.: London, UK, 1990.
34. Szwalec, A.; Lasota, A.; Kędzior, R.; Mundała, P. Variation in heavy metal content in plants growing on a zinc and lead tailings dump. *Appl. Ecol. Environ. Res.* **2018**, *16*, 5081–5094. [CrossRef]
35. Lehmann, C.; Rebele, F. Assessing the potential cadmium phytoremediation with *Calamagrostis epigejos*: A pot experiment. *Int. J. Phytorem.* **2004**, *6*, 169–183. [CrossRef] [PubMed]
36. Szöcs, E.; Schäfer, R.B. Ecotoxicology is not normal. A comparison of statistical approaches for analysis of count and proportion data in ecotoxicology. *Environ. Sci. Pollut. Res.* **2015**, *22*, 13990–13999. [CrossRef] [PubMed]
37. Miguel, B.; Edell, A.; Edson, Y.; Edwin, P. A phytoremediation approach using *Calamagrostis ligulata* and *Juncus imbricatus* in Andean wetlands of Peru. *Environ. Monit. Assess.* **2013**, *185*, 323–334. [CrossRef] [PubMed]
38. Dmuchowski, W.; Gozdowski, D.; Brągoszewska, P.; Baczewska, A.H.; Suwara, I. Phytoremediation of zinc contaminated soils using silver birch (*Betula pendula* Roth). *Ecol. Eng.* **2014**, *71*, 32–35. [CrossRef]
39. Bielecka, A.; Królak, E. Selected features of Canadian goldenrod that predispose the plant to phytoremediation. *Ecol. Eng.* **2019**, *20*, 88–93. [CrossRef]
40. Muthusaravanan, S.; Sivarajasekar, N.; Vivek, J.S.; Paramasivan, T.; Naushad, M.; Prakashmaran, J.; Gayathri, V.; Al-Duaij, O.K. Phytoremediation of heavy metals: Mechanisms, methods and enhancements. *Environ. Chem. Lett.* **2018**, *16*, 1339–1359. [CrossRef]
41. Awa, S.H.; Hadibarata, T. Removal of heavy metals in contaminated soil by phytoremediation mechanism: A review. *Water Air Soil Pollut.* **2020**, *231*, 47. [CrossRef]
42. Pandey, V.C.; Singh, K.; Singh, R.P.; Singh, B. Naturally growing *Saccharum munja* on the fly ash lagoons: A potential ecological engineer for the revegetation and stabilization. *Ecol. Eng.* **2012**, *40*, 95–99. [CrossRef]
43. Pandey, V.C. Suitability of *Ricinus communis* L. cultivation for phytoremediation of fly ash disposal sites. *Ecol. Eng.* **2013**, *57*, 336–341. [CrossRef]
44. Da Conceição Gomes, M.A.; Hauser-Davis, R.A.; de Souza, A.N.; Vitória, A.P. Metal phytoremediation: General strategies. genetically modified plants and applications in metal nanoparticle contamination. *Ecotoxicol. Environ. Saf.* **2016**, *134*, 133–147. [CrossRef]
45. Sarwar, N.; Imran, M.; Shaheen, M.R.; Ishaque, W.; Kamran, M.A.; Matloob, A.; Rehim, A.; Hussain, S. Phytoremediation strategies for soils contaminated with heavy metals: Modifications and future perspectives. *Chemosphere* **2017**, *171*, 710–721. [CrossRef]
46. Liu, L.; Li, W.; Song, W.; Guo, M. Remediation techniques for heavy metal-contaminated soils: Principles and applicability. *Sci. Total Environ.* **2018**, *633*, 206–219. [CrossRef] [PubMed]
47. Maiti, S.K.; Nandhini, S. Bioavailability of metals in fly ash and their bioaccumulation in naturally occurring vegetation: A pilot scale study. *Environ. Monit. Assess.* **2006**, *116*, 263–273. [CrossRef] [PubMed]
48. Desai, M.; Haigh, M.; Walkington, H. Phytoremediation: Metal decontamination of soils after the sequential forestation of former opencast coal land. *Sci. Total Environ.* **2019**, *656*, 670–680. [CrossRef] [PubMed]
49. Sebastiani, L.; Scebba, F.; Tognetti, R. Heavy metal accumulation and growth responses in poplar clones Eridano (*Populus deltoides* × *maximowiczii*) and I-214 (*P.* × *euramericana*) exposed to industrial waste. *Environ. Exp. Bot.* **2004**, *52*, 79–88. [CrossRef]
50. Hassinen, V.; Vallinkoski, V.M.; Issakainen, S.; Tervahauta, A.; Kärenlampi, S.; Servomaa, K. Correlation of foliar MT2b expression with Zn and Zn concentrations in hybrid aspen (*Populus tremula* × *tremuloides*) grown in contaminated soil. *Environ. Pollut.* **2009**, *157*, 922–930. [CrossRef]
51. De Oliveira, V.H.; Tibbett, M. Tolerance, toxicity and transport of Cd and Zn in *Populus trichocarpa*. *Environ. Exp. Bot.* **2018**, *155*, 281–292. [CrossRef]
52. Joshi, R.C. Fly ash-production, variability and possible complete utilization. In Proceedings of the Indian Geotechnical Conference—2010, GEOtrendz, Mumbai, India, 16–18 December 2010; pp. 103–111. Available online: https://gndec.ac.in/~{}igs/ldh/conf/2010/articles/v012.pdf (accessed on 29 April 2022).
53. Armesto, L.; Merino, J.L. Characterization of some coal combustion solid residues. *Fuel* **1999**, *78*, 613–618. [CrossRef]
54. Levandowski, J.; Kalkreuth, W. Chemical and petrographical characterization of feed coal, fly ash and bottom ash from the Figueira Power Plant, Paraná, Brazil. *Int. J. Coal Geol.* **2009**, *77*, 269–281. [CrossRef]
55. Maiti, D.; Pandey, V.C. Metal remediation potential of naturally occurring plants growing on barren fly ash dumps. *Environ. Geochem. Health* **2021**, *43*, 1415–1426. [CrossRef]
56. Pandey, V.C.; Sahu, N.; Singh, D.P. Physiological profiling of invasive plant species for ecological restoration of fly ash deposits. *Urban For. Urban Green.* **2020**, *54*, 126773. [CrossRef]
57. Pandey, V.C. Assisted phytoremediation of fly ash dumps through naturally colonized plants. *Ecol. Eng.* **2015**, *82*, 1–5. [CrossRef]

58. Malhotra, S.; Mishra, V.; Karmakar, S.; Sharma, R.S. Environmental predictors of indole acetic acid producing *Rhizobacteria* at fly ash dumps: Nature-based solution for sustainable restoration. *Front. Environ. Sci.* **2017**, *5*, 59. [CrossRef]
59. Glick, B.R. Using soil bacteria to facilitate phytoremediation. *Biotechnol. Adv.* **2010**, *28*, 367–374. [CrossRef] [PubMed]

Review

Clean Water Production Enhancement through the Integration of Small-Scale Solar Stills with Solar Dish Concentrators (SDCs)—A Review

Mohd Fazly Yusof [1], Mohd Remy Rozainy Mohd Arif Zainol [1,2,*], Andrei Victor Sandu [3,4,5,*], Ali Riahi [1], Nor Azazi Zakaria [1], Syafiq Shaharuddin [1], Mohd Sharizal Abdul Aziz [6], Norazian Mohamed Noor [7], Petrica Vizureanu [3,8], Mohd Hafiz Zawawi [9] and Jazaul Ikhsan [10]

1 River Engineering and Urban Drainage Research Centre (REDAC), Universiti Sains Malaysia, Nibong Tebal 14300, Penang, Malaysia; redac07@usm.my (M.F.Y.); redac_aliriahi@usm.my (A.R.); redac01@usm.my (N.A.Z.); redacsyafiq@usm.my (S.S.)
2 Department of Civil Engineering, Universiti Sains Malaysia, Nibong Tebal 14300, Penang, Malaysia
3 Faculty of Materials Science and Engineering, Gheorghe Asachi Technical University of Iasi, 61 D. Mangeron Blvd., 700050 Iasi, Romania; peviz@tuiasi.ro
4 Romanian Inventors Forum, St. P. Movila 3, 700089 Iasi, Romania
5 National Institute for Research and Development in Environmental Protection INCDPM, Splaiul Independentei 294, 060031 Bucharest, Romania
6 Department of Mechanical Engineering, Universiti Sains Malaysia, Nibong Tebal 14300, Penang, Malaysia; msharizal@usm.my
7 Faculty of Civil Engineering Technology, Universiti Malaysia Perlis, Arau 01000, Perlis, Malaysia; norazian@unimap.edu.my
8 Technical Sciences Academy of Romania, Dacia Blvd 26, 030167 Bucharest, Romania
9 Department of Civil Engineering, Universiti Tenaga Nasional, Kajang 43000, Selangor, Malaysia; Mhafiz@uniten.edu.my
10 Department of Civil Engineering, Universitas Muhammadiyah Yogyakarta, Yogyakarta 55183, Indonesia; jazaul.ikhsan@umy.ac.id
* Correspondence: ceremy@usm.my (M.R.R.M.A.Z.); sav@tuiasi.ro (A.V.S.)

Citation: Yusof, M.F.; Zainol, M.R.R.M.A.; Sandu, A.V.; Riahi, A.; Zakaria, N.A.; Shaharuddin, S.; Aziz, M.S.A.; Mohamed Noor, N.; Vizureanu, P.; Zawawi, M.H.; et al. Clean Water Production Enhancement through the Integration of Small-Scale Solar Stills with Solar Dish Concentrators (SDCs)—A Review. *Sustainability* 2022, 14, 5442. https://doi.org/10.3390/su14095442

Academic Editor: Omar I. Abdul-Aziz

Received: 29 March 2022
Accepted: 29 April 2022
Published: 30 April 2022

Publisher's Note: MDPI stays neutral with regard to jurisdictional claims in published maps and institutional affiliations.

Copyright: © 2022 by the authors. Licensee MDPI, Basel, Switzerland. This article is an open access article distributed under the terms and conditions of the Creative Commons Attribution (CC BY) license (https://creativecommons.org/licenses/by/4.0/).

Abstract: The conventional solar still, as a water treatment technique, has been reported to produce water at a low working temperature where various thermal resistance pathogens could survive in their distillate. In this work, the reviews of previous research on the quality of water produced by passive solar stills and their productivities in initial basin water temperatures were first presented and discussed. The next review discussed some recent studies on the performances of small-scale solar stills integrated with SDCs (with and without sun-tracking systems (STSs)) to observe the operating temperatures from early hours until the end of operations, daily water yield, and cost per liter. Based on these findings, it was revealed that SDCs with STSs indicated an instant increase in the absorber water temperature up to 70 °C at the starting point of the experiments in which this temperature range marked the unbearable survival of the pathogenic organisms and viruses, particularly the recent SARS-CoV-2. Furthermore, disinfection was also observed when the absorbers' water temperature reached beyond the boiling point until the end of operations. This indicates the effectiveness of SDCs with STS in reflecting a large amount of sun's rays and heat to the small-scale absorbers and providing higher operating absorbers temperatures compared to immobile SDCs. Daily productivities and costs per liter of the SDCs with STSs were found to be higher and lower than those of the other previous passive and active solar stills. Therefore, it is recommended that small-scale absorbers integrated with SDCs and STS can be used as a cost-effective and reliable method to produce hygienic pathogen-free water for the communities in remote and rural areas which encounter water scarcity and abundant annual bright sunshine hours.

Keywords: solar distiller; water temperature; pathogens removal; rural areas; sun-tracking system; cost-effective water production; water scarcity; SARS-CoV-2

1. Introduction

The components of the environment and rich sources on Earth [1] which are essential for humans include water [2], plant life [3], and animal life [4]. However, 3% of water sources are fresh water and only 0.30% of that fresh water is surface water which is accessible by humans [5]. Of the total available water on earth, 97% is salty [5–8], whereby its consumption can cause health problems to human beings, such as hypertension, stomach upset, and stroke effects [9]. Freshwater shortages are now affecting more than three billion people around the world as the amount of fresh water available per capita has dropped by a fifth over two decades [10] due to the impacts of climate change, global population growth, and industrial developments which have resulted in increasing freshwater demand across the strips of the globe [10–13]. It has become more difficult than before to obtain safe, potable water for a healthy life [14]. The vast problems caused by the lack of potable water and the transmission of waterborne diseases have been reported in some parts of the world, particularly involving communities in rural areas [15–21], which have generated public health distress. Around one in four people suffered from lack of safely managed drinking water in their homes in 2020 [15]. A report by WHO/UNICEF in 2020 stated that 81% of the world population have access to safe drinking water and about 1.6 billion people will need to survive without hygienic drinking water by 2023 [15]. Based on the surveys from 45 developing countries, 82% of people who lack access to safe clean water reside in rural communities, while the rest live in urban areas; meanwhile, 140 million hours are spent daily by millions of women and children living in villages to collect water from distant and often polluted sources, such as groundwater and natural water resources, for their day-to-day water consumption [15,16]. The drinking of water from the above contaminated water sources pose health risks to the villagers if consumed without any further purification [15–21]. Various types of pathogens—categorized as bacteria, viruses, fungi, and protozoa—which are found in environmental sources (such as water bodies) are risky and could lead to various diseases for living orgams, particularly human bodies. Waterborne diseases—such as cholera, dracunculiasis, infectious hepatitis, typhoid, bacillary dysentery, paratyphoid, colibacillosis, giardiasis, salmonellosis, filariasis, cryptosporidiosis, and amoebiasis—are mostly transmitted in contaminated fresh water due to pathogenic microorganisms in water sources from flowing rivers, groundwater, and runoff water from rooftops, as mostly consumed by the rural communities in Africa, India, and East Asia [20,21]. Skin keratosis is caused by high concentrations of arsenic in groundwater as reported in some rural areas of India, Pakistan, Nepal, and Bangladesh [17,19]. Many more bacterial and viral diseases can be caused by contaminated water [21]. Infection by viruses in untreated water—such as astrovirus, rotavirus, norovirus, and hepatitis A and B viruses—can result in a higher rate of mortality for vulnerable groups such as children, the elderly, and pregnant women, in which 6.3% of all causes of death in the world are attributed to the consumption of unsafe water and inadequate sanitation [22]. It was estimated that every year, one million people in developing countries could die due to contact with waterborne diseases [20–22]. In countries such as India, Bangladesh, Pakistan, and Nepal, there are certain regions where the arsenic concentration is more than 10 times that of the WHO drinking water quality standards of 0.01 mg/L [19]. In India alone, nearly 100 million people are at a health risk due to arsenic-rich water [19]. Some of them are suffering from arsenic-related diseases, such as skin burning and irritation, blackening of skin, paralytic attacks, and early greying of hair. Thus, the quality of drinking water has to be considered when evaluating the role of water in public health [20,22]. Reportedly, there are a number of water treatment methods for Arsenic removal from raw water. Oxidation reduction, adsorption, coagulation and precipitation, ion-exchange, membrane techniques and biological treatments, vapor compression distillation, reverse osmosis, and electro dialysis are some examples of desalination techniques which have been developed and tested by various researchers [17]. However, these methods require electricity and the observation of certain performance parameters on a steady basis; they also produce hazardous waste that restricts the sustained performance of these technologies, especially in rural areas which have restricted access to electricity and skilled manpower [17]. Over 10,000 desalination plants which exist in the world produce about 18.93 million cubic meters of treated water a day [23]. The required electricity for the above desalination methods is generated from coal and fossil fuel combustion as input energy which has contributed to the discharging of hazardous greenhouse gas emissions and increased global temperatures, thus leading to climate change and threatening the lives of millions of people on earth [24–26]. Meanwhile, some countries located in the middle east (e.g., Qatar, Lebanon, Iran, Jordan, Kuwait, Saudi Arabia, and UAE), South-East Asia (e.g., India, Bangladesh, Nepal, and some parts of China), and Africa (e.g., Libya, Egypt, Algeria, Sudan, Mali, Niger, and Nigeria), which are home to nearly a fourth of the world's population, have been facing extremely high levels of water stress and crisis recently [27]. All of these countries enjoy high levels of average daily solar irradiance and receive about 3000 bright sunshine hours annually [28]. Considering this, the application of solar energy to treat the contaminated surface or groundwater and the production of clean potable water can be used as an alternative to aid in eliminating water scarcity and stress issues for the local rural communities of the aforementioned countries.

Solar distillation still utilizing only solar energy is one of the most reported cost-effective, environmentally friendly, and sustainable water treatment technologies to supply high-quality drinking water which are safe from poor water sources for rural, remote, and coastal communities who lack access to other water treatment options [17,21,29–35]. Hence, the main aim of this work is to review the recent studies on performances of solar dish concentrators (SDCs) with different configurations which are integrated with the small-scale conventional solar stills (small absorbers) and a sun-tracking system. Specifically, the objectives of this work are to evaluate the capability of SDCs in the studies by: (1) eliminating the waterborne pathogens, bacteria, as well as SARS-CoV-2 virus from the absorbers water at the initial stages of the experimental work by achieving the initial absorber water temperatures at about 70 °C instantaneously; (2) disinfecting the water absorbers by increasing the water temperature to the boiling point and even at higher rates in continuous durations in the experiments; and (3) enhancing the production of hygienic, pathogen-free, and cost-effective fresh water for communities in remote and rural areas.

2. Performance of Passive Solar Desalination Still for Water Treatment

Solar stills are closed containers with different designs and configurations which are mainly comprised of basin/bed to keep the contaminated water and a transparent cover of the condensation to allow the sun's rays pass through it and heat the basin water [32–36] (Figure 1).

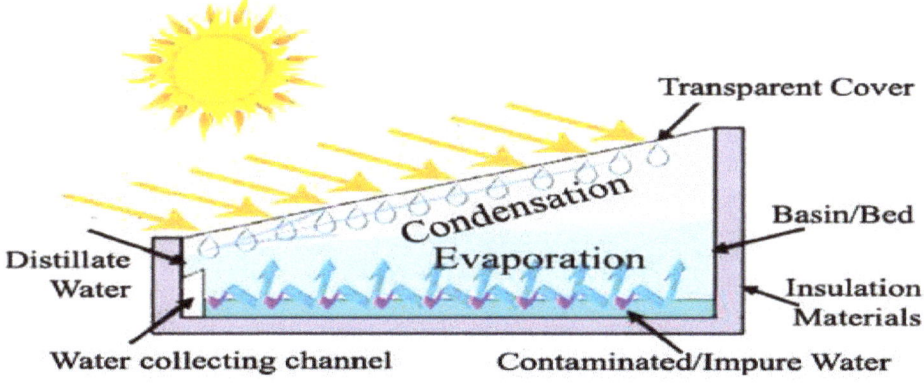

Figure 1. Sketch of a single slope passive solar still [36].

The basic process of the hydrological cycle—namely, evaporation and condensation phenomena—occurs inside a solar still between the surface water of the basin and inner cover of the solar still in order to produce clean water [37]. Solar distillation stills are categorized as passive and active solar stills [38]. The operation of passive solar stills depends greatly on the available direct solar irradiance to heat the basin water, while active solar stills are the similar to passive stills which are incorporated with additional external heat sources and receive direct solar irradiances [30,38–42]. The daily productivity of passive conventional solar stills (CSS) was investigated with different configurations in some countries, such as Malaysia [30,31,43], Saudi Arabia [44], India [45–48], Japan [49], Egypt [50,51], Jordan [52,53], Turkey [54] and Nigeria [21]; these productivity values were generally low—i.e., less than 5 L/m^2—due to failure in obtaining high basin water temperature which resulted in low evaporation rates and thus, low amounts of water production. Several researchers analyzed the quality of water produced by passive solar stills. In 2003, Hanson et al. [29] designed, fabricated, and studied the performance of a passive trapezoidal-shaped single basin single slope solar still in Southern New Mexico, USA in order to evaluate the treatment of samples of local tap water, brackish ground water, geothermal ground water, and diluted raw sewage. As proved in their study, 99% of non-volatile contaminants (such as salinity, total dissolved solids (TDS), total hardness (Caco3), electrical conductivity (EC), nitrate, fluoride), and 99.9% of *E.coli* and fecal coliform bacteria were successfully removed from the studied raw waters using the aforementioned passive solar distiller, and therefore it was concluded that the solar still produced high-quality hygienic drinking water [29]. In another study in Malaysia, lake water samples were treated using two passive glass (GSS) and polythene film (PSS) cover solar stills [30]. It was observed in the study that through both PSS and GSS (Table 1), the quality parameters of pH, TDS, salinity, nitrate, nitrite, iron, turbidity, and EC after the experiment were recorded within the acceptable ranges of WHO standards for drinking water [55]. Hence, the use of PSS was proposed as the economical means of production of healthy potable water for the benefit rural communities [30].

Table 1. Performances of several passive solar stills after the treatments of contaminated surface water [30], groundwater [17], and seawater [31] sample as recommended for the rural community consumption.

Water Quality Parameters	PSS [30]	GSS [30]	SSSB [17]	TrSS [31]	WHO Standards for Drinking Water [55]
pH	6.51	6.53	7.14	7.7	6.5–8.0
Total dissolved solids (TDS) mg/L	95	28	45	7.52	600
Total Arsenic (mg/L)	—	—	≤0.01	—	0.01
Salinity (ppt)	0.1	0	Na	0.006	<0.25
Nitrate (mg/L)	0.6	0.4	0.74	—	<50
Nitrite (mg/L)	0.03	0.01	Na	—	<0.05
Fluoride (mg/L)	—	—	0.02	—	1.5
Chloride (mg/L)	—	—	10.99	—	250
Hardness (mg/L)	—	—	33.81	—	200
Iron (mg/L)	0.03	0.02	0.00	—	0.3
Sulfate (mg/L)	—	—	0.72	—	250
Turbidity (NTU)	1.37	0.92	Na	—	<5
Electrical conductivity (EC) (µS/cm)	52.5	15.66	Na	11.6	<250

In a study, a single-slope single-basin (SSSB) passive solar still was designed and constructed, and then its performance for groundwater treatment was investigated in a rural community area affected by high arsenic levels in India, namely Kaudikasa village [17]. It was perceived in the study that the parameters of pH, TDS, total arsenic, nitrate, fluoride, chloride, hardness, iron, sulphate, and total coliform after conducting the experiment using SSSB [17] conformed with the WHO drinking water guideline ranges [55], as given in Table 1. In another study in Malaysia [31], seawater samples were treated using a low-cost passive triangular solar still (TrSS), and the results showed that the quality parameters of pH, salinity, TDS, and EC were also in compliance with the WHO standards of drinking water [55] as in Table 1. The distillate water produced by the solar distillers is deficient in minerals and fluoride concentration and therefore, some minerals and fluoride salts may be added to the distillate [17] to be in accordance with the current requirements as per drinking water quality standards which stated 1.5 mg/L in WHO, 2008 [55] as the requirement of fluoride so that the produced distilled water can be consumed as potable water without negatively affecting health. However, some recent studies [36,56–61] expressed their concern that working water temperatures in passive solar stills play an important role for the viability of various viruses and pathogens in the distillate due to their transmission through vapor in solar stills. This is because water vapor was observed at an extensive range of temperatures, and solar stills were able to produce distilled water even at low working temperatures [36]. With various modifications of passive solar stills, their maximum water temperature can reach up to 70 °C, and the temperature is considerably higher in active solar stills due to the use of different external heat sources, such as solar collectors, pre-heating, etc. [62]. However, in a study conducted by Parsa et al., the initial working water temperature in the early experimental hours using most passive and active solar stills was usually low, which was observed between 20 °C and 50 °C [61]. In another study conducted by Parsa S.M. [36], most solar stills, including passive and active solar stills [31,39,40,42–44,46–51,53,63–73], had the productivity at low working temperatures. The passive solar stills tested in Malaysia [31,43], Saudi Arabia [44], India [46–48,64,65,70,71,73], and Japan [49], had their initial productivity in basin water temperatures of 32, 35, 37, 35, 34, 33, 49.2, 18, 19.3, 25, 39, and 20 °C, respectively and most the active solar stills investigated in Malaysia [39,40], India [42,63,66,68,71,72], Saudi Arabia [44], Egypt [50,51], Jordan [53], Oman [67], and Iran [69] had their early water production in water temperatures of 47, 49, 25.5, 18.9, 9.25, 25, 26.6, 49, 48, 36, 25, 34.6, and 21.6 °C, respectively (Table 2) [36]. Generally, these results were obtained in the beginning of their experiments at early morning hours, with exposure to the low rate of solar radiation intensity [36]. In one study, the concentration of biological colonies in the distillate water produced by a passive stepped solar still was extremely high [56]; while, in another study, the presence of *E. coli* was noticed in the water produced by a passive plastic type solar still [57]. Another study reported the capability of various pathogens of *E. coli*, *Klebsiella pneumoniae*, and *Enterococcus faecalis* in transferring via vapor in a solar still [58]. The transmission rate of *E. coli* in water temperatures in the 30–35 °C range was found to be higher than *Enterococcus faecalis*, while the transmission rate of *Enterococcus faecalis* was higher than *E.coli* at the 40–45 °C and 50–55 °C temperature ranges [58]. As a thermally resistant pathogen, *Enterococcus faecalis* was able to survive in water with temperature up to 65 °C [58]. It was recommended that exposing all parts of solar stills to sunlight with a high rate of radiation intensity throughout the experiment is also important to prevent the growth of bacteria and pathogens in the produced water by solar stills [58–60]. However, this recommendation is not completely practical due to some parts of solar stills possibly failing to catch the solar intensity in early experimental hours (usually in the morning), and the presence of pathogens in the productivity of solar stills seems to be unavoidable [36].

Table 2. Initial produced water by some passive and active solar stills corresponded to their basin water temperature [36].

Solar Still Type	Modified/Incorporated with	Basin Water Temperature of the Distiller (°C)	Initial Productivity (L/m^2) of the Distiller Corresponded to the Basin Water Temperature	Countries/ Year of Experiment	Ref.
Passive, double slope	Polythene film cover and black painted Perspex sheet basin	32 °C	0.01 L/m^2	Malaysia/2014	[31]
Active, double slope	Photovoltaic modules-AC heater	47 °C	0.138 L/m^2	Malaysia/2016	[39]
Active, double slope	A photovoltaic module-DC heater	49 °C	0.32 L/m^2	Malaysia/2019	[40]
Active, single slope	Hybrid PV/T with cover cooling method	25.5 °C	0.08 L/m^2	India/2018	[42]
Passive, double slope	Black soil heat absorption materials	35 °C	0.048 L/m^2	Malaysia/2015	[43]
Passive, double slope	Black painted basin	37 °C	0.15 L/m^2	Saudi Arabia/2012	[44]
Active, double slope	Two immersed AC water heaters	49 °C	0.50 L/m^2	Saudi Arabia/2012	
Passive, single slope	Fin and sand as heat storing materials	35 °C	0.05 L/m^2	India/2008	[46]
Passive, single slope	Marble pieces in basin	34 °C	0.035 L/m^2	India/2017	[47]
Passive, single slope	Glass cover with 4 mm thickness	33 °C	0.04 L/m^2	India/2016	[48]
Passive, tubular shape	Polythene film cylindrical cover	20 °C	0.02 L/m^2	Japan, 2012	[49]
Passive, single slope		39 °C	0.07 L/m^2	Egypt/2012	[50]
Active, single slope	Vacuum tube collector and stepped basin	48 °C	0.15 L/m^2		
Active, single slope	A photovoltaic module-Rotating shaft	36 °C	0.05 L/m^2	Egypt/2005	[51]
Active, pyramid shape	A photovoltaic module-DC fan	25 °C	0.06 L/m^2	Jordan/2012	[53]
Active, single slope	Hybrid PV/T and flat plate collector (FPC)	18.9 °C	0.08 L/m^2	India/2018	[63]
Passive, single slope	Porous absorber and carbon foam	49.2 °C	0.10 L/m^2	India/2018	[64]
Passive, double slope	Multi wicks heat storage materials	18 °C	0.062 L/m^2	India/2017	[65]
Active, single slope	PV/T	15 °C	0.04 L/m^2	India/2010	[66]
Active, single slope	Inverted absorber	34.6 °C	0.06 L/m^2	Oman/2011	[67]
Active, single slope	Hybrid PV/T and heat exchanger	9.25 °C	0.0014 L/m^2	India/2018	[68]
Active, single slope	Reflectors	21.6 °C	0.0017 L/m^2	Iran/2021	[69]
Passive, single slope		19.3 °C	0.03 L/m^2	India/2006	[70]
Passive, single slope		12.2 °C	0.007 L/m^2	India/2006	[71]
Active, single slope	Flat plate collector (FPC)	25 °C	0.016 L/m^2		
Active, double slope	Flat plate collector (FPC)	26.6 °C	0.032 L/m^2	India/2011	[72]
Passive, single slope	Micro coated and nano-ferric oxide particles in basin	39 °C	0.13 L/m^2	India/2020	[73]

Nowadays, another worldwide concern, as reported by Parsa S.M. [36], was the presence of SARS-CoV-2 in the environment [74–78] which is able to survive in various water bodies with 4 °C temperature, room temperature of 20–25 °C, and hot temperature of 33–37 °C for 14, 7, and 1–2 days, respectively [75]. However, it was also reported that the novel coronavirus is unable to survive more than 30 min at temperatures within the range of

50–70 °C [74,75]. As recently noted, the water temperature in the basin of solar stills is one the most crucial factors affecting the viability of waterborne pathogens and SARS-CoV-2 in basin water, vapor, and distillate of the solar stills [36]. Thus, the risk of transmitting some pathogens and SARS-CoV-2 is higher in the produced water by the solar stills if the productivity occurs at low initial water temperatures, i.e., within the ranges of 20–25 °C and 33–37 °C [36] (Table 2). It is recommended that the best solution for treating water using solar stills and preventing the transmission of pathogens and viruses is by increasing the initial temperatures of water in the basin of solar stills instantaneously to 70 °C, and then to the boiling point (100 °C). Next, the boiling point temperature is maintained until the end of the experiment by integrating the external heat sources (such as external solar heat collectors) to the small-scale conventional passive solar stills (small-scale CSS) called absorbers or (boilers) with low water capacity. This will help to avoid the transmission of waterborne pathogens and the viruses, particularly SARS-CoV-2, in the vapor and solar still productivity during the pandemic. There are two types of conversion modes which are incorporated into the passive solar distillation stills to enhance the water production; the first mode is the solar flat plate collectors' approach and the second mode is the application of solar dish concentrator (SDC) [79]. In one study, the former type was used to increase the solar still basin water temperature up to 100 °C; while the second one—which is composed of an SDC, a focal absorber, and a sunlight tracking system—was used to enhance the freshwater production of the passive solar desalination approaches by increasing the temperatures of their boiler water to more than 100 °C [79]. As reported in the study, the thermal efficiency of the SDC system is higher than the efficiency of the flat plate collector (FPC) system as the receiver area of the SDC losses less heat temperature compared to the area of the FPC [79]. As mentioned previously, due to the recent concerns regarding the existence of waterborne pathogens and SARS-CoV-2 virus in the solar still vapors and distillate which are produced at low basin water temperature [36], one of the best alternatives is through the immediate increase in the initial basin water temperature of small-scale CSS or absorber above 70 °C in the early stages of the experiment. Other than that, it is also recommended that the absorber water temperature can be increased to be higher than the boiling point in order to remove the bacteria and viruses in the boiler. These methods can be employed by integrating the CSS with the SDC and the sun-tracking system. As reported by several studies, a disinfection process occurs during the continuous boiling process using the SDC system with the explosion of the solution to the solar radiation ultraviolet waves [80–83]. The solar thermal parabolic dish concentrators were also noted as one of the most cost-effective paths for renewable energy to displace fossil fuels [84] which can be employed in producing freshwater for the rural communities. The reason for incorporating a sun-tracking system to the SDC was to increase the solar energy density at the focal point of the dish by reflecting most of the sunlight onto the solar still through absorber. This approach could lead towards achieving higher water temperatures [80–82,85,86], compared to the immobile solar reflectors which was reported to be unable to increase the basin water temperature up to the boiling point and thus increase the distillate yield significantly [87,88]. To ensure the specified accuracy and smoothness of the SDC surface, Sinitsyn, S. et al., 2020 proposed a method of fan-shaped geometric parquetting of the surface of a parabolic concentrator [89] and Panchenko, V., 2021 stated that the overall efficiency of a solar module increased and the uniform illumination was provided by using a composite concentrator (SDC) by concentrating the solar radiation on the surface of the module [90].

2.1. Description of Solar Dish Concentrator (SDC)

Generally, an SDC is a parabolic-shaped device which is covered with mirror strips to reflect and focus on the radiation of the sun towards a receiver or absorber mounted on the focal point of the parabolic dish, as depicted in Figure 2 [82]. A dual-axis direct current (DC) sun tracker system is required to maintain the orientation of the dish towards the sun [79–82,85,86]. As shown in Figure 2, the parabolic dish is characterized by the parameters of an aperture area, acceptance angle, rim angle, focal length, intercept factor, and the absorber area [91]. The curvature area of the dish that receives the sun's rays and reflects them to the absorber is called the aperture area (Figure 2). The acceptance angle (θ_{lim}) is defined as the angular limit to which the direction of the sun passes from point A to point B, and its rays deviate from the curvature, reflect on, and still touch the bottom of the absorber that is mounted on the focal point (Figure 2), where point A and point B symbolize the position of the sun in the sky. In order to use the sun-tracking system, the acceptance angle from point A to point B must always be equal to 0° [82]. The rim angle (φ_r) is the angle between the edge of the dish and the center of dish curvature from the focal point (absorber) (Figure 2); meanwhile, the intercept factor (γ) is defined as the ratio of the solar energy intercepted (cut off) by the absorber to the total energy that is reflected by the parabola of the SDC [82].

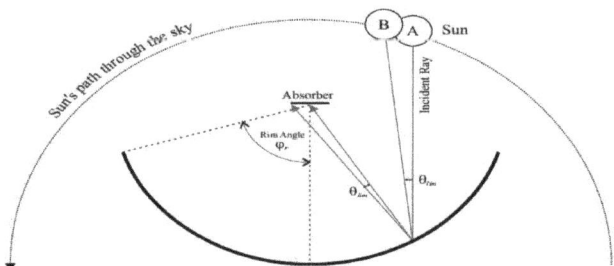

Figure 2. Acceptance and rim angles of an SDC [82].

The mathematical general equation for calculating the solar dish/parabolic concentrator (SDC) profile and the focal length (*f*) of the parabola of SDC was described by Johnston et al., 2003 [92] and Chaichan M.T. and Kazem H.A., 2015 [93] which is shown in Equation (1) when the coordinates of the parabola vertex (the point at the intersection of the parabola and its line of symmetry) is equal to (0, 0) (Figure 3).

$$y = x^2/(4f) \tag{1}$$

where *y* and *x* are the depth and radius of the SDC parabola, respectively, and *f* is the parabolic focal length (Figure 3) [92,93].

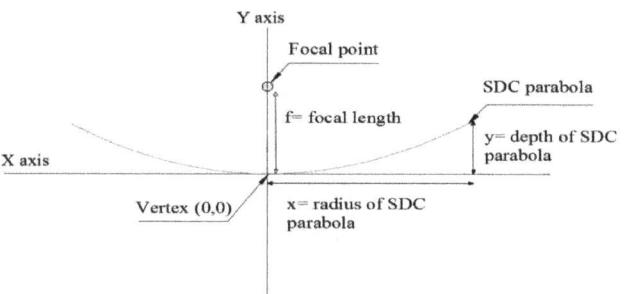

Figure 3. Sketch of an SDC parabola [92,93].

Maximized basin water temperature up to the boiling point and minimized thermal losses are among the advantages of SDCs compared to other heat sources which are coupled with passive solar distillation stills [94,95]. To achieve the above objectives, the quality of the SDC depends heavily on the quality of the reflecting surface; it is recommended that the surface be made using aluminum and stainless steel sheets for ensuring cost effectiveness and durability, as well as accuracy of the machining surface [94,95]. The solar still basin should be designed with small surfaces to ease its mounting at the focal point of the SDC for absorbing most of the reflected sunlight [79–82,85,86,94–96] as received from the optical concentration from tracking of the sun. As stated in a study, the typical temperature of the small-scale absorber integrated with the SDC and a dual-axis sun-tracking system ranges from 100 °C to 1500 °C [96].

2.2. Recent Findings on SDCs Integrated with Solar Stills

2.2.1. SDCs Integrated with Solar Stills and the Sun-Tracking System (STS)

The design and installation of an SDC that is integrated with a mini single-slope, air-tight solar still is called a 'modified receiver' (boiler) (Figure 4). This approach was presented and tested in a study [80] for the purpose of brackish water desalination under the climate in Egypt, in which the SDC performance was compared experimentally with the performance of a simple CSS. The dish-shaped concentrator (made of aluminum as a point-focus collector with an aperture diameter, depth, and focal length of 100, 20, and 40 cm, respectively) was selected and covered with highly reflective glass mirror strips (with 0.004 m of thickness) to reflect the intensity of the incoming solar insolation into the boiler located at the dish focal point. A tracking system was applied to the SDC to track the sun on two axes by using a 36 VDC tracking motor to move the SDC into the calculated positions. This was carried out throughout the day to maintain the focus of the sun's rays to the boiler in improving its water temperature, thermal efficiency, and distillate yield (Figure 4) [80]. The whole tracking system was powered by a 15 W amorphous silicon solar photovoltaic module, charge controller, battery, and inverter (Figure 4) [80]. Brackish water was preheated by a black hose which was exposed to solar irradiation throughout the day and supplied to both trapezoidal-shaped boiler and CSS. The boiler with a basin surface area of 0.046 m^2 had small dimensions (with the length, width, height from back, and front sides of 27, 16, 17, and 12 cm, respectively) to admit most of the reflected sunlight. The trapezoidal-shaped CSS with the basin area of 0.5 m^2 (with the high side and low side basin walls of 44 and 15 cm, respectively) was covered with a 30° inclined glass sheet (Figure 4). The experiments were conducted for nine hours, from 9:00 a.m. to 6:00 p.m. All of the boiler surfaces which were exposed to the sunlight (mainly UV) received some radiation from above as direct solar radiation, and most radiation of the sun's rays was reflected from the SDC surface to the sides and bottom surface of the boiler.

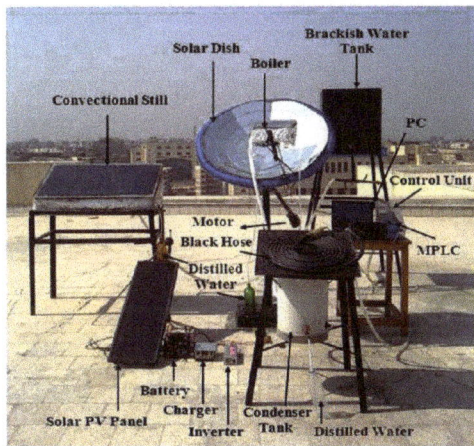

Figure 4. Experimental set-up photograph of the CSS and the solar still coupled with an SDC under Egypt's climate [80].

This process appeared to prevent the growth of bacteria and waterborne pathogens that could cause contamination of the distilled water [58–60] throughout the experiment.

However, this process is not completely practical in any other types of passive and active solar stills (CSS) which involve early experimental hours (usually in the morning) (Figure 5) [36]. The study results also showed that the water temperatures in the boiler were approximately 36–42 °C higher than the water temperatures in the basin of the CSS throughout the experiment (Figure 5). This occurred due to the additional concentrated sun's rays hitting the sides and bottom of the boiler as received from the SDC, which were about 230 w/m^2 at 9:00 a.m. in the first hour of the experiment, 100 w/m^2 at 12:00 p.m., and 600 w/m^2 at 5:00 p.m. in the day (Figure 5); it was also because the direct solar radiation intensity received from the top condensing surface of the boiler [80]. Meanwhile, the CSS only received direct intensities of solar radiation from its condensing cover which is located at the top (Figure 4). As can be observed from Figure 5, the starting (initial) brackish water temperature of the boiler, which was integrated with the SDC without preheating by the black hose, immediately reached 70 °C due to absorbing additional rate of solar intensities as reflected from the SDC. This rate was much higher than the temperature of the initial basin water of the CSS (which was recorded at about 45 °C) and solar stills in other studies [30,31,42–54,63–73], as presented in Table 2. The boiler water temperature increased to above 80 °C within an hour, reached approximately 105 °C at 11:00 a.m., and then continued to generate steam with a temperature of 105 °C for a three-and-a-half hour period until 2:30 p.m. Meanwhile, the maximum basin water temperature of the CSS was recorded at 63 °C at 12:00 p.m. (Figure 5) under similar climatic conditions. The produced steam from the boiler moved to a cylindrical tank, referred to as the 'condenser unit', that was filled with cold water which would be converted to fresh water droplets and collected in a graded container (Figure 4) [80].

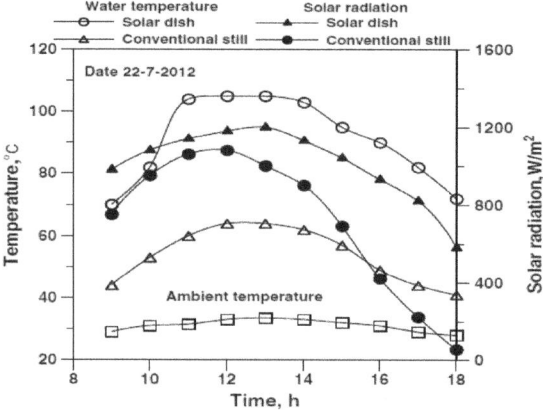

Figure 5. Diurnal hourly variations of direct and reflected intensities of solar radiation and water temperatures in the boiler and CSS and ambient temperature without pre-heating technique [80].

The study results showed that by integrating the SDC and small-scale solar still (boiler), the water temperature in the boiler drastically increased to above 70, 80, and 105 °C in the early experimental hours. This approach is seen as the most effective in removing any available waterborne pathogens, bacteria, and viruses—particularly SARS-CoV-2—from the boiler water and produced vapors while preventing the transmission of those impurities into the distillate (Figure 5). This is because the range of temperatures from 50 °C to 70 °C is the limit of the viability of waterborne pathogens and novel coronavirus, as recommended by several studies [74,75]. It was also stated in the study that the increased temperature of brackish water in the boiler by the SDC enhanced the amount of daily fresh water production, ranging from 0.65 and 0.55 L/h at 11:00 a.m. and 4:00 p.m., respectively (Figure 6) [80]. The boiler with SDC produced maximum water of 6.7 L/m^2 in a nine-hour period, while the CSS produced only 1.5 L/0.5 m^2 within the same period [80].

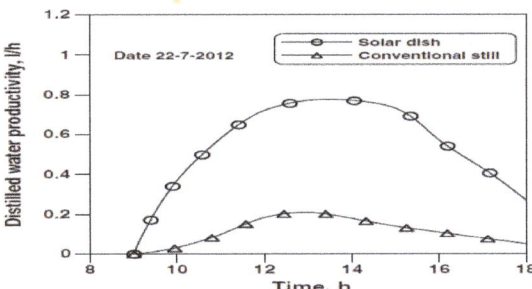

Figure 6. Hourly values of distilled water production of CSS and the boiler coupled with SDC without the preheating method [80].

In another study, a stand-alone point-focus parabolic solar still (PPSS) coupled with the sun-tracking system and a small-scale passive solar still as absorber or (boiler), was designed and fabricated for purification of the seawater and brackish water in Tehran, Iran (Figure 7) [81]. Salt with different masses (from 10 g to 40 g, with 5 g intervals) was dissolved into each kg of water sample before being fed into the boiler [81].

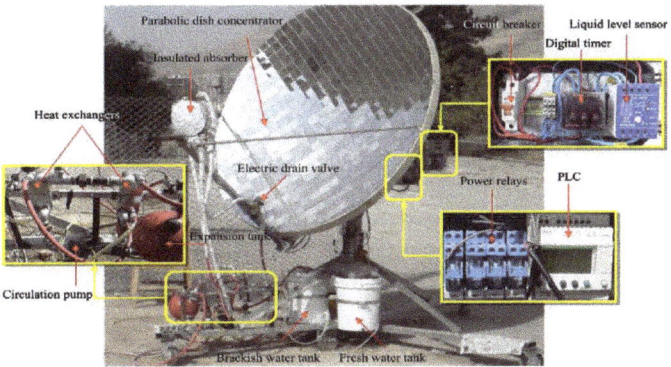

Figure 7. Photograph of the experimental set-up of an auto sun-tracking system SDC-solar still in Tehran, Iran [81].

The above-developed stand-alone system (PPSS) is comprised of several items, including a parabolic dish concentrator, a boiler mounted at the focal point of the dish collector, two plate heat exchangers (to condense the steam generated in the boiler and increase the brackish water temperature before entering the boiler, or the preheating process) and a brackish water level controller in the absorber (Figure 7) [81]. A programmable logic controller (PLC) was used to control the tracking motors to drive the SDC in two axes for tracking the sun based on the calculated positions [81] (Figure 7). The boiler had a receiving surface, which was as small as 0.031 m^2 and made of CK45 steel alloy, and the black chrome was coated at its bottom side to increase the absorptivity of the reflected sun's radiations [81]. The reflective area of the SDC was 3.142 m^2 with the aperture diameter of 2 m and the focal length of 0.693 m, and it was covered with silver-backed glass segments of 0.002 m thickness [81].

The study results showed that the initial boiler wall temperature (T_s) increased abruptly to about 70 °C (Figure 8) due to the reflection of solar radiation into the small-scale boiler [81]. All parts of the boiler were exposed to the sunlight in the early hours of the experiment, and the SDC-boiler system performed to produce water at temperatures higher than 70 °C, which is operative to prevent transmission of pathogens, bacteria, and viruses from the brackish water into the vapors and condensed vapors [74,75]. The boiler wall temperature increased drastically from 70 °C to about 100 °C in a short period of 30 min after the experiment started at 8:00 a.m. (Figure 8). Then it was maintained to reach above 100 °C (boiling point) for the remainder of the seven-hour experiment.

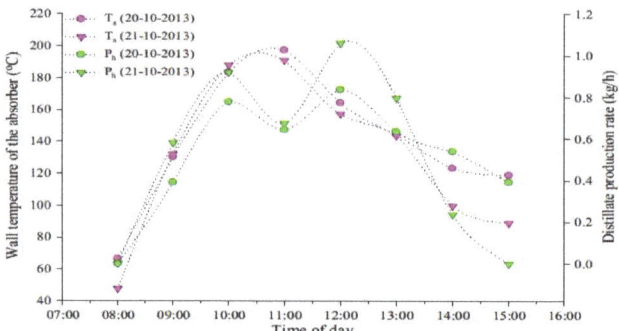

Figure 8. Hourly variations of temperature of absorber/or boiler wall of the PPSS versus the distillate production of the PPSS [81].

The maximum total daily water production (P_d) was reported on 18 and 22 October 2014 with values of 5.02 and 5.11 kg per 7 h, respectively. Meanwhile, the PPSS system was exposed to higher average solar radiation intensities ((I_b)$_{ave}$, more than 630 W/m^2) in these two days, causing the boiler wall temperatures (T_s)$_{ave}$ to reach maximum average values of about 140 °C and 150 °C, respectively. The highest average daily efficiency was reported as 36.7% on 22 October 2014 due to the highest average solar insolation, average absorber wall temperature, and total daily productivity [81]. However, it was reported that the average temperatures of air (T_{air})$_{ave}$, wind speed (V_w)$_{ave}$, and the salinity rates fed in water samples did not affect the daily water production considerably [81]. Water quality parameters of total dissolved solids (TDS) and electrical conductivity (EC) were also measured for feeding salt water into the boiler and discharging brine and distilled water from the boiler after the desalination process for the seven experimental days. As reported, the values of TDS and EC ranges for the distilled water produced by PPSS were the lowest and fell within the acceptable ranges of WHO standards for drinking water purposes [81]. The annual water production of the proposed PPSS system was calculated and stated as 2422.40 kg, while the cost of 1 kg of distilled freshwater produced by the system with an SDC coupled to a boiler was analyzed and reported as USD 0.012; under the Tehran climate, this cost was stated as sufficiently low and cost-effective for rural householders [81]. It was recommended that the photovoltaic modules can be employed as a useful alternative to supply power for the electrical components of the PPSS system, instead of consuming electricity directly in order to reduce the direct electricity consumption per kilogram as well as the operating costs of USD 6187.40 per year for the production of freshwater [81].

In another theoretical and experimental work [82], an SDC which was made from recyclable materials and coupled with a sun-tracking system and a boiler (evaporator) were designed, installed, and experimented for ground water and sea water desalination under the Brazil climate, as depicted in Figure 9. A recycled satellite dish antenna made from galvanized steel with two different aperture diameters (height of 68 cm and width of 62 cm) was selected, mirrored via an electrostatic chroming method, and then used as an SDC in the study [82].

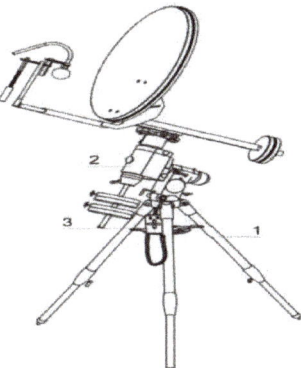

Figure 9. Sketch of a solar desalination unit of SDC coupled with a sun-tracking system [82].

Based on the experiment results as shown in Figure 10, an intermediate focal point in a focal region of the SDC which was determined between two different focus points of the reflected sun's rays was achieved at the best focal length of 51.5 cm [82]. A two-axis sun-tracking system was mounted on a steel tripod (1) and powered by two motors (2) which rotated the SDC in 64 steps per revolution with 1.8° in each step programmed through the control (3) (Figure 9) [82].

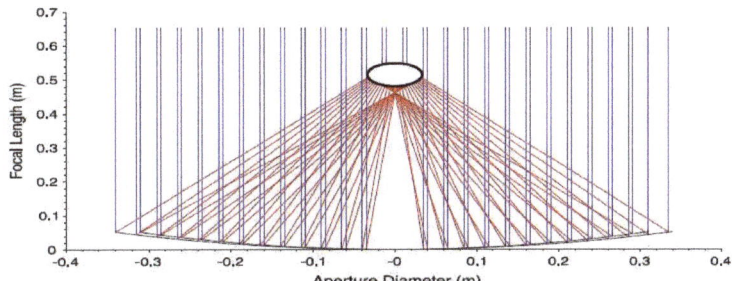

Figure 10. Simulation of the sun's rays (colored lines) as reflected from the two curvatures of the SDC had different diameters from the absorber which is located at the focal length of 51.5 cm [82].

The study's experiments were conducted from 9:00 a.m. to 4:00 p.m. for two months (September and October) [82]. As illustrated in Figure 11, a borosilicate glass sphere-shaped absorber called an evaporator absorber with the area of 0.1182 m^2 and the storage of 100 mL was filled with crushed basalt and coated with a matte black paint mounted at the focal point of the SDC. Samples of ground water and sea water with similar sea salt concentrations from 0% to 4% were pumped into the absorber from the storage tank (1). Next, the sun's rays' reflections were focused onto the sphere-shaped boiler to heat the sea water. Then, water vapor from the boiler passed through the copper tube with a length of 30 cm for the first phase of the condensing process (2) and was directed from a 1.5 m silicon tube for the second phase of condensation method (3) into the graduated container to store the produced water (4) [82].

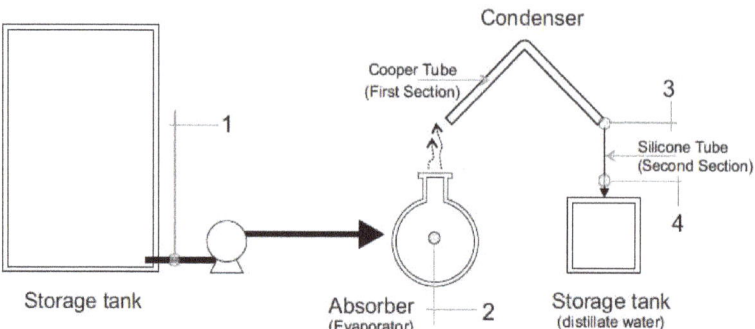

Figure 11. Processes of pumping brackish water into the absorber (1), evaporated from the absorber (2), and condensed in copper (3) and silicon (4) tubes [82].

As shown in Figure 12, two disk-shaped aluminum specimens, Specimen 1 and Specimen 2 (painted matte black with the effective areas of 0.1611, 0.1108 m^2, respectively) which were located at the focal region of the SDC acted as solar radiation absorbers and were tested theoretically and experimentally to determine their dynamic heating temperatures [82]. The intercept factors (γ) of Specimens 1 and 2 were experimentally investigated, analyzed, and recorded at 48.64 and 33.45 % respectively, which indicated that these factors were dependent on their diameters [82].

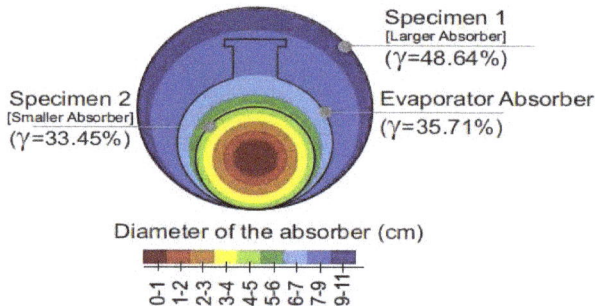

Figure 12. Relationship between absorbers' diameters and their intercept factors [82].

The temperature of the smaller absorber (Specimen 2) reached the maximum value of 319 °C in 840 s which was maintained until 1800 s, while the larger absorber (Specimen 1) experienced the maximum temperature of 198 °C at 1800 s which lasted until the end of experiment, i.e., at 3500 s. These results are shown in Figure 13a,b) [82]. It was also reported that the average boiling point temperature of the third absorber (called the 'evaporator absorber', with an optical efficiency of 0.273 and intercept factor of 35.71%) increased from 98.10 °C without sea salt concentration to 99.66 °C with 4% of salt concentration [82] during the desalination experimental works. This result indicated that a disinfection process occurs during the continuous boiling process with the explosion of the solution to the solar radiation ultraviolet waves [83]. It was observed during the study's experiments that all the three absorbers received ultraviolet waves (UV) of the sun's rays from their top sides and as reflected from the parabola of the SDC [82]. Thus, the reviews have proven the feasibility of using SDCs coupled with smaller absorbers of Specimen 2 and evaporator with the sun-tracking system in the study [82] for removing bacteria, waterborne pathogens, and viruses since the high initial temperatures of the absorber water were achieved. The highest yield of 4.95 kg/m^2 day of distilled water was attained under the average solar irradiances of 791 W/m^2 without adding salt in the sample [82].

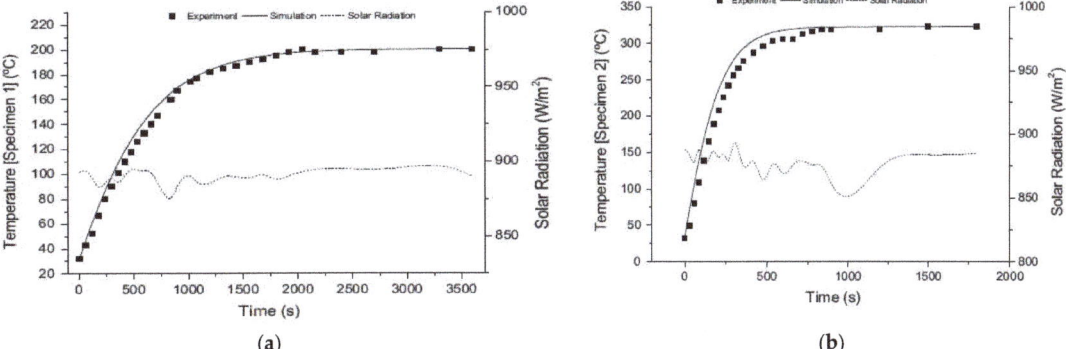

Figure 13. Experimental and simulated dynamic heating of Specimens 1 (**a**) and 2 (**b**) versus the values of solar radiation [82].

In a study conducted by Chaichan M.T. and Kazem H.A. in Baghdad, a solar distiller (absorber/receiver) was integrated with an SDC to heat the saline water (Figure 14) [93]. Then, hot water was transferred to a conical distiller by a heat exchanger to produce distilled water. The SDC had an aperture diameter of 1.5 m and a depth of 23 cm and the conical distiller was layered with paraffin wax, called 'PCM', as thermal energy storage material to expand the distillation process after daytime [93]. Aluminum foil was adhered to the parabola dish surface for reflecting the sunlight to the absorber.

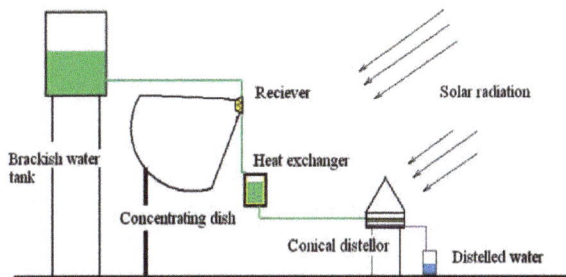

Figure 14. Sketch of the SDC with STS integrated with an absorber (receiver) with a conical distiller with PCM [93].

The experiments conducted for four cases consisted of the SDC without STS and PCM as Case 1, the SDC with STS and without PCM as Case 2, the SDC without STS and with PCM as Case 3, and the SDC with STS and PCM as Case 4. Using PCM with STS (Case 4) gave the highest temperatures compared to other three cases, especially for the period after 2:00 p.m. (Figure 15). However, the obtained temperature for Case 4 did not reach the boiling point of brackish water as the reflecting layer adhered to the surface of SDC was made of aluminum foils and had lower sun's rays reflectivity compared to the mirror. It can be seen that the water temperatures of the brackish water reached beyond 65 °C at about 2:00 p.m., and reduced significantly after 2:00 p.m. following the decrease in solar radiation intensity. Although the initial working temperatures of the absorber in Case 4 were between 10 °C and 40 °C in the early hours of the experiment (Figure 15). However, in the results obtained from the experiments conducted in [80–82], the initial temperature of the absorber was above 65 °C and it increased drastically beyond the boiling point immediately in a short period of time due to using glass mirror as the covered layer of the SDC surface. This can be resulted from covering the layer of the surface of the SDC with aluminum foil which has a lower solar radiation reflectivity as compared to mirror strips. Thus, it seems that the SDC layered with aluminum foil is unable to increase the absorber water temperature considerably (Figure 15) in order to remove bacteria, waterborne pathogens, and viruses due to the resulting low initial water temperatures in the absorber. Thus, it can be seen from the above results that the reflectivity of the cover layer of the SDC surface has a vital role in increasing the initial temperature of the brackish water in the absorber significantly.

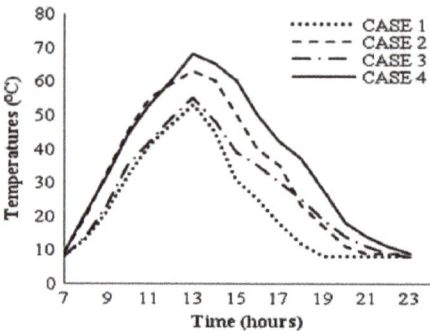

Figure 15. Variations of brackish water temperature versus time for the four studied cases [93].

In another study conducted by Bahrami et al., 2019 in Yasouj University, Iran, an SDC with an aperture diameter of 2.0 m integrated with an STS to reflect the solar radiation into an evaporator tank mounted on its focal point with a focal length of 1.4 m was designed, installed, and tested to desalinate saltwater (Figure 16) [97]. The evaporator had a base area of 0.2×0.2 m and saltwater in the range of 1.0 to 10 kg was fed into the evaporator during the experiment and maintained with the use of a float level controller (Figure 16).

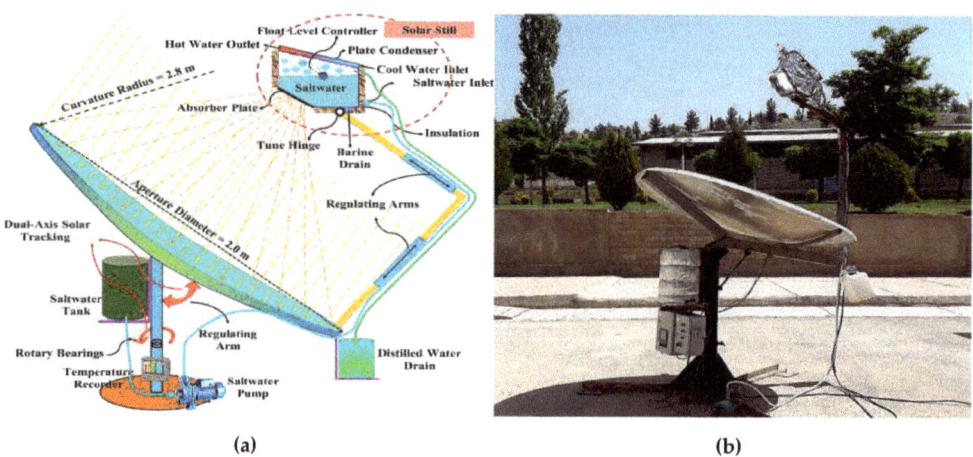

Figure 16. (a) Detailed sketch and (b) photograph of the experimental set up of the SDC and evaporator performed in Iran [97].

It has been stated by Bahrami et al. that the total amount of the produced distilled water increased from 11.5 to 50 kg by increasing the aperture diameter of the SDC from 1.5 to 3.0 m, respectively, and it increased twice while the optical efficiency of SDC increased from 0.5 to 0.8. The amount of produced water was also increased by more than double when the reflectivity of the evaporator base decreased from 0.7 to 0.4 [97]. They have also reported that the saltwater in the evaporator boils at earlier time for an SDC with larger aperture diameter. An SDC with an aperture diameter of 3.0 m was able to boil 8 kg of saltwater with a salinity rate of 30 (g salt/kg water) after about 20 min, while this took about 40 min for an SDC with a diameter of 2.0 m [97]. As depicted in Figure 17, the SDC with a diameter of 2.0 m was reported to boil 6.15 kg of saltwater with a salinity rate of 20 (g/kg) in the evaporator in a period of 1.0 h when the distillation process started from 11:40 a.m. and maintained the boiling point until 3:30 p.m. This highlighted that an SDC with larger aperture diameter [97] and smaller absorber area [82] is capable of reflecting more of the sun's rays to the evaporator (absorber) to reach the highest initial temperature and the boiling point in a shorter period.

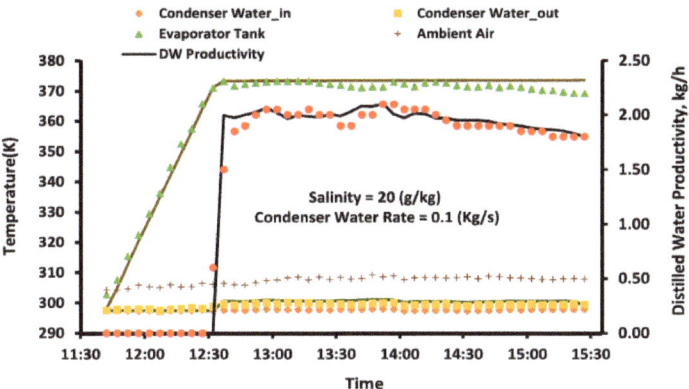

Figure 17. Hourly variation of saltwater temperature in the evaporator tank and the distilled water production of the SDC with aperture diameter of 2.0 m [97].

2.2.2. SDCs Integrated with Solar Stills without the Sun-Tracking System (STS)

In India, several studies investigated the performance of passive solar stills heated by SDCs and additional phase-change materials (PCM) in the still's basin using the cover cooling techniques, without employing the sun-tracking system [98,99]. In one of the studies [98], two passive single-slope solar stills (SSSS) were designed and fabricated, whereby each was mounted on a focal point of a fixed SDC (Figure 18) and stored the heat at their basins using six PCM copper balls filled with paraffin wax (Figure 19a,b); meanwhile, the cold water flow technique was employed at the top cover of one of the solar stills to improve the condensation rate. A black painted hemispherical copper bowl (with a diameter of 0.22 m and a thickness of 4 mm) was separately attached to each basin bottom of the passive SSSS mounted on the focal point of the SDC, which acted as receivers of the sun's rays' reflections to heat the basins water. Six hollow copper balls (each with a thickness of 1.2 mm, as in Figure 19a,b) filled with paraffin wax were used in the absorber of each solar still as the PCM. The balls acted as a heat source for the absorber water to maintain its temperature during the afternoon—i.e., when the solar irradiances started to decrease—and then continued to produce fresh water after sunset [98].

The performance of each solar still was strongly dependent on the intensities of solar absorption by the hemispherical copper bowl absorber from the concentrator, and the PCM balls located in the basin [98]. The temperatures of initial basin water temperatures in the early hours of the experiments with the solar stills with PCM and SDC using the top cover cooling techniques (with water flow rates of 40, 50, 60, 80, and 100 mL/min) were observed at 40, 43, 47, 47, and 48 °C at 9:00 a.m. and 56, 56, 56, 57, and 56 °C at 10:00 a.m. respectively; meanwhile, the temperatures recorded were 43 °C and 56 °C at 9:00 and 10:00 a.m., respectively, for the experiments without any water flow on the top cover [98].

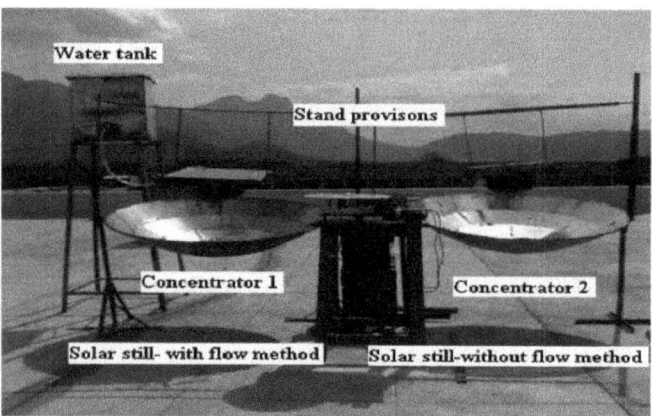

Figure 18. Photograph of two solar stills coupled two SDCs, with and without the cover cooling method [98].

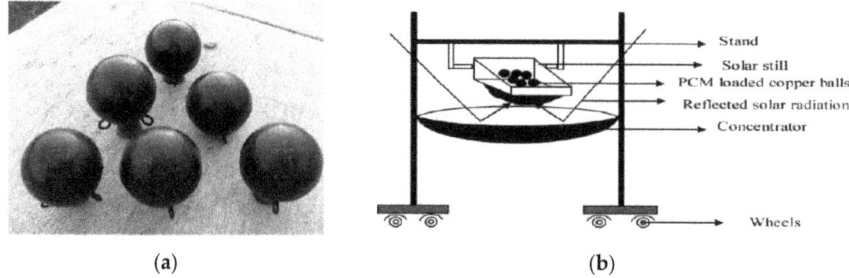

(a)   (b)

Figure 19. (a) Photograph of PCM copper balls used in a hemispherical SSSS; (b) sketch of a hemispherical SSSS with PCM balls in bowl-shaped copper basin while receiving the sun's rays from a fixed SDC [98].

As depicted in Figure 18, most parts of the solar still in the study included the top cover, basin bottom, and the sides which were exposed to the UV of solar radiation in the experiment from morning to evening [98], ranging from 580 to 1050 W/m^2. However, flowing water on the top cover and water droplets on the inner side of the solar still's cover reduced the inputs of the solar radiation to the basin water. Furthermore, as can be observed, there was a lack of coating of the mirrored layer on the SDC surface and no system to track the directions of the SDC and solar still towards the sun and to use the solar still to absorb the reflected sun's rays at a larger scale. As stated by Arunkumar et al., 2015, there were lower initial (ranging from 40 °C to 56 °C) and maximum (ranging from 92 °C to 88 °C) basin water temperatures and lower total yield (ranging from 3.557 to 3.80 L/m^2.day) in the solar stills throughout the experiment [98], compared to the use of solar stills with SDC, sun tracker system, and mirrored surfaces in other studies [80–82]. Meanwhile, as noted in other studies [36,74,75], it seems infeasible to produce water under low basin water temperatures with the use of solar stills coupled SDC and without the sun-tracking system [98], particularly in terms of removing bacteria, waterborne pathogens, and viruses due to the resulting low initial water temperatures (ranging 40 °C to 56 °C) in the basins of solar stills.

Another experimental study in India designed and fabricated a triple-basin solar distiller (TBSS) mounted on a focal point of an SDC [99]. Without engaging a sun-tracking system, it was heated by heat storing materials comprised of four triangular hollow fins filled with river sand (RS) and charcoal (CHAR) in the basins of the distiller which were exposed to the direct solar irradiances [99]. As depicted in Figure 20, a cover cooling (CC) approach using water with different flow rates (20 to 40 mL/s with the intervals of 5 mL/s) was also employed to decrease the still cover temperature and increase the condensation rate [99].

Figure 20. Photograph of a triple basin solar distillation still coupled with an immobile SDC [99].

As shown in Figure 21, the TBSS performed as an absorber whereby its three basins consisted of three basins (lower, middle, and upper basins) acting as an evaporator in the study [99]. Meanwhile, a trapezoidal-shaped glass casing (made of 4 mm thick window glass) was used as the condenser cover and placed at the top of the SDC (with a focal length of 50 cm) with no sun-tracking system to absorb the reflected sun's rays and direct solar intensities. The SDC had a diameter of 1.25 m and was made from a polished aluminum sheet with a thickness of 1 mm. The TBSS evaporator had an overall size of 0.3 × 0.36 m with a height of 0.33 m, while the three basins had a vertical gap of 0.12 m from each other to allow the water vapor to be directed into the inner surface of the condensing cover, as illustrated in Figure 21 [99]. The TBSS cover was constructed with the size of 0.4 × 0.46 m^2, heights of 0.4 m and 0.47 m at two different sides, and a 10° incline at the top. A plastic pipe with a diameter of 0.032 m and length of 0.46 m was punctured at regular intervals and then installed at the top of the outer surface of the condensing cover in order to cool the cover and maintain a uniform flow of water that was pumped over the outer glass of the condensing cover surface [99].

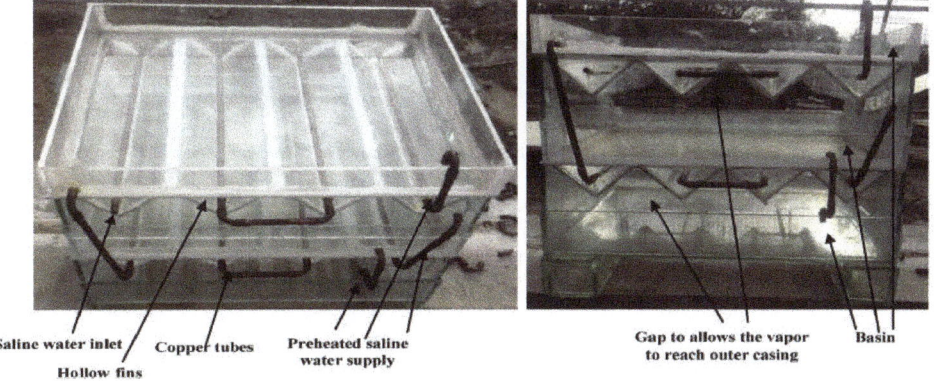

Figure 21. Photograph of a triple basin with triangular hollow fins [99].

The water temperatures in solar still basins of the TBSS, which was filled with charcoal and coupled with SDC without the sun-tracking system, were found to be 36 °C and

57 °C at the early experimental hours of experiment—i.e., at 9:00 a.m. and 10:00 a.m., respectively. As a result, about 0.30 kg/m² water was produced in the first hour of the experiment, as illustrated in Figure 22 [99]. These values are within the critical ranges for the transmission of pathogens and viruses in the produced water, as reported by several studies [36,74,75]. However, all parts of the TBSS were exposed directly to the reflected UV radiation of the sun. Hence, the competency of the TBSS which was coupled with SDC without a sun-tracking system as used in the study [99] seems to be impractical to remove bacteria, waterborne pathogens, and viruses. This was due to the water production at low basin water temperatures—i.e., ranged between 36 °C to 57 °C—as stated in other studies [36,74,75].

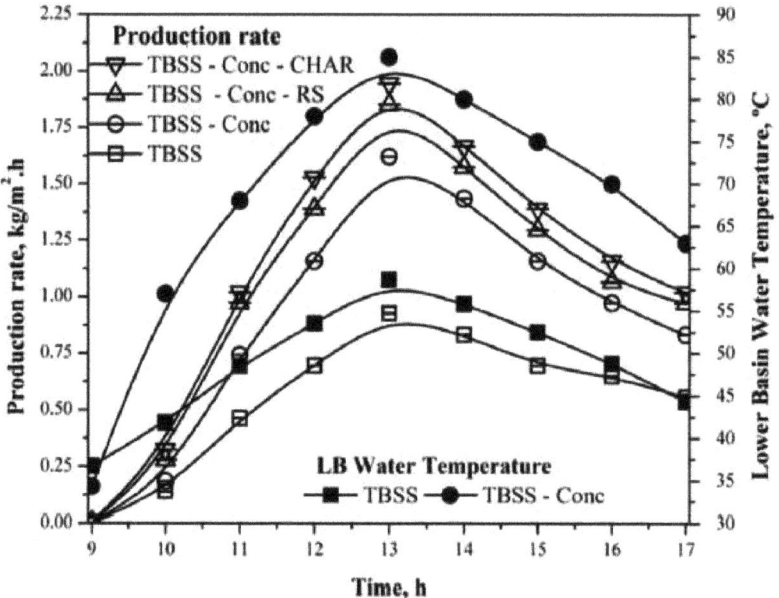

Figure 22. Hourly water temperatures and productivities of the TBSS (filled with and without heat storage materials in the basins) were affected by the use of SDC [99].

2.3. Cost Per Liter (USD) of Small-Scale Passive Solar Stills (Absorbers) Integrated with SDCs

In order to evaluate the economic benefits of passive solar stills (absorbers) integrated with SDCs for the remote and rural communities, it is essential to consider the cost per liter of the SDC distillation systems and their comparison against other passive and active solar stills. Previous studies revealed that the cost per liter (USD) of the small-scale solar stills (absorbers) coupled with SDCs and the sun-tracking system were USD 0.028 and 0.012 [80,81] and without the sun-tracking system were USD 0.0085 and 0.084 [98,99], respectively. As indicated in Table 3, these values were lower than the cost per liter of some conventional passive, and also active, solar stills used in several other studies [40,46,100–105]. Subsequently, the maximum water yields of the solar stills with both SDC and the sun-tracking system [80] and without the tracking system [99] were found to be higher than the maximum water production of passive and active solar stills tested in other studies [40,46,100–105] (Table 3).

Table 3. Cost per liter (USD) and maximum daily yield of the passive solar absorbers coupled with SDCs and the sun-tracking system in comparison with some other passive and active solar stills.

Types of Solar Still	Maximum Daily Water Production (L/m^2)	Cost per Liter (USD)
CSS with SDC with sun-tracking system and vapor condensing technique, Egypt [80]	6.70	0.028
CSS with SDC with sun-tracking system (PPSS) and vapor condensing method, Iran [81]	3.56	0.012
CSS with SDC with sun-tracking system and water heater/PV modules, Egypt [106]	13.63	0.25
SDC integrated with an evaporator and solar tracking system, Iran [97]	6.5	NA
Triple-basin solar still with SDC, charcoal in basins and cover cooling method without sun-tracking system, India [99]	16.94	0.084
SSSS with SDC, PCM balls, and cover cooling method without sun-tracking system, India [98]	3.80	0.0085
Conventional passive solar still, Egypt [80]	3.00	0.048
Conventional passive solar still, PSS, Malaysia [40]	3.21	0.015
Passive solar still (PSS) coupled with a PV module-DC heater, ACSS, Malaysia [40]	4.36	0.045
Single slope passive solar still, Pakistan [104]	3.25	0.063
Single slope hybrid (PV/T) active solar still, India [102]	1.91	0.14
Passive solar still coupled with a flat plate collector, Jordan [100]	4.69	0.103
Fin-type passive solar still, India [46]	4.00	0.054
Passive solar still with wick and fin in the basin, India [101]	4.06	0.065
Stepped passive solar still with fins and sponges in the basin, India [103]	3.03	0.064
Passive solar still with a shallow solar pond, Egypt [105]	4.65	0.08

3. Discussion

The limit of water temperatures ranging from 50 °C to 70 °C was reported as appropriate for the viability of some waterborne pathogens, bacteria, and viruses—particularly SARS-CoV-2—in water bodies [74,75]. The vast public health concern was pertaining to the existence of the aforementioned impurities during the pandemic, particularly SARS-CoV-2, in distilled water produced by passive and active solar stills, recently highlighted in previous studies [31,36,39,40,42–44,46–51,53,63–73], in which the solar stills were found to be able to generate the distillate in low initial operating water temperature. As can be observed from the reviews, using SDCs coupled with small-scale passive solar stills (i.e., absorbers or boilers) [80–82,97] and the sun-tracking systems could lead to drastic and instant increases in the initial water temperatures in the boilers until above 70, 80, and 105 °C in the early experimental hours. This was due to the high rates of the reflected sun's rays and heat from the SDC's mirrored surfaces onto the boiler outer surfaces, which is recommended as one of the most effective ways for removing any available waterborne pathogens, bacteria, and viruses—particularly SARS-CoV-2—from the absorber water in order to prevent the transmission of those impurities into the distillate.

However, initial basin water temperatures in the absorbers—which are coupled with SDCs, but without the sun-tracking systems as experimented in several studies [98,99]—were lower than 50 °C in the beginning of the experiments. Such conditions are an important factor for the viability and survival of water borne pathogens and viruses in the basins water and distillates, as noted by other studies [36,56–60].

Furthermore, as seen from the experimental works on the SDCs integrated with absorbers and sun tracking devices [80–82,97], all parts of the absorbers were exposed to the sunlight (mainly ultraviolet waves (UV)) and received direct radiation at the top surfaces and the reflected sun's radiation at the bottom and sides from the parabola surface of the SDCs throughout the experiments. Exposing all parts of the solar stills to the sun's rays is an efficient technique to prevent the growth of bacteria and pathogens in the distillate [58–60]. However, this method was not completely practical in the use of any other types of passive and active solar stills because the sun's rays were only received from the top condensing cover surfaces in the early experimental hours [21,29,31,39,40,42–44,46–51,53,63–73].

It was also stated in another theoretical and experimental study [80–82] that the absorbers with smaller surfaces areas and lower water capacity have experienced greater water temperatures, as compared to those with larger surfaces areas [97–99], when the SDCs and the boiler were used under the hourly sunlight periods. The water temperature of the small-scale absorbers coupled with SDCs and the sun-tracking systems as reported in several studies [80–82] increased drastically from about 70 °C to above 100 °C (i.e., the boiling point). The maximum values were achieved at 105, 150.7, and 319 °C, respectively, within a few minutes in the early morning after the daily experiments began, and then the condition was maintained for several hours until the evening. This indicated that a disinfection process occurred during the continuous boiling processes in the absorbers due to the explosion of the solution onto the solar radiation ultraviolet waves [83]. It was obtained from the results of the above studies [80–82,93,97] that an SDC with largest aperture diameter, greatest optical efficiency, and reflectivity with STS integrated with an absorber with smallest area and lowest reflectivity had a vital role in increasing the initial temperature of the brackish/saline water in the absorber around 70 °C, maintaining the water temperature beyond the boiling point and enhancing the amount of distillate significantly.

On the other hand, other studies [87,88,98,99] reported that solar stills with immobile solar reflectors were unable to significantly improve basin water temperature to reach the boiling point.

Furthermore, as reported in several studies [80–82], small-scale absorbers coupled with SDCs and dual-axis sun-tracking systems had better performance and were more effective in obtaining higher productivities with lower cost per liter, compared to passive and active solar stills investigated by others [40,46,100–105]. This was due to the resulting higher average water temperatures of the absorbers. Nevertheless, solar stills integrated with immobile SDCs and heat storage materials in their basins [98,99] had higher water productivities and lower costs per liter compared to the mobile SDCs—distillation systems [80–82]. Despite this, low initial absorber temperatures of the absorbers are highlighted as a public health concern in terms of preventing the transmission of pathogens and viruses into the distillate.

4. Conclusions

Based on the above reviews and discussions, SDCs with mirrored surfaces and sun-tracking systems were seen as capable of increasing the initial water temperature of the integrated small-scale absorbers until exceeding 70 °C. Furthermore, continuous increase in the absorbers' wall temperatures beyond the boiling point until the end of the operation is also recommended as another efficient technique to demolish the waterborne pathogens and viruses, especially SARS-CoV-2, at the same time to prevent transmitting these impurities to the produced water during the pandemic. Smaller scale absorbers were found to be more effective in terms of the SDC's surfaces' ability to absorb more heat from the reflected sun's rays, compared to those with larger areas. SDCs with and without the sun-tracking systems (STS) produced greater amounts of freshwater at a lower cost compared to the other previous passive and active solar stills. An SDC with larger aperture diameter, greater optical efficiency, and reflectivity with the STS integrated with an absorber with smaller area and lower reflectivity was perceived to be more operative in increasing the initial temperature of the brackish/saline water in the absorber around 70 °C, maintaining the water temperature at the boiling point during sunshine hours and enhancing the amount of distillate significantly. SDCs with the STS were more effective than the immobile and non-sun tracking SDCs in terms of obtaining higher operating absorber temperatures. Therefore, SDCs which are integrated with small-scale absorbers and sun-tracking systems are recommended as a cost-effective and reliable alternative of an impure water treatment system that can produce hygienic and pathogen-free fresh water, particularly during the SARS-CoV-2 pandemic, for the benefit of the communities in remote and rural areas—including those located in the Middle East, South-East Asia, and Africa—which are suffering from water scarcity and have abundant annual bright sunshine hours.

Author Contributions: M.F.Y. and M.R.R.M.A.Z. wrote the original draft of the manuscript; A.R., N.A.Z., A.V.S., S.S., M.S.A.A., M.H.Z., P.V., M.R.R.M.A.Z., N.M.N., and J.I. edited the manuscript, data curation, validation, and prepared the technical aspects of the paper. All authors have read and agreed to the published version of the manuscript.

Funding: This research was funded by the "Ministry of Higher Education (MOHE), Malaysia", grant number "FRGS/1/2021/TK0/USM/02/17" and by TUIASI Internal Grants Program (GI_Publications/2021), financed by the Romanian Government. APC was funded by the "River Engineering and Urban Drainage Research Centre (REDAC)".

Institutional Review Board Statement: Not applicable.

Informed Consent Statement: Informed consent was obtained from all subjects in the study.

Data Availability Statement: The data presented in this study are available on request from the corresponding author.

Acknowledgments: The authors greatly appreciate the support and funding provided by the Ministry of Higher Education (MOHE), Malaysia.

Conflicts of Interest: The authors declare no conflict of interest.

References

1. Gleick, P.H. *Water in Crisis: A Guide to the World's Freshwater Resources*; Oxford University Press Pacific: Oxford, NY, USA, 1993.
2. Holdsworth, J. 18—Authorised EU health claims for water. In *Foods, Nutrients and Food Ingredients with Authorised Eu Health Claims*; Sadler, M., Ed.; Woodhead Publishing Series in Food Science, Technology and Nutrition; Woodhead Publishing: Sawston, UK, 2014; pp. 373–395.
3. Gutierrez, J.R.; Whitford, W.G. Chihuahuan desert annuals: Importance of water and nitrogen. *Ecology* **1987**, *68*, 2032–2045. [CrossRef] [PubMed]
4. Rastogi, S.C. *Essentials of Animal Physiology*, 4th ed.; New Age International Publishers: New Delhi, India, 2008; Available online: https://www.vet-ebooks.com/essentials-of-animal-physiology-4th-edition/ (accessed on 9 January 2022).
5. Science Learning Hub—Pokapū Akoranga Pūtaiao. Earth's Water Distribution. Available online: https://www.sciencelearn.org.nz/images/802-earth-s-water-distribution (accessed on 20 December 2021).
6. Cech, T.V. *Principles of Water Resources: History, Development, Management, and Policy*, 3rd ed.; John Wiley & Sons: Hoboken, NJ, USA, 2010.
7. Kumar, K.V.; Bai, R.K. Performance study on solar still with enhanced condensation. *Desalination* **2008**, *230*, 51–61. [CrossRef]
8. Tiwari, G.N.; Singh, H.N.; Tripathi, R. Present status of solar distillation. *Sol. Energy* **2003**, *75*, 367–373. [CrossRef]
9. Vineis, P.; Chan, Q.; Khan, A. Climate change impacts on water salinity and health. *J. Epidemiol. Glob. Health.* **2011**, *1*, 5–10. [CrossRef] [PubMed]
10. Guardian News & Media Limited. The Guardian for 200 Years. Available online: https://www.theguardian.com/environment/2020/nov/26/more-than-3-billion-people-affected-by-water-shortages-data-shows (accessed on 26 November 2020).
11. Tabrizi, F.F.; Sharak, A.Z. Experimental study of an integrated basin solar still with a sandy heat reservoir. *Desalination* **2010**, *253*, 195–199. [CrossRef]
12. El-Ghonemy, A.M.K. Water desalination systems powered by renewable energy sources: Review. *Renew. Sustain. Energy Rev.* **2012**, *16*, 1537–1556. [CrossRef]
13. Sampathkumar, K.; Arjunan, T.V.; Pitchandi, P.; Senthilkumar, P. Active solar distillation—A detailed review. *Renew. Sustain. Energy Rev.* **2010**, *14*, 1503–1526. [CrossRef]
14. Arunkumar, T.; Raj, K.; Dsilva Winfred Rufuss, D.; Denkenberger, D.; Tingting, G.; Xuan, L.; Velraj, R. A review of efficient high productivity solar stills. *Renew. Sustain. Energy Rev.* **2019**, *101*, 197–220. [CrossRef]
15. United Nations. *WHO/UNICEF Joint Monitoring Program (JMP) for Water Supply, Sanitation and Hygiene—Progress on Household Drinking Water, Sanitation and Hygiene 2000–2020*; IMI-SDG6 SDG 6 Progress Reports; United Nations: San Francisco, CA, USA; Available online: https://www.unwater.org/publications/who-unicef-joint-monitoring-program-for-water-supply-sanitation-and-hygiene-jmp-progress-on-household-drinking-water-sanitation-and-hygiene-2000-2020/ (accessed on 1 July 2021).

16. World Health Organization. *Global Costs and Benefits of Drinking-Water Supply and Sanitation Interventions to Reach the MDG Target and Universal Coverage*, 1st ed.; WHO Press: Geneva, Switzerland, 2012; pp. 1–67.
17. Jasrotia, S.; Kansal, A.; Kishore, V.V.N. Application of solar energy for water supply and sanitation in Arsenic affected rural areas: A study for Kaudikasa village, India. *J. Clean. Prod.* **2012**, *37*, 389–393. [CrossRef]
18. World Health Organization. *Guidelines for Drinking-Water Quality, Incorporating First Addendum Vol. 1. Recommendations*, 3rd ed.; WHO Press: Geneva, Switzerland, 2008.
19. Bhattacharya, P.; Welch, A.H.; Stollenwerk, K.G.; McLaughlin, M.J.; Bundschuh, J.; Panaullah, G. Arsenic in environment: Biology and chemistry. *Sci. Total Environ.* **2007**, *379*, 109–120. [CrossRef]
20. Mahendra, P.; Yodit, A.; Angesom, H.; Sumitra, P.; Vijay, J.J. Public Health Hazards Due to Unsafe Drinking Water. *Air Water Borne Dis.* **2018**, *7*, 1000138.
21. Adio, S.A.; Osowade, E.A.; Muritala, A.O.; Fadairo, A.A.; Oladepo, K.T.; Obayopo, S.O.; Fase, P.O. Solar distillation of impure water from four different water sources under the southwestern Nigerian climate. *Drink. Water Eng. Sci.* **2021**, *14*, 81–94. [CrossRef]
22. Pruss-Ustun, A.; Bos, R.; Gore, F.; Bartram, J. *Safer Water, Better Health: Costs, Benefits and Sustainability of Interventions to Protect and Pro Mote Health*; World Health Organization: Geneva, Switzerland, 2008.
23. Cengel, Y.A.; Boles, M.A. *Thermodynamics: An Engineering Approach*, 8th ed.; McGraw Hill: New York, NY, USA, 2015; p. 712.
24. Goosen, M.; Mahmoudi, H.; Ghaffour, N. Overview of renewable energy technologies for freshwater production. In *Renewable Energy Applications for Freshwater Production*; CRS Press: London, UK; IWA Publishing: London, UK, 2012; pp. 25–77.
25. Kalogirou, S.A. Concentrating solar power plants for electricity and desalinated water production. In Proceedings of the World Renewable Energy Congress, Linköping, Sweden, 8 May 2011. [CrossRef]
26. Kalogirou, S.A. *Solar Energy Engineering: Processes and Systems*; Academic Press: San Diego, CA, USA, 2014.
27. UAE. World Resources Institute's Aqueduct Water Risk Atlas August 2019 Report. Available online: https://www.thenationalnews.com/uae/environment/uae-water-resources-under-extreme-stress-new-report-finds-1.895660 (accessed on 14 January 2022).
28. Solar Power—Open Source Learning. Available online: https://mediawiki.middlebury.edu/wikis/OpenSourceLearning/images/5/53/Sunshine_Map.jpg (accessed on 14 January 2022).
29. Hanson, A.; Zachritz, W.; Stevens, K.; Mimbela, L.; Polka, R.; Cisneros, L. Distilate water quality of a single-basin solar still: Laboratory and field studies. *Sol. Energy* **2004**, *76*, 635–645. [CrossRef]
30. Riahi, A.; Yusof, K.W.; Isa, M.H.; Mahinder Singh, B.S.; Mustaffa, Z.; Ahsan, A.; Ul Mustafa, M.R.; Sapari, N.; Zahari, N.A.M. Potable water production using two solar stills having different cover materials and fabrication costs. *Environ. Prog. Sustain. Energy* **2018**, *37*, 584–596. [CrossRef]
31. Ahsan, A.; Imteaz, M.; Thomas, U.A.; Azmi, M.; Rahman, A.; Nik Daud, N.N. Parameters affecting the performance of a low cost solar still. *Appl. Energy* **2014**, *114*, 924–930. [CrossRef]
32. Malik, M.A.S.; Tiwari, G.N.; Kumar, A.; Sodha, M.S. *Solar Distillation*; Pergaman Press: Oxford, UK.; London, UK, 1982.
33. Tiwari, G.N.; Madhuri. Effect of water depth on daily yield of the still. *Desalination* **1987**, *61*, 67–75. [CrossRef]
34. Akash, B.A.; Mohsen, M.S.; Nayfeh, W. Experimental study of the basin type solar still under local climate conditions. *Energy Convers. Manag.* **2000**, *41*, 883–890. [CrossRef]
35. Toure, S.; Meukam, P. A numerical model and experimental investigation for a solar still in climatic conditions in Abidjan (Côte d'Ivoire). *Renew. Energy* **1997**, *11*, 319–330. [CrossRef]
36. Parsa, S.M. Reliability of thermal desalination (solar stills) for water/wastewater treatment in light of COVID-19 (novel coronavirus "SARS-CoV-2") pandemic: What should consider? *Desalination* **2021**, *512*, 115106. [CrossRef]
37. Muthu Manokar, A.; Prince Wiston, D.; Kabeel, A.E.; El-Agouz, S.A.; Sathyamurthy, R.; Arunkumar, T.; Madhu, B.; Ahsan, A. Integrated PV/T solar still-A mini-review. *Desalination* **2018**, *435*, 259–267. [CrossRef]
38. Dwivedi, V.K.; Tiwari, G.N. Experimental validation of thermal model of a double slope active solar still under natural circulation mode. *Desalination* **2010**, *250*, 49–55. [CrossRef]
39. Riahi, A.; Wan, Y.K.; Mahinder Singh, B.S.; Isa, M.H.; Olisa, E.; Zahari, N.A.M. Sustainable potable water production using a solar still with photovoltaic modules-AC heater. *Desalination Water Treat.* **2016**, *57*, 14929–14944. [CrossRef]
40. Riahi, A.; Zakaria, N.A.; Isa, M.H.; Yusof, K.W.; Mahinder Singh, B.S.; Mustaffa, Z.; Takaijudin, H. Performance investigation of a solar still having polythene film cover and black painted stainless steel basin integrated with a photovoltaic module–direct current heater. *Energy Environ.* **2019**, *30*, 1521–1535. [CrossRef]
41. Manchanda, H.; Kumar, M. Study of water desalination techniques and a review on active solar distillation methods. *Environ. Prog. Sustain. Energy* **2018**, *37*, 444–464. [CrossRef]
42. Kumar, B.P.; Winston, D.P.; Pounraj, P.; Manokar, A.M.; Sathyamurthy, R.; Kabeel, A.E. Experimental investigation on hybrid PV/T active solar still with effective heating and cover cooling method. *Desalination* **2018**, *435*, 140–151. [CrossRef]
43. Riahi, A.; Yusof, K.W.; Mahinder Singh, B.S.; Olisa, E.; Sapari, N.B.; Isa, M.H. The performance investigation of triangular solar stills having different heat storage materials. *Int. J. Energy Environ. Eng.* **2015**, *6*, 385–391. [CrossRef]
44. Al-Garni, A.Z. Productivity enhancement of solar still using water heater and cooling fan. *J. Sol. Energy Eng. Trans. ASME* **2012**, *134*, 031006. [CrossRef]

45. Phadatare, M.K.; Verma, S.K. Influence of water depth on internal heat and mass transfer in a plastic solar still. *Desalination* **2007**, *217*, 267–275. [CrossRef]
46. Velmurugan, V.; Deenadayalan, C.K.; Vinod, H.; Srithar, K. Desalination of effluent using fin type solar still. *Energy* **2008**, *33*, 1719–1727. [CrossRef]
47. Panchal, H.; Patel, P.; Patel, N.; Thakkar, H. Performance analysis of solar still with different energy-absorbing materials. *Int. J. Ambient. Energy* **2017**, *38*, 224–228. [CrossRef]
48. Panchal, H. Performance investigation on variations of glass cover thickness on solar still: Experimental and theoretical analysis. *Technol. Econ. Smart Grids Sustain. Energy* **2016**, *1*, 7. [CrossRef]
49. Ahsan, A.; Imteaz, M.; Rahman, A.; Yusuf, B.; Fukuhara, T. Design, fabrication and performance analysis of an improved solar still. *Desalination* **2012**, *292*, 105–112. [CrossRef]
50. Kabeel, A.E.; Khalil, A.; Omara, Z.M.; Younes, M.M. Theoretical and experimental parametric study of modified stepped solar still. *Desalination* **2012**, *289*, 12–20. [CrossRef]
51. Abdel-Rehim, Z.S.; Lasheen, A. Improving the performance of solar desalination systems. *Renew. Energy* **2005**, *30*, 1955–1971. [CrossRef]
52. Tarawneh, M.S.K. Effect of water depth on the performance evaluation of solar still. *Jordan J. Mech. Ind. Eng.* **2007**, *1*, 23–29.
53. Taamneh, Y.; Taamneh, M.M. Performance of pyramid-shaped solar still: Experimental study. *Desalination* **2012**, *291*, 65–68. [CrossRef]
54. Aybar, H.S.; Assefi, H. Simulation of a solar still to investigate water depth and glass angle. *Desalination Water Treat.* **2009**, *7*, 35–40. [CrossRef]
55. World Health Organization. *Guidelines for Drinking-Water Quality*, 3rd ed.; WHO Press: Geneva, Switzerland, 2008.
56. Balladin, D.A.; Headley, O.; Roach, A. Evaluation of a concrete cascade solar still. *Renew. Energy* **1999**, *17*, 191–206. [CrossRef]
57. KIkuchi, S.; Oyoda, H.T.; Akami, A.T.; Himada, S.S.; Oba, M.O.; Ekiyama, T.S. Simple solar still using solar energy and compost heat for family use. *J. Arid L. Stud.* **2012**, *21*, 207–210.
58. Ayoub, G.M.; Dahdah, L.; Alameddine, I.; Malaeb, L. Vapor-induced transfer of bacteria in the absence of mechanical disturbances. *J. Hazard. Mater.* **2014**, *280*, 279–287. [CrossRef]
59. Ayoub, G.M.; Dahdah, L.; Alameddine, I. Transfer of bacteria via vapor in solar desalination units. *Desalination Water Treat.* **2015**, *53*, 3199–3207. [CrossRef]
60. Malaeb, L.; Ayoub, G.M.; Al-Hindi, M.; Dahdah, L.; Baalbaki, A.; Ghauch, A. A biological, chemical and pharmaceutical analysis of distillate quality from solar stills. *Energy Procedia* **2017**, *119*, 723–732. [CrossRef]
61. Parsa, S.M.; Rahbar, A.; Javadi, Y.D.; Koleini, M.H.; Afrand, M.; Amidpour, M. Energy-matrices, exergy, economic, environmental, exergoeconomic, enviroeconomic, and heat transfer (6E/HT) analysis of two passive/active solar still water desalination nearly 4000 m: Altitude concept. *J. Clean. Prod.* **2020**, *261*, 121243. [CrossRef]
62. Kalbasi, R.; Esfahani, M.N. Multi-effect passive desalination system, an experimental approach. *World Appl. Sci. J.* **2010**, *10*, 1264–1271.
63. Pounraj, P.; Prince Winston, D.; Kabeel, A.E.; Praveen Kumar, B.; Manokar, A.M.; Sathyamurthy, R.; Christabel, S.C. Experimental investigation on Peltier based hybrid PV/T active solar still for enhancing the overall performance. *Energ. Conver. Manag.* **2018**, *168*, 371–381. [CrossRef]
64. Arunkumar, T.; Kabeel, A.E.; Raj, K.; Denkenberger, D.; Sathyamurthy, R.; Ragupathy, P.; Velraj, R. Productivity enhancement of solar still by using porous absorber with bubble-wrap insulation. *J. Clean. Prod.* **2018**, *195*, 1149–1161. [CrossRef]
65. Pal, P.; Yadav, P.; Dev, R.; Singh, D. Performance analysis of modified basin type double slope multi–wick solar still. *Desalination* **2017**, *422*, 68–82. [CrossRef]
66. Kumar, S.; Tiwari, A. Design, fabrication and performance evaluation of a hybrid photovoltaic/ thermal (PV/T) double slope active solar still. *Energy Convers. Manag.* **2010**, *51*, 1219–1229. [CrossRef]
67. Dev, R.; Abdul-Wahab, S.A.; Tiwari, G.N. Performance study of the inverted absorber solar still with water depth and total dissolved solid. *Appl. Energy* **2011**, *88*, 252–264. [CrossRef]
68. Joshi, P.; Tiwari, G.N. Energy matrices, exergo-economic and enviro-economic analysis of an active single slope solar still integrated with a heat exchanger: A comparative study. *Desalination* **2018**, *443*, 85–98. [CrossRef]
69. Sohani, A.; Hoseinzadeh, S.; Berenjkar, K. Experimental analysis of innovative designs for solar still desalination technologies; an in-depth technical and economic assessment. *J. Energy Storage.* **2021**, *33*, 101862. [CrossRef]
70. Tripath, R.; Tiwari, G.N. Effect of water depth on heat and mass transfer in a solar still: In summer climate condition. *Desalination* **2006**, *217*, 267–275.
71. Tripathi, R.; Tiwari, G.N. Thermal modelling of passive and active solar stills for different depths of water by using the concept of solar fraction. *Sol. Energy* **2006**, *80*, 956–967. [CrossRef]
72. Singh, G.; Kumar, S.; Tiwari, G.N. Design, fabrication and performance evaluation of a hybrid photovoltaic thermal (PVT) double slope active solar still. *Desalination* **2011**, *277*, 399–406. [CrossRef]
73. Balachandran, G.B.; David, P.W.; Mariappan, R.K.; Kabeel, A.E.; Athikesavan, M.M.; Sathyamurthy, R. Improvising the efficiency of single-sloped solar still using thermally conductive nano-ferric oxide. *Environ. Sci. Pollute. Res.* **2020**, *27*, 32191–32204. [CrossRef]

74. Chin, A.W.H.; Chu, J.T.S.; Perera, M.R.A.; Hui, K.P.Y.; Yen, H.L.; Chan, M.C.W.; Peiris, M.; Poon, L.L.M. Stability of SARS-CoV-2 in different environmental conditions. *Lancet Microbe* **2020**, *1*, 10. [CrossRef]
75. Chan, K.H.; Sridhar, S.; Zhang, R.R.; Chu, H.; Fung, A.Y.F.; Chan, G.; Chan, J.F.W.; To, K.K.W.; Hung, I.F.N.; Cheng, V.C.C.; et al. Factors affecting stability and infectivity of SARS-CoV-2. *J. Hosp. Infect.* **2020**, *106*, 226–231. [CrossRef]
76. Yu, L.; Peel, G.K.; Cheema, F.H.; Lawrence, W.S.; Bukreyeva, N.; Jinks, C.W.; Peel, J.E.; Peterson, J.W.; Paessler, S.; Hourani, M.; et al. Catching and killing of airborne SARS-CoV-2 to control spread of COVID-19 by a heated air disinfection system. *Mater. Today Phys.* **2020**, *15*, 100249. [CrossRef]
77. Wang, D.; Sun, B.C.; Wang, J.X.; Zhou, Y.Y.; Chen, Z.W.; Fang, Y.; Yue, W.H.; Liu, S.M.; Liu, K.Y.; Zeng, X.F.; et al. Can masks be reused after hot water decontamination during the COVID-19 pandemic? *Engineering* **2020**, *6*, 1115–1121. [CrossRef]
78. Ahmed, W.; Bertsch, P.M.; Bibby, K.; Haramoto, E.; Hewitt, J.; Huygens, F.; Gyawali, P.; Korajkic, A.; Riddell, S.; Sherchan, S.P.; et al. Decay of SARS-CoV-2 and surrogate murine hepatitis virus RNA in untreated wastewater to inform application in wastewater-based epidemiology. *Environ. Res.* **2020**, *191*, 110092. [CrossRef]
79. Chaouchi, B.; Zrelli, A.; Gabsi, S. Desalination of brackish water by means of a parabolic solar concentrator. *Desalination* **2007**, *217*, 118–126. [CrossRef]
80. Omara, Z.; Eltawil, M.A. Hybrid of solar dish concentrator, new boiler and simple solar collector for brackish water desalination. *Desalination* **2013**, *326*, 62–68. [CrossRef]
81. Gorjian, S.; Ghobadian, B.; Tavakkoli Hashjin, T.; Banakar, A. Experimental performance evaluation of a stand-alone point-focus parabolic solar still. *Desalination* **2014**, *352*, 1–17. [CrossRef]
82. Prado, G.O.; Vieira, L.G.M.; Damasceno, J.J.R. Solar dish concentrator for desalting water. *Sol. Energy* **2016**, *136*, 659–667. [CrossRef]
83. McGuigan, K.G.; Conroy, R.M.; Mosler, H.-J.; du Preez, M.; Ubomba-Jaswa, E.; Fernandez-Ibanez, P. Solar water disinfection (SODIS): A review from bench-top to roof-top. *J. Hazard. Mater.* **2012**, *235–236*, 29–46. [CrossRef]
84. Aljabair, S.; Habeeb, L.J.; Ali, A.M. Experimental analysis of parabolic solar dish with radiator heat exchanger receiver. *J. Eng. Sci. Technol.* **2020**, *15*, 437–454.
85. Li, L.; Kecskemethy, A.; Arif, A.F.M.; Dubowsky, S. A Novel Approach for Designing Parabolic Mirrors Using Optimized compliant Bands. In Proceedings of the International Design Engineering Technical Conference and Computers and Information in Engineering, IDETC, Washington, DC, USA, 28 August 2011.
86. Aliman, O.; Daut, I.; Isa, M.; Adzman, M.R. Simplification of sun tracking mode to gain high concentration solar energy. *Am. J. Appl. Sci.* **2007**, *4*, 171–175. [CrossRef]
87. Singh, S.K.; Bhatnagar, V.P.; Tiwari, G.N. Design parameters for concentrator assisted solar distillation system. *Energy Convers. Manag.* **1996**, *37*, 247–252. [CrossRef]
88. Riffat, S.; Mayere, A. Performance evaluation of v-trough solar concentrator for water desalination applications. *Appl. Therm. Eng.* **2013**, *50*, 234–244. [CrossRef]
89. Sinitsyn, S.; Panchenko, V.; Kharchenko, V.; Vasant, P. Oprimization of parquetting of the concentrator of photovoltaic thermal module. In *Intelligent Computing and Optimization. ICO 2019. Advances in Intelligent Systems and Computing*; Vasant, P., Zelinka, I., Weber, G.W., Eds.; Springer: Cham, Switzerland, 2020; Volume 1072.
90. Panchenko, V. Photovoltaic thermal module with paraboloid type solar concentrators. *Int. J. Energy Optim. Eng. (IJEOE)* **2021**, *10*, 23. [CrossRef]
91. Garg, H.P.; Prakash, J. *Solar Energy Fundamentals and Applications*, 1st ed.; McGraw Hill: Delphi, Greece, 2000.
92. Johnston, G.; Lovegrove, K.; Luzzi, A. Optical performance of spherical reflecting elements for use with paraboloidal dish concentrators. *Sol. Energy* **2003**, *74*, 133–140. [CrossRef]
93. Chaichan, M.T.; Kazem, H.A. Water solar distiller productivity enhancement using concentrating solar water heater and phase change material (PCM). *Case Stud. Therm. Eng.* **2015**, *5*, 151–159. [CrossRef]
94. Nuwayhid, R.Y.; Mrad, F.; Abu-Said, R. The realization of a simple solar tracking concentrator for university research applications. *Renew. Energy* **2001**, *24*, 207–222. [CrossRef]
95. Kaushika, N.D.; Reddy, K.S. Performance of a low cost solar paraboloidal steam generating system. *Energy Convers. Manag.* **2000**, *41*, 713–726. [CrossRef]
96. Born, M.; Wolf, E. *Principles of Optics*, 5th ed.; Pergamon Press: Oxford, NY, USA, 1975.
97. Bahrami, M.; Avargani, V.M.; Bonyadi, M. Comprehensive experimental and theoretical study of a novel still coupled to a solar dish concentrator. *Appl. Therm. Eng.* **2019**, *151*, 77–89. [CrossRef]
98. Arunkumar, T.; Denkenberger, D.; Velraj, R.; Sathyamurthy, R.; Tanaka, H.; Vinothkumar, K. Experimental study on a parabolic concentrator assisted solar desalting system. *Energy Convers. Manag.* **2015**, *105*, 665–674. [CrossRef]
99. Srithar, K.; Rajaseenivasan, T.; Karthik, N.; Periyannan, M.; Gowtham, M. Standalone triple basin solar desalination system with cover cooling and parabolic dish concentrator. *Renew. Energy* **2016**, *90*, 157–165. [CrossRef]
100. Badran, A.A.; Al-Hallaq, A.A.; Eyal Salman, I.A.; Odat, M.Z. A solar still augmented with a flat plate collector. *Desalination* **2005**, *172*, 227–234. [CrossRef]
101. Velmurugan, V.; Gopalakrishnan, M.; Raghu, R.; Srithar, K. Single basin solar still with fin for enhancing productivity. *Energy Convers. Manag.* **2008**, *49*, 2602–2608. [CrossRef]

102. Kumar, S.; Tiwari, G.N. Life cycle cost analysis of single slope hybrid (PV/T) active solar still. *Appl. Energy* **2009**, *86*, 1995–2004. [CrossRef]
103. Velmurugan, V.; Kumaran, S.S.; Prabhu, V.N.; Srithar, K. Productivity enhancement of stepped solar still performance analysis. *Therm. Sci.* **2008**, *12*, 153–163. [CrossRef]
104. Ali Samee, M.; Mirza, U.K.; Majeed, T.; Ahmad, N. Design and performance of a simple single basin solar still. *Renew. Sustain. Energ. Rev.* **2007**, *11*, 543–549. [CrossRef]
105. El-Sebaii, A.A.; Ramadan, M.R.I.; Aboul-Enein, S.; Salem, N. Thermal performance of a single-basin solar still integrated with a shallow solar pond. *Energy Convers. Manag.* **2008**, *49*, 2839–2848. [CrossRef]
106. Kabeel, A.E.; Dawood, M.M.K.; Ramzy, K.; Nabil, T.; Elnaghi, B.; Elkassar, A. Enhancemnet of single solar still integrated with solar dishes: An experimental approach. *Energy Convers. Manag.* **2019**, *196*, 165–174. [CrossRef]

Article

Comprehensive Energy Consumption of Elevator Systems Based on Hybrid Approach of Measurement and Calculation in Low- and High-Rise Buildings of Tropical Climate towards Energy Efficiency

Jia Hui Ang [1], Yusri Yusup [1], Sheikh Ahmad Zaki [2,*], Ali Salehabadi [1] and Mardiana Idayu Ahmad [1,*]

1. Environmental Technology Division, School of Industrial Technology, Universiti Sains Malaysia, Penang 11800, Malaysia; jiahui_93@live.com (J.H.A.); yusriy@usm.my (Y.Y.); alisalehabadi@usm.my (A.S.)
2. Department of Mechanical Precision Engineering, Malaysia-Japan International Institute of Technology, Universiti Teknologi Malaysia, Kuala Lumpur 54100, Malaysia
* Correspondence: sheikh.kl@utm.my (S.A.Z.); mardianaidayu@usm.my (M.I.A.)

Abstract: Rapid population growth and urbanization contribute to an ever-increasing global energy demand, of which the building sector accounts for one-third. The increasing average height and density of buildings escalate the need for vertical transportation, expanding elevator usage and energy needs. This phenomenon accounts for a significant amount of the total building energy use, necessitating a study of elevator system energy consumption. This study aimed to analyze the energy consumption and carbon emissions of elevator systems in low- and high-rise buildings towards energy-efficient estimations. A comprehensive analysis was performed based on a hybrid approach of measurement and calculation using a formula and reference values derived from previous studies. Four buildings were selected and thoroughly studied, representing the low- and high-rise categories. Data were collected based on on-site sampling and observation, as well as information from the building management offices. The mechanical parameters of the elevator system in each building and operational factors in terms of speed, number of trips, load, travel distance, and time were studied. In this analysis, the energy consumption calculation was performed according to International Standard ISO 25745. Annual carbon emissions were calculated in accordance with the USA EPA and IPCC guidelines. The elevator energy efficiency class was determined based on daily energy consumption. It was found from this study that the annual energy consumption of an elevator system is positively correlated to an elevator's daily energy consumption. The annual carbon emissions of the elevator systems are dependent on increasing annual energy consumption, which is also connected to building height indirectly. The low-rise buildings showed better energy efficiency compared to the high-rise buildings due to lower travel distance, less trips, and fewer floors. The annual number of trips, travel distances, and energy consumption had an effect on the energy efficiency of the elevator systems in this study.

Keywords: energy consumption; greenhouse gas emissions; energy efficiency; elevator; buildings

1. Introduction

World energy consumption is increasing on an annual basis in tandem with high population expansion and urbanization. Many countries across the world contribute to the upward trend. For instance, in 2020, the energy consumptions in Europe, the Commonwealth of Independent States (CIS), North America, Latin America, Asia, the Pacific, Africa, and the Middle East were 1689 Million tonnes of oil equivalent (Mtoe), 1015 Mtoe, 2327 Mtoe, 758 Mtoe, 5955 Mtoe, 152 Mtoe, 809 Mtoe, and 803 Mtoe, respectively [1]. Global energy demand is projected to increase 28 to 30% between 2015 and 2040, from 575 quadrillion British Thermal Unit (Btu) to 736 quadrillion Btu [2,3]. Among various energy sectors, buildings account for a significant percentage of the overall global energy consumption, up to 20.1% to 40% [4–6]. In Malaysia, with a tropical climate, the energy consumption of the building sector contributes up to 48% to 54% of the total electricity usage in the country [5]. Buildings have substantial energy needs throughout their entire life cycles, from the construction phase, operation phase, and maintenance phase, until the demolition phase [7]. In many European countries, residential buildings contribute more to building energy consumption compared to commercial buildings [8–11]. In Malaysia, the energy consumption of commercial buildings outweighs residential buildings [5,12]. Specifically, the nonresidential buildings consume more energy per area (kWh/m^2) than the residential buildings [13]. Commercial buildings have an alarmingly high energy consumption. This is contributed to by the increased number of heating, ventilation, and air conditioning (HVAC) systems in commercial office buildings compared to residential buildings in tropical climate countries [14]. Occupant behaviour was also identified as a factor in these buildings' energy consumption [15].

Additionally, when average building height and density increase, vertical transportation demand within the building rises, leading to a surge in elevator usage. As a result, the energy consumption of the building continues to rise significantly with increasing building height [16,17]. According to a previous report by Al-kodmany [18], more than seven billion elevator trips are made every single day in the world. This adds to elevators' high energy consumption, which may account for 2% to 40% of the total building energy consumption. In most cases, the elevator systems in commercial office buildings meet more stringent criteria (e.g., higher rated load, rated speed, and number of cars) than the elevator systems in residential buildings, owing to increased elevator traffic and requirements [19]. Elevator systems with better specifications and more traffic in commercial office buildings consume more energy than elevator systems in residential buildings. This increased traffic in commercial office buildings is a result of tenants' access to entrances and exits and their interfloor travels [20].

The number of elevator cars and their configuration are determined by the number of floors in a building, as well as the population density on each floor [21]. Elevator energy consumption is highly dependent on elevator car and shaft characteristics, motor type, control system, auxiliary system, elevator traffic, and population density in the building [22,23]. Geared and gearless traction roped elevators are both less energy-intensive than conventional elevators and are commonly used in mid-rise and high-rise buildings, respectively. In contrast, hydraulic elevators are more energy-intensive and are typically used in low-rise buildings.

1.1. Determination of Elevator Energy Consumption

There are five common methods used to determine the energy consumption of an elevator system, which are: (i) calculation from first principles [24–26]; (ii) calculation using formulas and reference values derived from previous studies [27,28]; (iii) measurement [29,30]; (iv) a hybrid method of measurement and calculation using formulas and reference values derived from previous studies [31,32]; and (v) modelling and simulation [33–35].

At an early time, Kirchenmayer [25] and White [26] both studied the energy consumption of traction and hydraulic elevators using a method of calculation based on first principles [33]. Both studies reported that the traction-type elevator was a more energy-efficient mode of transport. This method is low-cost and easily implemented. However, this method does not distinguish between elevator load and speed, as well as building usage. Thus, this method is suitable to be used for any investigations that aim to focus on elevator energy efficiency technology instead of elevator energy consumption quantitative analysis. The measurement method refers to direct measurement by installing an energy meter on the elevator system, which provides the most accurate elevator energy consumption value. This method enables continuous monitoring of transient elevator energy consumption parameters, which are current and voltage, according to the desired sampling rate and average time specified. To produce a high-quality outcome, precise and accurate tools should be employed. This method is the most expensive, yet it provides the most precise and immediate findings. It is vital for the validation of alternative methods of determining elevator energy consumption.

The hybrid method comprises measurement and calculation using formulas and reference values derived from previous studies, involving measurement and secondary data for further analysis of calculation and extrapolation. This method entails measuring field measurements of energy consumption at specific parameters, such as speed, load, direction, and traffic load. The energy consumption is estimated for various parameter values using relevant and appropriate equations. For instance, data from measurements of loads of 500 kg and 600 kg can be used to predict the energy consumption of a load of 700 kg. Energy assessment and classification schemes, such as Verein Deutscher Ingenieure (VDI) 4707-1:2009 and the International Organization for Standardization (ISO) 25745-2:2015, are established examples of this method [36–38]. These scientific experiments give standard reference values in a variety forms, including energy models, categorization tables, and elevator energy profiles, to be used in different cases with listed conditions.

This method was used in a study by Hu et al. [39] to compare the effectiveness of variable voltage variable frequency (VVVF) drives in the energy efficiency of typical elevator systems, which are hydraulic and traction, following the VDI 4707 method. Bannister et al. [31] collected data on the system and technology within a specific building, developing empirical correlations of these data to the energy consumption of the elevator in that building in order to anticipate the elevator energy consumption in an office building over 3000 m^2. The predictive benchmark equation produced in this study can be used in future studies categorized under the hybrid method comprising measurement and calculation using formulas and reference values derived from previous studies. On the other hand, Tukia et al. [32] studied and compared two approaches to predict annual elevator energy consumption from short-term measurement data. In their study, daily consumption measurement data were used to derive the equations for annual elevator energy projection. In the study, a simpler method that worked based on linear extrapolation of the annual consumption based on measurement data was used. Another method used in the study took into account seasonal influences on elevator usage, making it more accurate but necessitating a greater level of data detail. The latter method's results were shown to be accurate when compared to the actual measurement of annual elevator energy consumption, as well as to the estimated results based on VDI 4707-1:2009 and ISO 25745-2:2015.

According to existing studies in the literature, the estimation and calculation method based on short-time measurement may be used to represent long-time measurement, as both results are similar [40]. The cost is much lower than that of the method of pure measurement but higher than that of the method of calculation. It is less expensive than modelling and simulation, but it requires more time and has a lesser degree of comprehensiveness due to its lower complexity. As such, it is well suited for the preliminary monitoring of a specific elevator in order to develop subsequent improvement. Despite this, the assumptions may deviate from the real case due to spatial and temporal differences. Moreover, it does not cover more detailed and short-time energy consumption estimation that is needed for the development of an efficient elevator control system.

1.2. Current Situation, Research Gaps, and the Aim of the Current Study

The literature reports that energy consumption in buildings accounts for around one-third of overall consumption and contributes an equal share of carbon dioxide emissions [41,42]. To obtain a better grasp of the existing body of information in this field, studies on elevator energy consumption and its association with carbon emissions are beneficial towards effective energy management in buildings. This enables better monitoring of the energy performance of an elevator system and the energy-saving awareness of users, resulting in more targeted solutions for excessive elevator energy consumption. Despite this, there are few studies on elevator system energy consumption, particularly in tropical climate countries, compared to total building energy consumption and other high-energy-consuming systems in buildings, such as mechanical ventilation air-conditioning (MVAC) and lighting, as well as building structure and design [43–49]. For example, in Malaysia's hot and humid environment, the majority of researchers have concentrated on the energy consumption and efficiency of a structure or entire building, the cooling system, and the electrical appliances for thermal comfort [5,45,50–55]. There is a lack of studies examining elevator system energy usage and efficiency in relation to carbon emissions. Furthermore, there are a very limited studies available on this topic that employ the hybrid method comprising measurement and calculation using formulas and reference values derived from previous works.

Therefore, considering this research gap, this study was carried out to analyze the energy consumption of elevator systems in low- and high-rise buildings in tropical climates towards energy-efficienct estimation. The study uses a detailed hybrid approach utilizing a combination of measurements and calculation methods. The new contributions of this paper are a contribution towards energy analysis that also takes into account the elevator system type, characteristics, and operating parameters, such as elevator speed, number of trips, loads, travel distance, and time, as well as elevator energy aspects in terms of usage category and elevator powers, in the addition to carbon emissions estimation. The mechanical specifications and operational variables of each building's elevator system were evaluated. In view of demographic trends and the increasing need for convenience, the number of elevator systems is expected to increase. More urbanization in developing countries and a growing awareness of issues of accessibility due to a growing population, particularly in a tropical climate, stimulates the need for more advanced and efficient systems. Thus, this study contributes to the current pool of data to improve the energy efficiency of elevator systems in the building sector, especially in tropical climates.

2. Methods

This study covers field data collection and calculations based on a formula and reference values from previous studies, as well as analysis based on the established standards and past studies. The methods of this study are explained in the following subsections.

2.1. Characteristics of the Selected Buildings and the Elevator System

Most of the features of low-rise, mid-rise, and high-rise buildings are the same in different countries, except for a lesser number and a smaller size of windows in tropical climates compared to temperate climates [56]. The most common type of elevator system used in the existing and current buildings in tropical climates is the traction-type elevator [57,58]. Four low- and high-rise buildings with common building characteristics and elevator system specifications were selected in this study to represent the typical low- and high-rise buildings in the Malaysian tropical climate. The characteristics of the four selected buildings were collected from the building management offices.

The building dimensions (in rough form), building floor plan (detailed, exact shape), and function and operating lift of different floors for high-rise residential apartment (HA), low-rise residential apartment (LA), high-rise commercial office (HO), and low-rise commercial office (LO) buildings are shown in Figures 1–4(a–c), respectively. Also, the respective characteristics of the buildings and elevator cars are summarized in Appendices A–D.

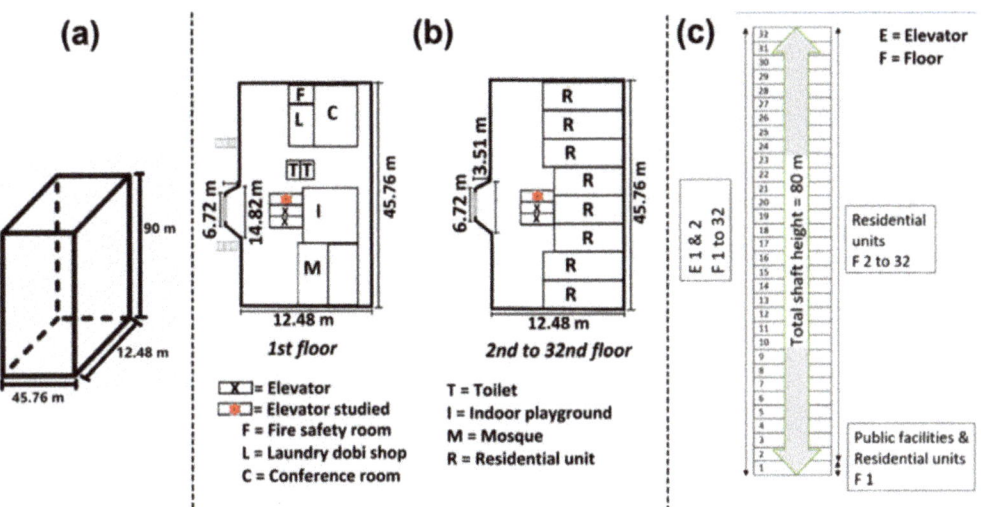

Figure 1. HA (**a**) building dimensions, (**b**) building floor plan, and (**c**) function and operating lift of different floors.

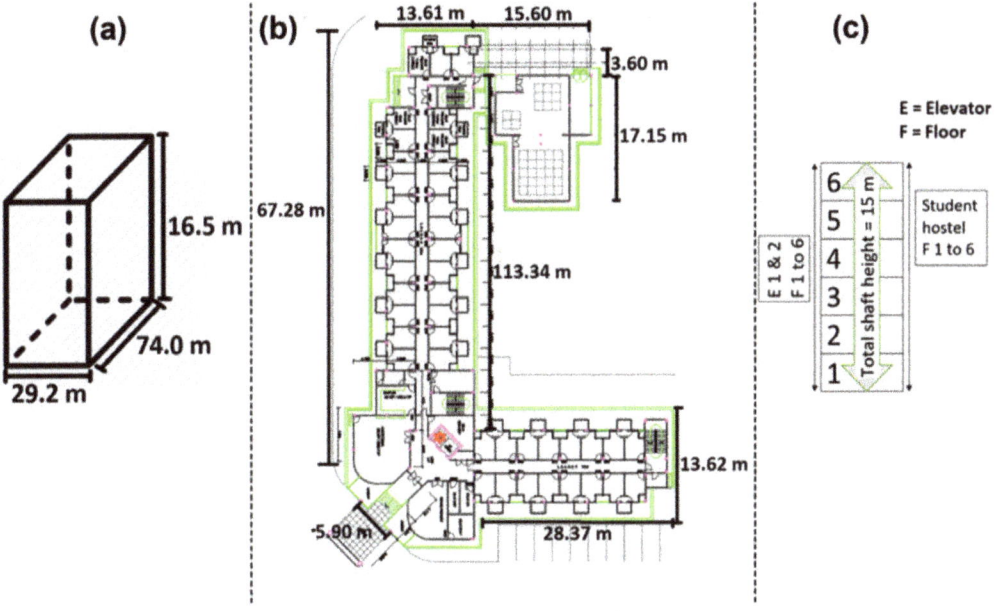

Figure 2. LA (**a**) building dimensions, (**b**) building floor plan, and (**c**) function and operating lift of different floors.

Sustainability 2022, 14, 4779

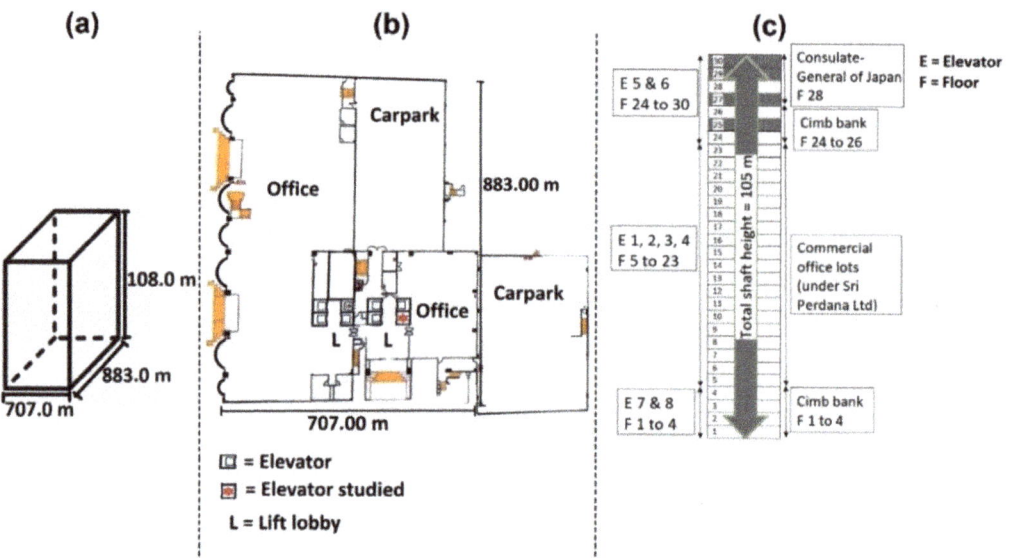

Figure 3. HO (**a**) building dimensions, (**b**) building floor plan, and (**c**) function and operating lift of different floors. Shaded floors are vacant during the study.

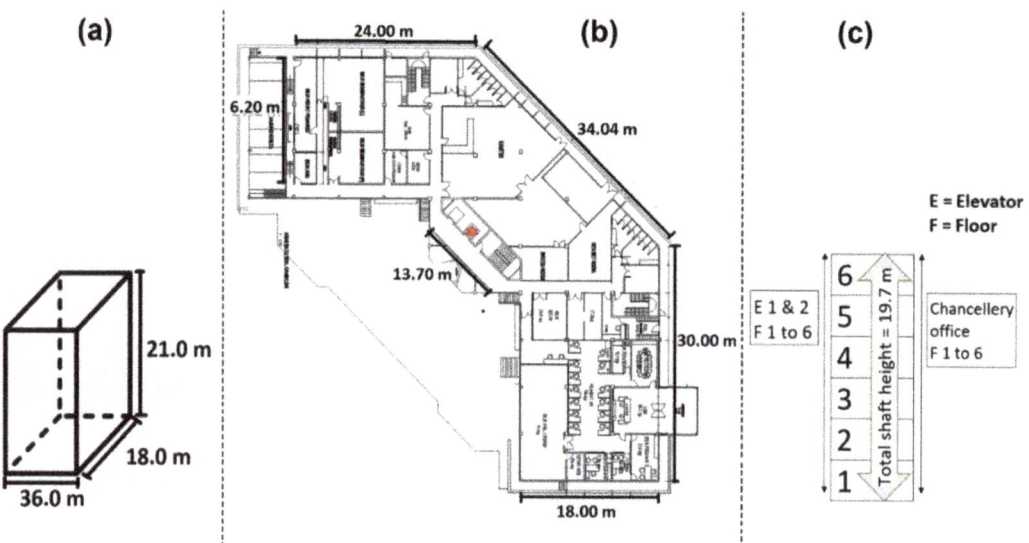

Figure 4. LO (**a**) building dimensions, (**b**) building floor plan, and (**c**) function and operating lift of different floors.

2.2. Field Data Collection

Based on this study's conditions, a hybrid approach of measurement and calculation was used to estimate the energy consumption of the elevator systems in the selected buildings. The annual energy consumption of the elevator systems was estimated based on the elevator traffic and operation power following guidelines from ISO 25745 [38]. The elevator traffic was determined based on observation at the elevator site for two weeks to include the variations among weekdays and weekends [35]. The elevator traffic was measured in terms of the number of trips travelled by an elevator, in which a single set of start and stop represented one trip. The average daily number of trips over two weeks was recorded and used for the calculation of each building [36,38]. On the other hand, the operation power was calculated from the data of running power, standby power, and idle power provided by the building management departments and (or) the elevator manufacturer company. The results served to evaluate energy efficiency [20,59].

2.3. Elevator Energy Consumption Analysis

The main formulas used in the calculation of the energy consumptions are stated in Equations (3)–(6). Further details of the calculations are shown in [32,36,38,60,61]:

$$E_d = E_r + E_{nr} \tag{1}$$

$$E_r = \frac{(n_d \times 0.5\, t_{av} \times R)}{3600} \tag{2}$$

$$E_{nr} = \frac{t_{nr}}{100}(P_{id}R_{id} + P_{st5}R_{st5} + P_{st30}R_{st30}) \tag{3}$$

$$E_a = E_d + D \tag{4}$$

where, E_d, E_a, E_r, and E_{nr} are the daily energy consumption of the elevator system, the annual energy consumption of the elevator system, the daily running energy of the elevator system, and the daily nonrunning energy of the elevator system in kWh, respectively. Moreover, n_d is the number of trips per day, D is the number of days operated, t_{av} is the average trip time the elevator travelled (s), R is the rated power of the elevator (kW), and t_{nr} is the nonrunning time of the elevator system (s). In all the above equations, P_{id} and R_{id} are the idle power (W) and the time ratio of the idle phase to the nonrunning phase, respectively. P_{stx} and R_{stx} (x = 5, 30) are the first "x" minutes of standby power (W), and the time ratio of the first "x" minutes standby phase to the nonrunning phase, respectively.

2.4. Elevator Energy Efficiency Analysis

The efficiency class of daily energy consumption was determined using the equations shown in Table 1 taken from the ISO 25745-2:2015 guidelines [38]. Class A shows the highest energy efficiency, while class G shows the lowest energy efficiency. In the equations in Table 1, the parameters accounted for in the energy efficiency evaluation are as follows: Q is the rated load of the elevator car (kg), n_d is the daily number of trips, S_{av} is the average distance travelled (m), and t_{nr} is the nonrunning time of the elevator system (s).

Table 1. Energy efficiency class of elevator systems based on ISO 25745 standards [38]. Reproduced with permission from ISO, ISO 25745-2:2015 guidelines; published by ISO, 2015.

Energy Efficiency Class	Energy Consumption Per Day (Wh)
A	$0.72 \times Q \times n_d \times S_{av}/1000 + (50 \times t_{nr})$
B	$1.08 \times Q \times n_d \times S_{av}/1000 + (100 \times t_{nr})$
C	$1.62 \times Q \times n_d \times S_{av}/1000 + (200 \times t_{nr})$
D	$2.43 \times Q \times n_d \times S_{av}/1000 + (400 \times t_{nr})$
E	$3.65 \times Q \times n_d \times S_{av}/1000 + (800 \times t_{nr})$
F	$5.47 \times Q \times n_d \times S_{av}/1000 + (1600 \times t_{nr})$
G	$5.47 \times Q \times n_d \times S_{av}/1000 + (1600 \times t_{nr})$

S_{av} is the average displacement per day.

2.5. Elevator Carbon Emissions Analysis

According to the United States Environmental Protection Agency (US EPA) and the Intergovernmental Panel on Climate Change (IPCC) guidelines, carbon emissions were calculated by multiplying the energy consumption with the carbon emission factor [62–64]. Since carbon dioxide is the most significant greenhouse gas (GHG), this study focused only on the carbon dioxide emissions from anthropogenic activities in building sectors [65].

The grid carbon dioxide emissions factor of Malaysia was obtained from the Sustainable Energy Development Authority (SEDA) of Malaysia in accordance with the revised 1996 IPCC Guidelines for National Greenhouse Gas Inventories and 2006 IPCC Guidelines for National Greenhouse Gas Inventories [66–69]. In this study, the carbon emissions of the elevator in each building were calculated using Equation (7), where E_a is the annual energy consumption of the elevator, and 0.639 tCO_2/MWh is the latest grid carbon dioxide emissions factor of Malaysia (updated in 2016) [67].

$$\text{Carbon emissions (tCO}_2) = E_a(\text{MWh}) \times 0.639 \left(\frac{\text{tCO}_2}{\text{MWh}}\right) \quad (5)$$

3. Results and Discussion

3.1. Energy Consumption and Energy Efficiency of the Elevator Systems

Table 2 shows the usage, usage category, running power, idle power, standby power 5 min, standby power 30 min, and nonrunning power, as well as daily and annual energy consumption of the elevator systems in the selected buildings based on the present study. In accordance with ISO 25745-2:2015, the elevator usage category was categorized based on the counted number of trips of the elevator and was indicated by an integer from 1 to 6 with increasing usage intensity or frequency.

Table 2. Daily usage, usage category, and energy consumption of the elevator systems in the selected buildings.

Elevator Energy Aspects	High-Rise Building		Low-Rise Building	
	Residential Apartment (HA)	Commercial Office (HO)	Residential Apartment (LA)	Commercial Office (LO)
No. of trips per day (trip)	340	504	176	589
Usage category	3	4	2	4
Running power, P_r (W)	13,000	13,000	4900	8500
Idle power, P_{id} (W)	1334.9	1334.9	1326.45	208
Standby power 5 min, P_{st5} (W)	252.6	252.6	190.25	120.1
Standby power 30 min, P_{st30} (W)	161.6	161.6	128.3	81.7
Nonrunning power, P_{nr} (W)	1749.1	1749.1	1645	409.8
Daily energy consumption (kWh)	49.81	52.52	5.35	29.68
Annual energy consumption (kWh)	14485	13,617.6	1656.48	6422.88

Based on Figure 5, the present study reported that the annual energy consumption of the elevator systems increased with the daily energy consumption of the elevator systems, as expected. This can be justified by the calculation methods of elevator annual energy consumption in the ISO and VDI standards. In the mentioned methods, the daily elevator traffic is assumed to be consistent throughout the year and is used to calculate the annual energy consumption by multiplication with the operating days of the elevator system throughout the year [37,38]. This is because office and residential buildings are occupied by mostly long-term tenants and operate as usual throughout the entire year.

The main contributing factors to elevator energy consumption are elevator usage and building height, which affects the total shaft height, and there are positive correlations between each of these variables, which are illustrated in Figure 6. In past and present studies, for similar-height buildings, the annual energy consumption of the elevator system has increased with the annual number of trips for the elevator system [70,71]. However, in the comparison between high-rise and low-rise buildings, the low-rise buildings showed a higher annual number of trips but lower annual energy consumption because the lower building height had a lower total shaft height (full travel distance), leading to a lower total distance travelled throughout the year. This phenomenon is shown in Figure 6, where the low-rise residential apartment and low-rise commercial offices had a high annual number of trips but lower annual energy consumption compared to the high-rise residential apartment and high-rise commercial offices. The high-rise buildings had a higher full travel distance for the elevators and heavier traffic due to a lower number of shared trips and high tenant density. Higher building height with a higher number of floors caused the tenants to be distributed more widely among the different floors, causing a fewer number of trips to be shared by more than one tenant [71]. Moreover, the higher tenant density in the high-rise residential apartment caused a higher difference in the working and school hours of different tenants in the high-rise residential apartment. The high-rise commercial office, which is a multi-tenant office, had lighter traffic compared to the low-rise commercial office, which is a single-tenant office with higher interfloor trips [72]. The low-rise office showed the highest number of trips among all four buildings in the study. This was due to the limitations of the building structure and elevator system of the low-rise office building, which is further elaborated in Section 3.3.

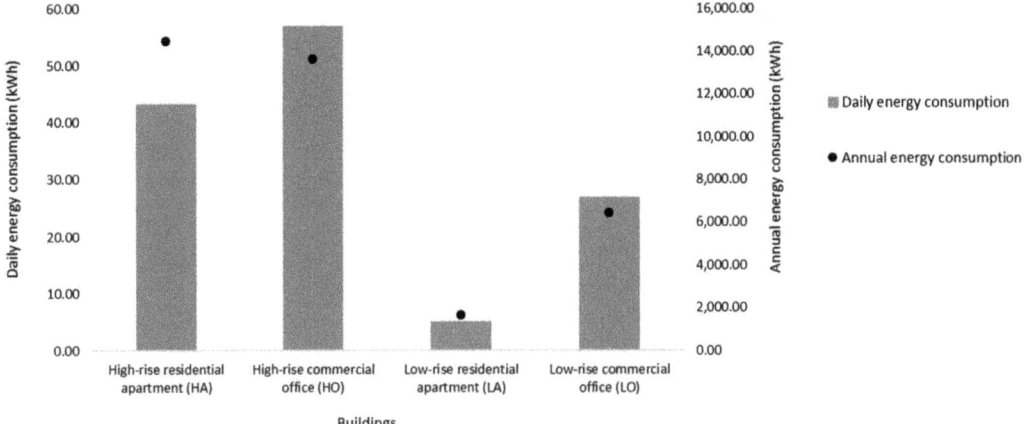

Figure 5. Daily energy consumption versus annual energy consumption of the elevator systems in the buildings in the present study.

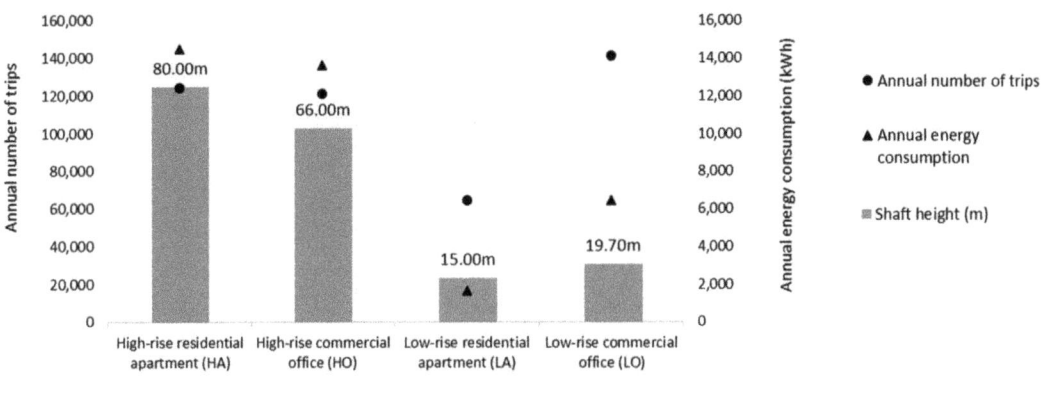

Figure 6. Annual number of trips versus annual energy consumption of the elevator system in buildings with different shaft heights based on the present study.

The data from existing studies are collected and shown in Tables 3 and 4. As expected, the elevator energy consumption of the high-rise buildings was higher than that of the low-rise buildings due to their higher full travel distance. Also, building use affected elevator usage and energy consumption. Commercial office buildings recorded a higher elevator energy consumption than residential apartment buildings. This was because the elevator usage of commercial buildings, such as offices, is generally higher than that of residential buildings, such as apartments, due to the higher incoming and outgoing population and activities. Considering the similar number of elevator cars of the elevator systems, the results from this study obey the theoretical results and the trends in past studies' results. Despite this, there are some noteworthy trends shown in Figure 6, in which the annual elevator energy consumption of the high-rise residential apartment building showed a slightly higher value than that of the high-rise commercial office buildings. This was because the high-rise commercial office building has a zoned elevator system with a higher number of cars servicing only specific floors, comprising six cars, compared to the high-rise residential apartment building with only three cars servicing all floors. A higher number of cars distributes the demand for trips travelled, decreasing the energy consumption of each car. Different elevators zones have different home landing or parking floors (resting terminals), enabling more efficient travel among the specific floors. For the cases with the same number of cars in the elevator systems, the low-rise commercial office building had a higher annual elevator energy consumption than the low-rise residential apartment building.

An analysis was conducted based on the elevator annual energy consumptions of past studies and the present study. The past study data are shown in the boxplot with blue markers illustrating the present study (Figure 7). All buildings, except the low-rise office building, had annual energy consumptions of an elevator system within the ranges of typical elevator annual energy consumption. This was due to the limitations of the building structure and elevator system of the low-rise office building, which is further elaborated in Section 3.3.

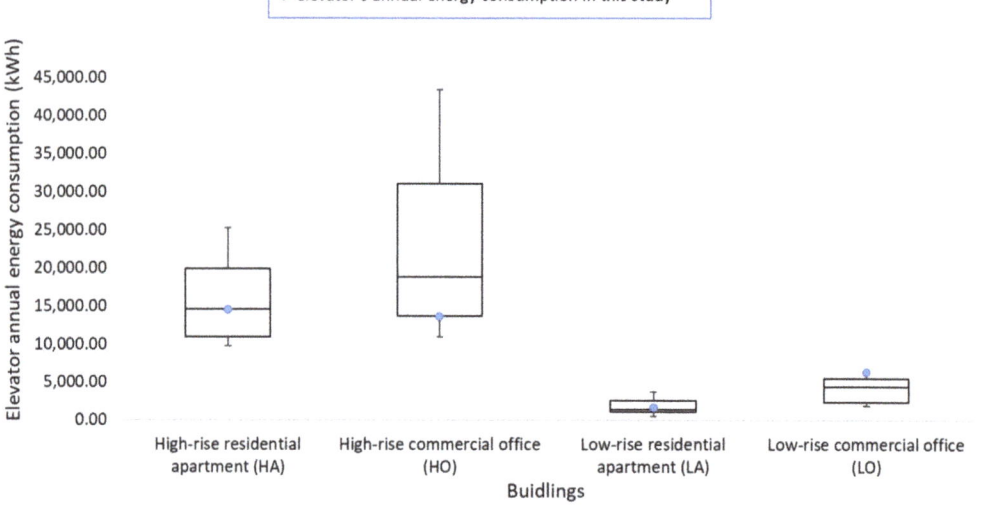

Figure 7. Annual energy consumptions of the elevator systems in high-rise and low-rise residential and office buildings based on the present study and past studies.

Table 3. Characteristics and annual energy consumption of the elevator systems in different heights of residential buildings based on past studies.

	Building Height (m) and Floors	Elevator Type	Parameters			Annual Energy Consumption (kWh)	Ref.
			Elevator Speed (m/s)	Load (kg)	Elevator Travel Power (kW)		
High-rise	136.5; 45	Gearless traction	3	1350	39.7	26,462.50	[73]
	230.6; 69 (The Met)	Gearless traction, regenerative, compact machine room, KONE Jump-Lift	4.0–6.0	2000	N/A	19,846.50	[74–76]
	180.0; 63 (Ashok tower)	Gearless traction, regenerative, compact machine room, KONE Mini-Space (old)	4	2000	N/A	10,878.00	[77,78]
Mid-rise	59.0; 16	Traction, gearless PMSM, non-regenerative	2.5	1500	3.8	5324.00	[32]
	50.0; 14	Gearless traction	1.5	1000	13	4350.00	[79]
	26.1; 7	Geared traction	1.6	750	7.2	4763.30	[80]
	29.0; 7	Gearless traction	1	630	5.3	3969.40	[80]
	23.7; 6	Geared traction	1	1000	8.5	5294.30	[80]
	14.0; 4	Geared traction	1	320	3.9	711.8	[80]
	14.0; 4	Geared traction	1	630	3.4	1126.60	[80]
	14.6; 4	Gearless traction	1	630	4.8	1219.10	[80]
	17.4; 4	Gearless traction	1	630	5.2	1274.90	[80]
	16.5; 4	Gearless traction	1	800	4.9	1408.90	[80]
Low-rise	11.2; 3	Hydraulic	0.6	320	9.5	2677.30	[80]
	13.4; 3	Hydraulic	0.6	500	12.6	2565.40	[80]
	12.4; 3	Hydraulic	1	500	4.7	1192.30	[80]
	11.6; 3	Geared traction	1	630	6	1481.00	[80]
	10.2; 2	Gearless traction	1.6	500	7.5	2329.40	[80]
	11.2; 5	Gearless traction	0.6	320	travel 8.93 mWh/(m·kg), standby 200W	2124.3	[81]
	14.0; 6	Gearless traction	1	630	4	950	[79]
	14.0; 6	Gearless traction	1	1000	6.1	953.8	[79]
	13.0; 6	Gearless traction	0.6	500	travel 91 Wh, idle 50 W, standby 31 W	511	[36]

Table 4. Characteristics and annual energy consumption of the elevator systems in different heights of office buildings based on past studies.

	Building Height (m) and Floors	Parameters					Ref.
		Elevator Type	Elevator Speed (m/s)	Load (kg)	Elevator Travel Power (W)	Annual Energy Consumption (kWh)	
High-rise	192.0; 48	Traction regenerative, MG-DC gearless	6.1	1360.8	N/A	47,815.0	[82]
	192.0; 48	Traction regenerative, Quattro-DC gearless	6.1	1360.8	N/A	23,725.0	[82]
	140.0; 35	Traction regenerative, MG-DC gearless	5.1	1360.8	N/A	33,507.0	[82]
	140.0; 35	Traction regenerative, Quattro-DC gearless	5.1	1360.8	N/A	13,833.5	[82]
Mid-rise	81.0; 19 (Moorhouse)	Gearless traction, KONE Mini-Space C7	4.0	2000.0	N/A	6900.0	[83]
	75.0; 21	Gearless traction	2.5	1600.0	Travel 170 Wh, idle 500 W, standby 120 W	12,306.0	[36]
	72.0; 18	Traction, VVVF, regenerative	2–3.0	1800.0	N/A	20,622.5	[84]
	76.0; 19	Traction regenerative, MG-DC gearless	4.1	1360.8	N/A	25,367.5	[82]
	76.0; 19	Traction regenerative, Quattro-DC gearless	4.1	1360.8	N/A	9198.0	[82]
	73.7; 18	Gearless traction	2.5	1500.0	19.3	23,494.3	[80]
	68.1; 17	Gearless traction	2.5	1500.0	27.1	42,705.0	[80]
	60.9; 15	Traction inductive AC non-regenerative	3.1	1135.0	N/A	38,106.0	[82]
	60.9; 15	Traction permanent magnet AC non-regenerative	3.1	1135.0	N/A	28,449.0	[82]
	60.9; 15	Traction SCR-DC regenerative	3.1	1135.0	N/A	27,798.0	[82]
	60.9; 15	Traction permanent magnet AC regenerative	3.1	1135.0	N/A	14,690.0	[82]
	51.9; 13	Gearless traction	1.6	2000.0	18.0	23,925.8	[80]
	42.9; 11	Gearless traction	2.0	3000.0	40.0	50,008.7	[80]

Table 4. *Cont.*

Building Height (m) and Floors		Parameters					Ref.
		Elevator Type	Elevator Speed (m/s)	Load (kg)	Elevator Travel Power (W)	Annual Energy Consumption (kWh)	
Low-rise	N/A	Traction	2.0	2000.0	19.0	17,700.0	[79]
	N/A	Traction	1.5	1000.0	13.0	4350.0	[40]
	27.5; 9	Gearless traction	1.0	630.0	4.2	3531.4	[80]
	19.6; 8	Gearless traction	1.5	1000.0	21.0	4388.9	[79]
	26.0; 7 (Five boats, Duisburg)	Gearless traction, KONE Mono-Space C5 with energy-efficient options activated	1.6	1000.0	N/A	5100.0	[83]
	18.0; 6	Non-regenerative geared traction	N/A	N/A	N/A	5800.0	[85]
	18.0; 6	Non-regenerative geared traction	N/A	N/A	N/A	6902.0	[85]
	18.0; 6	Regenerative geared traction	N/A	N/A	N/A	2441.0	[85]
	14.6; 4	Gearless traction	1.6	630.0	15.8	5010.7	[80]
	15.6; 4	Geared traction	1.0	800.0	6.7	2346.4	[80]
	11.8; 3	Geared traction	1.0	630.0	6.3	1956.0	[80]
	6.9; 2	Gearless traction	1.0	630.0	4.0	1827.9	[80]

The efficiency class of the daily energy consumption of an elevator had a positive correlation with the running and nonrunning energy consumption, but it also depended on the load factor, number of trips, and average displacement, as well as on running and nonrunning time [37,60]. Thus, an elevator system with good running and nonrunning energy performances did not necessarily have a good energy efficiency class. Based on the results of the present study, Table 5, the low-rise buildings showed relatively good energy efficiency (A and B) compared to the high-rise buildings (D and B). This can be justified by the lower distance to travel, as well as a smaller number of floors, that caused more passengers to share the same destination floor and pick-up floor, reducing the trip demand of the elevator system. Despite this, the high-rise commercial office recorded the same energy efficiency as the low-rise commercial office. This was due to the zoned elevator system of the high-rise office building and the limitations of the building structure and elevator system of the low-rise office building, as mentioned above. However, running and nonrunning energy consumption were the main factors of energy efficiency.

Table 5. Energy efficiency class of the elevator systems based on the present study.

Buildings	High-Rise Building		Low-Rise Building	
	Residential Apartment (HA)	Commercial Office (HO)	Residential Apartment (LA)	Commercial Office (LO)
Efficiency class	D	B	A	B

3.2. Carbon Emissions of the Elevator System

The annual carbon emissions were calculated from multiplication of the annual energy consumption of elevator with the constant grid carbon emission factor of Malaysia. Thus, the annual carbon emissions of the elevator system depended on only the annual energy consumption and increased with annual energy consumption (Figure 8). This was because carbon emissions are produced mainly from electricity generation, which is used to power the elevator operation. These results are supported by previous studies [86,87]. There is a positively correlated trend between growing urbanization and carbon emissions that are generated by the building sector. Carbon emissions depend on building height, which affects elevator usage. It was deduced that there is a positive correlation between energy consumption, building height, and carbon emissions.

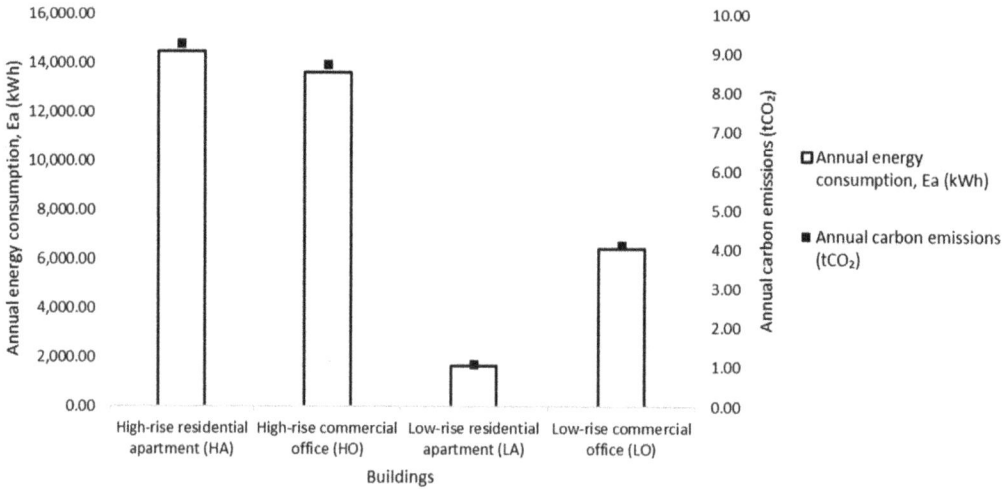

Figure 8. Annual energy consumption against the annual carbon emissions of the elevator systems.

In Figure 8, for the cases of the HA, HO, and LO buildings, the black markers indicating annual carbon emissions are further from the bar chart, indicating the annual energy consumption compared to that in the LA building. This can be justified by Equation (7), where elevator energy consumption is the only changing variable, and the emission factor is a fixed variable. Therefore, the only factor to be taken into account was the elevator energy consumption. The small value of elevator energy consumption in the LA building led to the small difference of the value between annual energy consumption and annual carbon emissions, and vice versa, for the HA, HO, and LO buildings. Therefore, the two graphs are more closely plotted for the LA building compared to the HA, HO, and LO buildings.

3.3. Limitations of the Studied Elevator Systems in the Buildings and Recommendations

Based on the present case study, lower efficiency was observed in the cases of high-rise residential apartments and the low-rise commercial office. An analysis of observation was performed by inspecting the elevator system and the building structure and operation. For the high-rise residential apartment, the elevator system did not have a zoned system. All three of the elevator cars travelled for the full shaft distance (covered all thirty-two service floors), making it less energy-efficient and operation-efficient. For the low-rise commercial office, resting terminals were on the ground floor, but its offices were mainly located from the second floor to the sixth floor, and its main entrance and car park were on the second floor. This caused the elevator to travel more from the ground floor resting terminal to the second floor to conduct passengers from the second floor to their destination floor and back to the ground floor resting terminal again. This issue greatly reduced the energy and operation efficiency of the elevator system. The second floor (where the main entrance and car park were located) should be set as the elevator resting terminal in order to avoid the unnecessary travel of the empty elevator car from the terminal to the main entrance and car park floor. Another alternative is the main entrance and car park should be built on the ground floor, which is the resting terminal of the elevator car.

4. Conclusions

The annual energy consumption, energy efficiency, and carbon emissions of the elevator systems in the selected low-rise and high-rise buildings in the Malaysian tropical climate were successfully evaluated. From the results, it was concluded that:

- The annual energy consumption of an elevator system had a positive correlation with the average daily energy consumption of the elevator system calculated on a weekly basis, since elevator traffic was relatively consistent weekly throughout the year. Low-rise buildings had a higher ratio of annual number of trips to annual energy consumption. The annual elevator energy consumption of the high-rise residential apartment building showed a slightly higher value than that of the high-rise commercial office buildings with an elevator zoning system;
- A higher number of cars distributing the demand for trips travelled decreased the energy consumption of each car. With the same number of cars in the elevator system, the low-rise commercial office building had a higher annual elevator energy consumption than the low-rise residential apartment building due to the higher incoming and outgoing tenants and passengers, as well as building structure limitations;
- Low-rise buildings showed relatively good energy efficiency compared to high-rise buildings;
- The usage, energy consumption, and carbon emissions had a positive correlation to each other.

The main challenge in this study was the assessment of elevator traffic for the elevator systems in the buildings during peak hours. This caused the risk of under- or overcounting the number of trips. As a result, future studies should look into a more advanced technology to enable a tracking system to replace manual counting, as well as auto-estimates of energy consumption, energy efficiency, and carbon emissions of the elevator systems. Consequently, the building management can have a better management plan to achieve the maximum energy efficiency and minimum emissions. Last but not least, this study contributed to the reference data of elevator energy consumption and carbon emissions to promote more similar, yet improved, versions of studies on different buildings in the future to facilitate energy efficiency and GHG mitigation efforts.

Author Contributions: Conceptualization, M.I.A., S.A.Z., A.S., J.H.A. and Y.Y.; investigation, J.H.A. and M.I.A.; methodology, J.H.A., S.A.Z., Y.Y. and M.I.A.; project administration, M.I.A., Y.Y. and S.A.Z.; resources, J.H.A., M.I.A., S.A.Z. and Y.Y.; visualization, J.H.A., Y.Y. and A.S.; writing—original draft, M.I.A., S.A.Z., A.S., J.H.A. and Y.Y.; writing—review and editing, M.I.A., S.A.Z., A.S. and Y.Y.; project administration, M.I.A., S.A.Z. and A.S.; funding acquisition, M.I.A. and S.A.Z. All authors have read and agreed to the published version of the manuscript.

Funding: The authors would like to thank Universiti Teknologi Malaysia and the Ministry of Education (MOE) for funding through the Fundamental Research Grant Scheme [FRGS/1/2019/TK07/UTM/02/5]. This work was also supported by the Universiti Sains Malaysia through funding related to this study (Publication Funding and 1001/PTEKIND/8014124).

Institutional Review Board Statement: Not applicable.

Informed Consent Statement: Not applicable.

Data Availability Statement: Not applicable.

Conflicts of Interest: The authors declare no conflict of interest.

Appendix A

Table A1. Characteristics of a studied building (high-rise residential apartment).

Use	Residential
Type	High-rise
Height (m)	Approximately 90
Shaft height (m)	80
Floor-to-floor height (m)	2.50
Number of blocks	4
Number of floors	32
Estimated number of units	247 per block, 8 per floor
Estimated built-up area (m^2)	Approximately 65 per unit, and 608.69 per block
Estimated total number of occupants (person)	1976
Number of elevator cars	3 per block

Table A2. Characteristics of a studied elevator car (high-rise residential apartment).

Type; brand	Gearless traction; Northern
Car dimension (m)	1.6 × 2.4 × 1.68
Rated load (kg); capacity	1150; 17 passengers
Rated speed (m/s)	1.50
Rated power (running power) * (kW)	13
Full travel distance (m)	80
Counterbalancing * (%)	50
Door operation time (open, remained open, close) (s)	12
Acceleration * (m/s^2)	1.00 (gearless traction type)
Jerk * (m/s^3)	1.20

* Assumptions based on the typical values of parameters for each type of standard passenger elevator.

Appendix B

Table A3. Characteristics of a studied building (low-rise residential apartment).

Use	Residential
Type	Low-rise
Building height (m)	16.50
Shaft height (m)	15
Floor-to-floor height (m)	2.50
Number of blocks	1
Number of floors	6
Estimated number of units	627
Estimated built-up area (m^2)	1615.1
Estimated total number of occupants (person)	627
Number of elevator cars	2

Table A4. Characteristics of a studied elevator car (low-rise residential apartment).

Type; brand	Geared traction; Otis
Car dimension (m)	1.37 × 2.10 × 1.40
Rated load (kg); capacity	750; 11
Rated speed (m/s)	0.65
Rated power (running power) * (kW)	4.90
Full travel distance (m)	15
Counterbalancing * (%)	50
Door operation time (open, remained open, close) (s)	9.13
Acceleration * (m/s^2)	0.80 (geared traction type)
Jerk * (m/s^3)	1.20

* Similar assumptions to Appendix A.

Appendix C

Table A5. Characteristics of a studied building (high-rise commercial office).

Use	Commercial
Type	High-rise
Height (m)	108
Shaft height (m)	66 out of 105 (accounting for only 19 occupied floors out of 28)
Floor-to-floor height (m)	3.50
Number of blocks	1
Number of floors	19 out of 30 (accounting for only 19 occupied floors out of 28)
Estimated number of units	94 out of 138 (accounting for only 19 occupied floors out of 28)
Estimated built-up area (m^2)	Office lot: from 65 to 278.71 Total building: 624,281
Estimated total number of occupants (person)	2070
Number of elevator cars	4 (for floor 5 to 28), 2 (for floor 1 to 4), 2 (for floor 24 to 28)

Table A6. Characteristics of a studied elevator car (high-rise commercial office).

Type; brand	Gearless traction; Northern
Car dimension (m)	1.78 × 2.20 × 1.60
Rated load (kg); capacity	900; 13 passengers
Rated speed (m/s)	1.60
Rated power (running power) * (kW)	13
Full travel distance (m)	66
Counterbalancing * (%)	50
Door operation time (open, remained open, close) (s)	4.17
Acceleration * (m/s^2)	1.00 (gearless traction type)
Jerk * (m/s^3)	1.2

* Similar assumptions to Appendix A.

Appendix D

Table A7. Characteristics of a studied building (low-rise commercial office).

Use	Office
Type	Low-rise
Building height (m)	21
Shaft height (m)	19.70
Floor-to-floor height (m)	2.74
Number of blocks	1
Number of floors	6
Estimated number of units	229
Estimated built-up area (m^2)	1148
Estimated total number of occupants (person)	458
Number of elevator cars	2

Table A8. Characteristics of a studied elevator car (low-rise commercial office).

Type; brand	Geared traction; Schindler
Car dimension (m)	1.55 × 2.24 × 1.55
Rated load (kg); capacity	1000; 15
Rated speed (m/s)	0.70
Rated power (running power) * (kW)	8.50
Full travel distance (m)	19.70
Counterbalancing * (%)	50
Door operation time (open, remained open, close) (s)	7.39
Acceleration * (m/s^2)	0.80 (geared traction type)
Jerk * (m/s^3)	1.20

* Similar assumptions to Appendix A.

References

1. Enerdata. World Energy Consumption Statistics. Available online: https://yearbook.enerdata.net/total-energy/world-consumption-statistics.html (accessed on 14 January 2020).
2. International Energy Agency. *World Energy Outlook 2017*; International Energy Agency: Paris, France, 2017; pp. 1–19.
3. U.S. Energy Information Administration (EIA). Emerging Asian Economies Drive the Increase in World Energy Use from 2015 to 2040. Available online: https://www.eia.gov/pressroom/releases/press448.php (accessed on 6 February 2022).
4. U.S. Energy Information Administration. Building Sector Energy Consumption. In *International Energy Outlook 2016*; U.S. Energy Information Administration (EIA): Washington, DC, USA, 2016; pp. 101–112.
5. Hassan, J.S.; Zin, R.M.; Majid, M.Z.A.; Balubaid, S.; Hainin, M.R. Building Energy Consumption in Malaysia: An Overview. *J. Teknol.* **2014**, *70*, 33–38. [CrossRef]
6. Ali, S.B.M.; Hasanuzzaman, M.; Rahim, N.A.; Mamun, M.A.A.; Obaidellah, U.H. Analysis of energy consumption and potential energy savings of an institutional building in Malaysia. *Alexandria Eng. J.* **2021**, *60*, 805–820. [CrossRef]
7. Chel, A.; Kaushik, G. Renewable energy technologies for sustainable development of energy efficient building. *Alexandria Eng. J.* **2018**, *57*, 655–669. [CrossRef]

8. Azari, R. *Life Cycle Energy Consumption of Buildings; Embodied + Operational*; Elsevier Inc.: Amsterdam, The Netherlands, 2019; ISBN 9780128117491.
9. Pérez-Lombard, L.; Ortiz, J.; Pout, C. A review on buildings energy consumption information. *Energy Build.* **2008**, *40*, 394–398. [CrossRef]
10. Department of Energy United States of America. Chapter 5—Increasing Efficiency of Building Systems and Technologies. In *Quadrennial Technology Review an Assessment of Energy Technologies and Research*; Department of Energy United States of America: Washington, DC, USA, 2015.
11. U.S. Energy Information Administration (EIA). *Monthly Energy Review January 2022*; U.S. Energy Information Administration: Washington, DC, USA, 2022; Volume 0035.
12. Aldhshan, S.R.S.; Abdul Maulud, K.N.; Wan Mohd Jaafar, W.S.; Karim, O.A.; Pradhan, B. Energy consumption and spatial assessment of renewable energy penetration and building energy efficiency in malaysia: A review. *Sustainability* **2021**, *13*, 9244. [CrossRef]
13. European Comission. Energy Use in Buildings. Available online: https://ec.europa.eu/energy/eu-buildings-factsheets-topics-tree/energy-use-buildings_en (accessed on 6 February 2022).
14. U.S. Energy Information Administration (EIA). Electricity Intensity of U.S. Homes and Commercial Buildings Decreases in Coming Decades. Available online: https://www.eia.gov/todayinenergy/detail.php?id=38332 (accessed on 17 April 2019).
15. Chen, S.; Zhang, G.; Xia, X.; Chen, Y.; Setunge, S.; Shi, L. The impacts of occupant behavior on building energy consumption: A review. *Sustain. Energy Technol. Assessments* **2021**, *45*, 101212. [CrossRef]
16. Godoy-Shimizu, D.; Steadman, P.; Hamilton, I.; Donn, M.; Evans, S.; Moreno, G.; Shayesteh, H. Energy use and height in office buildings. *Build. Res. Inf.* **2018**, *46*, 845–863. [CrossRef]
17. Ang, J.H.; Yusup, Y.; Zaki, A.; Salim, S.; Ahmad, M.I. A CFD Study of Flow Around an Elevator Towards Potential Kinetic Energy Harvesting. *J. Adv. Res. Fluid Mech. Therm. Sci. J.* **2019**, *59*, 54–65.
18. Al-kodmany, K. Tall Buildings and Elevators: A Review of Recent Technological Advances. *Buildings* **2015**, *5*, 1070–1104. [CrossRef]
19. Eninter Ascensores. What Determines the Number of Lifts a Building Needs? Available online: http://www.astarlifts.com/en/blog-lifts/lifts-lifts/what-determines-the-number-of-elevators-you-need-a-building (accessed on 17 April 2019).
20. Barney, G.; Al-Sharif, L. *Elevator Traffic Handbook: Theory and Practice*, 2nd ed.; Taylor & Francis: Abingdon-on-Thames, UK, 2016; ISBN 9781138852327.
21. Al-Sharif, L.; Seeley, C. The effect of the building population and the number of floors on the vertical transportation design of low and medium rise buildings. *Build. Serv. Eng. Res. Technol.* **2010**, *31*, 207–220. [CrossRef]
22. Hui, S.C.M.; Yeung, C. Analysis of standby power consumption for lifts and escalators. In Proceedings of the 7th Greater Pearl River Delta Conference on Building Operation and Maintenance—SMART Facilities Operation and Maintenance, Hong Kong, 6 December 2016; pp. 35–47.
23. Zubair, M.U.; Zhang, X. Explicit data-driven prediction model of annual energy consumed by elevators in residential buildings. *J. Build. Eng.* **2020**, *31*, 101278. [CrossRef]
24. Barney, G.C.; Loher, A.G. *Elevator Electric Drives: Concepts and Principles, Controls and Practice*; International Association of Elevator Engineers by Ellis Horwood: New York, NY, USA, 1990; ISBN 0132614626.
25. Kirchenmayer, G. *Energy Consumption of Elevators*; Elevator World, Inc.: Mobile, AL, USA, 1981; p. 48.
26. White, L.E. *Energy Consumption: Hydraulic Elevators and Traction Elevators*; Elevator World, Inc.: Mobile, AL, USA, 1984.
27. Barney, G. Energy Models for Lifts. In Proceedings of the 1st Symposium on Lift and Escalator Technologies, Northampton, UK, 29 September 2011; Volume 1, pp. 25–34.
28. Schroeder, J. Energy consumption and power requirements of elevators. In *Second Century of the Skyscraper*; Springer: Boston, MA, USA, 1988; Volume 34, pp. 621–627. [CrossRef]
29. Daniel, C.; Forth, J.; Arthur, W.; Eric, H.; Douglas, R.; Martin, H. Methods and Apparatus for Retrieving Energy Readings from an Energy Monitoring Device. U.S. Patent 7089089B2, 8 August 2006.
30. Chan, C.Y.B. *Elevator Drive Systems Energy Consumption Study Report—UBC Social Ecological Economic Development Studies (SEEDS)*; University of British Columbia: Vancouver, BC, Canada, 2012; p. 125.
31. Bannister, P.; Bloomfield, C.; Chen, H. Empirical prediction of office building lift energy consumption. In Proceedings of the Proceedings of the Building Simulation 2011, 12th Conference of International Building Performance Simulation Association, Sydney, Australia, 14–16 November 2011; pp. 2635–2642.
32. Tukia, T.; Uimonen, S.; Siikonen, M.L.; Hakala, H.; Donghi, C.; Lehtonen, M. Explicit method to predict annual elevator energy consumption in recurring passenger traffic conditions. *J. Build. Eng.* **2016**, *8*, 179–188. [CrossRef]
33. Al-Sharif, L. Lift Energy Consumption: General Overview (1974–2001). *Elev. World* **2004**, *52*, 61–67.
34. Al-Sharif, L.; Peters, R.; Smith, R. *Elevator Energy Simulation Model*; Elevator World, Inc.: Mobile, AL, USA, 2016.
35. Tukia, T.; Uimonen, S.; Siikonen, M.L.; Donghi, C.; Lehtonen, M. Modeling the aggregated power consumption of elevators—the New York city case study. *Appl. Energy* **2019**, *251*, 113356. [CrossRef]
36. Barney, G.; Lorente, A. Simplified Energy Calculations for Lifts Based on ISO/DIS 25745-2. In Proceedings of the Symposium on Lift and Escalator Technologies, Northampton, UK; The CIBSE Lifts Group: London, UK, 2013; Volume 3, pp. 10–19.

37. Association of German Engineers. *VDI 4707 Guideline—Lifts Energy Efficiency*; Association of German Engineers: Düsseldorf, Germany, 2007.
38. *ISO 25745-2:2015 (en)*; Energy Performance of Lifts, Escalators and Moving Walks—Part 2: Energy Calculation and Classification for Lifts (elevators). ISO: Geneva, Switzerland, 2015.
39. Hu, D.M.; Xu, B.; Yang, H.Y. VVVF controlled closed-circuit energy-saving hydraulic lift system. *J. Zhejiang Univ. Sci.* **2008**, *42*, 209.
40. Lehtonen, M.; Tukia, T. Determining and Modeling the Energy Consumption of Elevators. Master's Thesis, Aalto University, Espoo, Finland, 2014.
41. Jing, R.; Wang, M.; Zhang, R.; Li, N.; Zhao, Y. A study on energy performance of 30 commercial office buildings in Hong Kong. *Energy Build.* **2017**, *144*, 117–128. [CrossRef]
42. Mardiana, A.; Riffat, S.B. Building Energy Consumption and Carbon dioxide Emissions: Threat to Climate Change. *J. Earth Sci. Clim. Change* **2015**, *S3*, 1. [CrossRef]
43. Mirrahimi, S.; Mohamed, F.M.; Haw, L.C.; Lukman, N.I.; Fatimah, W.Y.M.; Aflaki, A.; Yusoff, W.F.M.; Aflaki, A.; Mohamed, M.F.; Haw, L.C.; et al. The effect of building envelope on the thermal comfort and energy saving for high-rise buildings in hot–humid climate. *Renew. Sustain. Energy Rev.* **2016**, *53*, 1508–1519. [CrossRef]
44. Tahir, M.Z.; Jamaludin, R.; Nasrun, M.; Nawi, M.; Baluch, N. Building energy index (BEI): A study of government office building in Malaysian public university. *J. Eng. Sci. Technol.* **2017**, *12*, 192–201.
45. Akashah, F.W.; Ali, A.S. *Analysis of Energy Trend of Office Building in Malaysia—Case Study in Universiti Pendidikan Sultan Idris (UPSI)*; University of Malaya: Kuala Lumpur, Malaysia, 2017.
46. Sadrzadehrafiei, S.; Mat, K.S.S.; Lim, C. Energy consumption and energy saving in Malaysian office buildings. *Models Methods Appl. Sci.* **2009**, *75*, 152–156.
47. Tee, B.T.; Yahaya, A.Z.; Breesam, Y.F.; Dan, R.M.; Zakaria, M.Z. Energy analysis for lighting and air-conditioning system of an academic building. *J. Teknol.* **2015**, *76*, 5536.
48. Saidur, R. Energy consumption, energy savings, and emission analysis in Malaysian office buildings. *Energy Policy* **2009**, *37*, 4104–4113. [CrossRef]
49. Sadrzadehrafiei, S.; Sopian, K.; Mat, S.; Lim, C.; Hashim, H.S.; Zaharim, A. Enhancing energy efficiency of office buildings in a tropical climate, Malaysia. *Int. J. Energy Environ.* **2012**, *6*, 209–310.
50. Taufiq, B.N.; Masjuki, H.H.; Mahlia, T.M.I.; Amalina, M.A.; Faizul, M.S.; Saidur, R. Exergy analysis of evaporative cooling for reducing energy use in a Malaysian building. *Desalination* **2007**, *209*, 238–243. [CrossRef]
51. Ossen, D.R.; Ahmad, M.H.; Madros, N.H. Impact of solar shading geometry on building energy use in hot humid climates with special reference to Malaysia. In Proceedings of the NSEB2005—SUSTAINABLE SYMBIOSIS, National Seminar on Energy in Buildings, UiTM, Subang Jaya, Malaysia, 10–11 May 2005; pp. 10–11.
52. Noranai, Z.; Kammalluden, M.N.; Tun, U.; Onn, H. Study of building energy index in Universiti Tun Hussein Onn Malaysia. *Acad. J. Sci.* **2012**, *1*, 429–433.
53. Saidur, R.; Hasanuzzaman, M.; Yogeswaran, S.; Mohammed, H.A.; Hossain, M.S. An end-use energy analysis in a Malaysian public hospital. *Energy* **2010**, *35*, 4780–4785. [CrossRef]
54. Tang, F.E. An Energy Consumption Study for a Malaysian University. *Int. J. Environ. Chem. Ecol. Geol. Geophys. Eng.* **2012**, *6*, 99–105.
55. Saidur, R.; Masjuki, H.H. Energy and associated emission analysis in office buildings. *Int. J. Mech. Mater. Eng.* **2008**, *3*, 90–96.
56. Bureau of Energy Efficiency. *Energy Conservation Building Code (ECBC)—Building Envelope for Hot & Dry Climate 2017*; Bureau of Energy Efficiency: New Delhi, India, 2017; p. 26.
57. Allied Market Research. *Traction Elevator Market by Order Type (Geared Traction Elevators and Gearless Traction Elevators) and Application (Residential and Commercial): Global Opportunity Analysis and Industry Forecast 2021–2030*; Allied Market Research: Portland, OR, USA, 2021.
58. Elevator Wiki. Traction Elevators. *Elevatorpedia* 2019; pp. 1–5. Available online: https://elevation.fandom.com/wiki/Traction_elevators (accessed on 6 February 2022).
59. Menezes, A.C.; Cripps, A.; Buswell, R.A.; Wright, J.; Bouchlaghem, D. Estimating the energy consumption and power demand of small power equipment in office buildings. *Energy Build.* **2014**, *75*, 199–209. [CrossRef]
60. ISO. Energy Performance of Lifts, Escalators and Moving Walks—Part 3: Energy Calculation and Classification of Escalators and Moving Walks. Available online: https://www.iso.org/obp/ui/#iso:std:iso:25745:-3:ed-1:v1:en (accessed on 10 December 2017).
61. Barney, G. Energy efficiency of lifts—Measurement, conformance, modelling, prediction and simulation. 2007. Available online: www.cibseliftsgroup.org/CIBSE/papers/Barney-on-energy%20efficiency%20of%20lifts.pdf (accessed on 6 February 2022).
62. U.S. Environmental Protection Agency. *Indirect Emissions from Purchased Electricity*; Environmental Protection Agency: Washington, DC, USA, 2018.
63. Intergovernmental Panel on Climate Change. *2006 IPCC Guidelines for National Greenhouse Gas Inventories*; IPCC: Geneva, Switzerland, 2006.
64. Malaysian Green Technology Corporation. *2017 Electricity Baseline for Malaysia*; Malaysian Green Technology Corporation: Selangor, Malaysia, 2017.
65. U.S. Environmental Protection Agency. Overview of Greenhouse Gases. Available online: https://www.epa.gov/ghgemissions/overview-greenhouse-gases (accessed on 6 January 2022).

66. Intergovernmental Panel on Climate Change. Precursors and Indirect Emissions. In *2006 IPCC Guidelines for National Greenhouse Gas Inventories*; IPCC: Geneva, Switzerland, 2006.
67. Sustainable Energy Development Authority Malaysia (SEDA). Average Carbon Dioxide Emission Factor of Malaysia. Available online: http://www.seda.gov.my/statistics-monitoring/co2-avoidance/ (accessed on 6 January 2022).
68. Intergovernmental Panel on Climate Change. *2019 Refinement to the 2006 IPCC Guidelines for National GHG Inventories*; IPCC: Geneva, Switzerland, 2019.
69. Intergovernmental Panel on Climate Change. *Revised 1996 IPCC Guidelines for National Greenhouse Gas Inventories*; IPCC: Geneva, Switzerland, 1996.
70. Patrão, C.; De Almeida, A.; Fong, J.; Ferreira, F. Elevators and Escalators Energy Performance Analysis. In Proceedings of the 2010 ACEEE Summer Study on Energy Efficiency in Buildings, Pacific Grove, CA, USA, 15–20 August 2010; pp. 53–63.
71. Siikonen, M.; Sorsa, J.; Hakala, H. Impact of Traffic on Annual Elevator Energy Consumption. In Proceedings of the ELEVCON 2010, Lucerne, Switzerland, 2–4 June 2010; Volume 18, pp. 344–352.
72. Hakala, H.; Siikonen, M.-L.; Tyni, T.; Ylinen, J. Energy-efficient elevators for tall buildings. In Proceedings of the Council on Tall Buildings and Urban Habitat, 6th World Congress, Melbourne, Australia, 26 February–1 March 2001; pp. 1–13.
73. Marsong, S. Analysis of Energy Consumption and Behavior of Elevator in a Residential Building. In Proceedings of the 5th International Electrical Engineering Congress, Pattaya, Thailand, 8–10 March 2017.
74. KONE Corporation. *KONE JumpLift*; KONE Corporation: Espoo, Finland, 2017.
75. CTBUH. The Met—The Skyscraper Center. Available online: http://skyscrapercenter.info/building/the-met/1134 (accessed on 7 February 2022).
76. Elevator Wiki Fandom. Kone JumpLift. Available online: https://elevation.fandom.com/wiki/Kone_JumpLift (accessed on 7 February 2022).
77. EMPORIS. Ashok Towers D, Mumbai 261721. Available online: https://www.emporis.com/buildings/261721/ashok-towers-d-mumbai-india (accessed on 19 September 2018).
78. KONE. *KONE MiniSpace*; KONE: Chertsey, UK, 2018.
79. Nipkow, J.; Schalcher, M. Energy consumption and efficiency potentials of lifts. *Hospitals* **2006**, *2*, 3.
80. VDI. *VDI4707 Technical Equipment for Buildings Manual*; VDI: Düsseldorf, Germany, 2008; Volume 5.
81. Zarikas, V.; Papanikolaou, N.; Loupis, M.; Spyropoulos, N. Intelligent Decisions Modeling for Energy Saving in Lifts: An Application for Kleemann Hellas Elevators. *Energy Power Eng.* **2013**, *5*, 31126. [CrossRef]
82. Vollrath, D. *Elevator Drives Energy Consumption & Savings*; Magnetek: New York, NY, USA, 2011; p. 16.
83. KONE Corporation. *KONE C-Series*; KONE Corporation: Espoo, Finland, 2008.
84. Nguyen, H.; Gaudet, E.W. Comparative Evaluation of Line Regenerative and Non-Regenerative Vector Controlled Drives for AC Gearless Elevators. In Proceedings of the Conference Record of the 2000 IEEE Industry Applications Conference, Thirty-Fifth IAS Annual Meeting and World Conference on Industrial Applications of Electrical Energy (Cat. No. 00CH37129), Rome, Italy, 8–12 October 2000; pp. 1431–1437.
85. Hackel, S.; Kramer, J.; Li, J.; Lord, M.; Marsicek, G.; Petersen, A.; Schuetter, S.; Sippel, J. *Proven Energy-Saving Technologies for Commercial Properties*; National Renewable Energy Laboratory: Golden, CO, USA, 2015; pp. 36–47.
86. US EPA. Sources of Greenhouse Gas Emissions. Available online: https://www.epa.gov/ghgemissions/sources-greenhouse-gas-emissions (accessed on 22 April 2019).
87. Herbert, R.B.; Malmström, M.; Ebenå, G.; Salmon, U.; Ferrow, E.; Fuchs, M. Quantification of abiotic reaction rates in mine tailings: Evaluation of treatment methods for eliminating iron- and sulfur-oxidizing bacteria. *Environ. Sci. Technol.* **2005**, *39*, 770–777. [CrossRef] [PubMed]

Article

Removal of Reactive Black 5 *Dye* by Banana Peel Biochar and Evaluation of Its Phytotoxicity on Tomato

Riti Thapar Kapoor [1], Mohd Rafatullah [2,*], Masoom Raza Siddiqui [3], Moonis Ali Khan [3] and Mika Sillanpää [4]

1. Amity Institute of Biotechnology, Amity University Uttar Pradesh, Noida 201 313, India; rkapoor@amity.edu
2. Division of Environmental Technology, School of Industrial Technology, Universiti Sains Malaysia, Penang 11800, Malaysia
3. Chemistry Department, College of Science, King Saud University, Riyadh 11451, Saudi Arabia; mrsiddiqui@ksu.edu.sa (M.R.S.); mokhan@ksu.edu.sa (M.A.K.)
4. Department of Biological and Chemical Engineering, Aarhus University, Nørrebrogade 44, 8000 Aarhus C, Denmark; mikasillanpaa@bce.au.dk
* Correspondence: mrafatullah@usm.my

Abstract: Removal of Reactive Black 5 (*RB5*) *dye* from an aqueous solution was studied by its adsorption on banana peel biochars (BPBs). The factors affecting *RB5 dye* adsorption such as pH, exposure time, *RB5 dye* concentration, adsorbent dose, particle size and temperature were investigated. Maximum 97% *RB5 dye* removal was obtained at pH 3 with 75 mg/L adsorbate concentration by banana peel biochars. Fourier transform infrared (FTIR) and scanning electron microscopy (SEM) were used to characterize the adsorbent material. The data of equilibrium were analyzed by Langmuir and Freundlich isotherm models. The experimental results were best reflected by Langmuir isotherm with maximum 7.58 mg/g adsorption capacity. Kinetic parameters were explored and pseudo-second order was found suitable which reflected that rate of adsorption was controlled by physisorption. Thermodynamic variables exhibited that the sorption process was feasible, spontaneous, and exothermic in nature. Banana peel biochar showed excellent regeneration efficiency up to five cycles of successive adsorption-desorption. Banana peel biochar maintained >38% sorption potential of *RB5 dye* even after five cycles of adsorption-desorption. The phytotoxic study exhibited the benign nature of BPB-treated *RB5 dye* on tomato seeds.

Keywords: banana peel biochar; reactive black 5; isotherm; kinetic; phytotoxicity; tomato

1. Introduction

The exponential growth in global population, industrialization, urbanization, and unskilled utilization of natural water resources has enhanced the requirements of freshwater. Due to the limited availability of freshwater resources, currently 1 billion people have no safe drinking water but this number may rise as world population is predicted to increase up to 10 billion by the year 2050 [1]. *Dyes* are widely used for the coloration of materials in textile, plastic, cosmetics, pharmaceutical, and paper industries [2–5]. Increasing demand for *dyes* in different industries has exacerbated the release of *dye* wastewater into the environment [6,7]. The textile industry, one of the big industries globally, utilizes synthetic *dyes* and approximately 8×10^5 tons of *dye* are produced per year [8,9]. Most of the textile industries are located in developing nations where they enhance employment opportunities and boost the economy of a country by foreign exchange earnings [10]. The textile industry consumes approximately 56% of total *dye* generated every year at a global level and releases >280,000 tons of *dyes* as industrial effluent which poses serious threat to the ecosystem [11,12]. The presence of a small amount of *dye* in water (<1 ppm) is quite visible and unacceptable [13,14]. Most of the industries do not follow effluent discharge

norms properly and they release large quantities of untreated or partially treated *dye* effluents in water resources, resulting pollution of the environment with a decrease in the availability of clean water [15,16]. The presence of *dyes* not only degrades the aesthetic value of water but also alters pH, BOD, and COD, enhances toxicity and turbidity, and reduces the sunlight penetration into water which leads to the deterioration of aquatic ecosystems [17,18]. *Dyes* show mutagenic and carcinogenic effects on animals and humans [11]. *Dye* adversely affects brain, kidney, liver, heart, respiratory, immune, and reproductive systems in humans [19]. Reactive *dyes* are utilized in fabric industries because of fast coloration, availability of reactive groups to form covalent bonds with different types of fibers, wide ranges of colors for printing and their permanent effect under a wide range of temperature [20]. According to Jozwiak et al. [21] around thirty percent of total coloring materials in the world are composed of reactive *dyes*. Reactive Black 5 (*RB5*), tetrasulphonated disazo *dye*, is used in fabric industries for coloring nylon and cotton stuff, etc. [22]. Reactive Black 5 and its intermediates are highly toxic in nature; however, no information is available in literature on the adverse effects of *RB5 dye* on animals, most of the reports revealed its deleterious impacts on human health. Some reports are available on Reactive red 120 which exerts genotoxic impact on *Catla catla* and damage DNA in the cells [23]. Reactive Black 5 exhibited adverse impacts on aquatic animals such as zebrafish embryo and digestive and central nervous system of humans [24,25]. The exposure to the *RB5 dye* causes allergy, skin irritation, nausea, bronchitis, confusion, high bold pressure, headache, cancer, etc., [26]. Removal of *dye* from industrial runoff is an arduous task due to stability of *dyes* against light, oxidation, temperature, and complex aromatic structure [27]. Different conventional methods such as coagulation, flocculation, electrochemical degradation, oxidation, membrane separation, ultra-filtration, microbial degradation, reverse osmosis, ion exchange, ozonation, and adsorption have been exploited to treat *dye* containing effluent [28–33]. Commercial adsorbents such as activated carbon, chitosan, graphene, and zeolite have been applied for removal of harmful contaminants from wastewater [34–36]. The Abovementioned wastewater treatment technologies are expensive, complex in operation, take a long operation time and are ineffective for *dye* removal on a large scale with sludge production at the end of the process.

Adsorption is considered as a significant treatment method because of its simple, affordable, and cost-effective nature and capacity to use locally available waste biomass [33]. Agricultural wastes such as wheat straw, cabbage and coconut waste, peanut husk, pumpkin seed hulls, mango seed husks, cashew nutshell, bamboo, spent tea leaves, walnut shells and orange peel have been used in preparation of carbon materials that can absorb *dye* from wastewater [37–40]. Biochar produced from agro-wastes are a rich source of bioavailable nitrogen, phosphorus, potassium, and magnesium [41]. Due to the high specific surface area and availability of functional groups on their surface, biochar can be used for the adsorption of contaminants. Banana (Family: Musaceae) plant grows around the year in the places with tropical climate. Banana peels are rich in carbon amount due to cellulose, hemicellulose, chlorophyll, and pectin presence and can be considered an excellent source of activated carbon [24]. The presence of functional groups such as hydroxyl, carboxyl, carbonyl, and amide groups on banana peel surface act as a binder in the adsorption process [24]. Banana peel has the potential to remove heavy metals, and pharmaceutical and phenolic compounds as reported by earlier workers [42,43]. Application of banana peel as an adsorbent removes contaminants from wastewater as well as solves the disposal problem of biowaste. In the earlier published work of Munagapati et al. [24], they used chemically modified banana peel powder for removal of *RB5 dye*. Their process was not eco-friendly as they utilized chemicals such as HCHO and HCOOH for modification/activation of banana peel biochar. They did not report the effect of chemically modified banana peel powder-treated *RB5 dye* solution on the growth of plants. In our study, we did not utilize any chemicals, so our process is environmentally benign. We also checked the impact of banana peel biochar-treated *RB5 dye* solution on germination and growth of tomato seeds. The irrigation of agricultural fields with *dye* contaminated water or industrial effluent inhibits crop growth and reduces land productivity [44,45]. In the present study, we wanted to evaluate the applicability of BPB-treated *RB5 dye* contaminated water for irrigation purposes. Wastewater after biochar treatment can be reused for various purposes and it can reduce the demand of fresh water. To the best of our knowledge, the phytotoxic effect of Reactive Black 5 *dye* and banana peel Biochar (BPB)-treated RB 5 *dye* solution on the development of tomato plants was not reported previously. Hence, the objective of the present investigation was to develop efficient and cost-effective adsorbent from banana peel for Reactive Black 5 *dye* removal from contaminated water and to analyze its impact on various growth parameters of tomato plants.

2. Materials and Methods

2.1. Preparation of Banana Peel Biochar and Proximate Analysis

Bananas were purchased from the local market of Noida, Uttar Pradesh, India. Banana peel was washed in tap water for 4–5 min, washed thrice with double deionized water to remove dust from its surface [46,47] as the presence of dust may provide different results of proximate analysis. Banana peel was cut into small pieces and dried at room temperature for 5 days to reduce moisture. The pyrolysis reactor was used for generation of biochars with nitrogen gas inside a pyrolyzer and the temperature was regulated by an electric heater. Dried banana peel was crushed and pyrolyzed at 500 °C for 3 h. Banana peel biochar was washed with lukewarm double deionized water and kept at 75 °C for 2 h in an electric oven to check micro-organisms growth [48]. The proximate analysis of BPB was conducted to confirm its stability for thermochemical transformation procedure. The proximate analysis was performed to verify ash, moisture, volatile material, and carbon contents.

2.2. Preparation of Dye Stock Solution

Reactive Black 5 *dye* was obtained from Sigma Aldrich (Mumbai, India). *RB5 dye* (1000 mg/L) stock solution was prepared with sterilized double deionized water which was used to prepare different concentrations as per requirements. The absorbance of *dye* was measured by UV-vis spectrophotometer (Shimadzu 1800, Kyoto, Japan) and the maximum absorbance (λmax) for *RB5 dye* was recorded at 597 nm (Table 1).

Table 1. Properties of *RB5 dye*.

Dye	RB5
Chemical name	Remazol Black B
Solubility	High solubility in water, easily forms covalent bonds with cellulosic fibers, resistant to sunlight and aerobic decomposition
Melting point	>300 °C
CAS number	17095-24-8
Dye type	Anionic *dye*
Appearance	Black colored powder
IUPAC Name	4-amino-5-hydroxy-3,6-bis[[4-(2-sulfonatooxyethylsulfonyl)phenyl]diazenyl]naphthalene-2,7-disulfonate
Empirical Formula	$C_{26}H_{21}N_5Na_4O_{19}S_6$
Molecular Weight	991.82 g/mol

Table 1. Cont.

Dye	RB5
Molecular Structure	(structure of RB5 dye)
λ$_{max}$	597 nm

2.3. Characterization of Banana Peel Biochar

Functional groups available on banana peel biochar (BPB) prior and afterwards *RB5 dye* sorption were analyzed by FTIR (Fourier-transform infrared spectroscopy) (Perkin Elmer 2000, Waltham, MA, USA) in wavenumber (400–4000 cm^{-1}) by utilizing KBr pellet method. Outer surfaces of BPB prior and afterwards *RB5 dye* sorption were observed by scanning electron microscopy (SEM) (Quanta FEG 650, Thermofisher, Beverly, CA, USA).

2.4. Batch Adsorption Experiments

The batch study was carried out to detect applicability of BPB as an adsorbent for removal of *RB5 dye*. The impact of various variables such as pH (3–11), adsorbent dose (0.2–1.0 g), size of particles (0–500 µm), concentration of *dye* (25–150 mg/L), temperature (25–65 °C), and contact time (30–180 min) were assessed at stirring speed 120 rpm for *RB5 dye* removal from aqueous solution by BPB. The experiment was conducted in six sets and in each set eighteen Erlenmeyer flasks were taken as triplicate, each flask with 100 mL of *RB5 dye* concentrations (25, 50, 75, 100, 125, and 150 mg/L) and 0.2, 0.4, 0.6, 0.8, 1, and 1.2 g BPB, respectively. The experiments were performed in triplicate. In the control set, no banana peel biochar was used in the *RB5 dye* containing flasks but in treatment, different concentrations of BPB were added in *RB5 dye* solution to study its *dye* sorption capacity. The UV-vis spectrophotometer (λmax = 597 nm) was utilized to detect concentration of *RB5* before and after BPB treatment and the absorption efficiency was measured by the following formula:

$$Removal\ of\ RB5\ dye = \frac{C_0 - C_t}{C_0} \times 100 \quad (1)$$

C_0 and C_t are initial and final *RB5 dye* concentrations in mg/L.

The isotherm models were applied for determination of sorption equilibrium. A total of 100 mL of *RB5 dye* (25–150 mg/L) solution were taken with different dosages of BPB to confirm the feasibility of the isotherm by comparing the adsorption potential. The Langmuir isotherm indicates that adsorption energy is constant over the adsorbate layer on the adsorbent surface at a constant temperature [49]. Langmuir equation is expressed as:

$$\frac{C_e}{q_e} = \frac{1}{q_e}K_L + \frac{C_e}{q_m} \quad (2)$$

where q_e (mg/g) = *RB5 dye* adsorbed at equilibrium, q_m (mg/g) = maximum *RB5* adsorbed, C_e = *dye* concentration at equilibrium (mg/L), K_L = Langmuir constant for binding ability of *RB5* on BPB.

The Freundlich isotherm illustrates distribution of *dye* molecule between BPB and solution at equilibrium. The isotherm defines an expanding inconsistency of active sites surface energy during adsorption and reduction in adsorption heat [50]. The Freundlich equation can be mentioned as:

$$ln q_e = ln K_F + \left(\frac{1}{n}\right) ln C_e \tag{3}$$

where Freundlich constants n = intensity of sorption and K_F = uptake capacity (n shows nature of process, $n < 1$ indicates chemisorption, $n > 1$ implies physisorption and $n = 1$ shows linear sorption).

The kinetic study determines the equilibrium time and rate of adsorption through adsorption modelling. Two kinetic models such as pseudo-first and second order were applied for rate constant calculation in sorption procedure. The pseudo-first order [51] and pseudo-second order [52] reaction mechanism were calculated by given equations:

$$\ln(q_e - q_t) = ln q_e - K_1 t \tag{4}$$

$$\frac{t}{q_t} = \frac{1}{k_2 q_e^2} + \frac{t}{q_e} \tag{5}$$

where q_e and q_t are RB5 dye adsorbed at equilibrium and time, k_1 = pseudo-first order adsorption rate constant (min^{-1}), and k_2 = pseudo-second order adsorption rate constant (g/mg.min).

2.5. Regeneration Analysis

The regeneration analysis was performed by the procedure of Kapoor and Sivamani [53]. In RB5 dye solution (100 mL) of 75 mg/L concentration, 0.8 g BPB was added and kept in shaking incubator under constant shaking condition at 32 °C for 45 min. Banana peel biochar containing dye was segregated by centrifugation. After filtration with Whatman no. 1 filter paper, the filtrate was analyzed by measuring absorbance for determination of dye content adsorbed by BPB. The blank sample (without BPB) was taken to compare the impact of BPB on RB5 dye removal compared with those samples in which BPB was added. After that, 0.1 g of RB5 dye containing biochar was placed at 50 °C for 6–7 h as drying may influence the sorption ability and microstructure of BPB [54] which was mixed with the desorbing solution. For the desorbing solution, 1N HCl and 1N NaOH reagents each were prepared in two different flasks separately and BPB was first washed with 1N HCl. Then the same biochar was washed with 1N NaOH and agitated at 180 rpm for 45 min. RB5 present in desorbing solution was recorded by UV-vis spectrophotometer. BPB isolated from desorbing solution was washed with sterilized distilled water three to four times to remove the desorbing solution and particles of BPB were kept at 50 °C for 9 h. The regeneration analysis was conducted up to five cycles for identification of re-applicability of used biochar. Dye desorption (%) was calculated by following formula [55]:

$$Desorption\ (\%) = Amount\ of \frac{RB5 desorbed}{Amount} of\ RB5\ adsorbed \times 100 \tag{6}$$

2.6. Evaluation of Phytotoxicity

Reactive Black 5 dye toxicity before and after treatment with BPB was tested on tomato seeds. A total of 9 test tubes, 9 petri plates, and 90 tomato seeds were taken for the phytotoxicity test. Seeds of tomato (Solanum lycopersicum L. variety Heera) were washed with tap water and the surface of seeds was sterilized with sodium hypochlorite solution (10% w/v) for 5 min for inhibition of microbial activities and cleaned again with sterilized distilled water. As the experiment was conducted in triplicate, nine test tubes were arranged and in each test tube, ten tomato seeds were kept in 10 mL of distilled water, RB5 dye solution (75 mg/L), and BPB-treated RB5 dye solution, respectively for 4 h according to the treatment. Soaking of the tomato seeds was performed for the activation of enzymes. Then, tomato seeds were transferred into sterilized petri plates (ten tomato seeds were placed in each petri plate) and the petri plates were kept in the seed germinator for 8 days under 87% relative humidity with 12 h photoperiod at 26 ± 2 °C. Three petri plates were taken for the control, three petri plates for *RB5 dye*, and three petri plates were taken for the BPB-treated *RB5 dye* solution. In the control set, distilled water was used for watering/irrigation of tomato seeds whereas two other sets were arranged in which tomato seeds of the second set were irrigated with *RB5 dye* solution (75 mg/L) and in the third set BPB-treated *RB5 dye* solution was used for watering of tomato seeds. *RB5 dye* concentration (75 mg/L) was selected for the evaluation of phytotoxic effects of *RB5 dye* on tomato seeds as maximum *dye* removal was obtained with this concentration in the batch study. Seed germination, length, and *vigor index* of 90 seedlings were measured in the control and treatment set of all the nine petri plates after 8 days of treatment [56]. The *germination percentage* and *vigor index* were analyzed by a given equation [57]:

$$\text{Germination (\%)} = \text{Total number of tomato seeds germinated} / \text{Total number of tomato seeds taken for germination} \times 100 \quad (7)$$

$$\text{Vigor index} = \text{Total length of seedling in mm} \times \text{germination percentage} \quad (8)$$

2.7. Estimation of Biochemical Components

The chlorophyll content was assessed in tomato seedlings through the Lichtenthaler [58] procedure. The total sugar and protein contents present in tomato seedlings were analyzed by the method of Hedge and Hofreiter [59] and Lowry et al. [60], respectively.

2.8. Statistical Analysis

Treatments with three replicates were arranged in a randomized block design. A randomized block design is an experimental design in which the experimental units are kept in groups called blocks. Data were analyzed by ANOVA and SPSS software and the treatment mean was calculated by DMRT at $p < 0.05$.

3. Results and Discussion

3.1. Proximate Analysis of Banana Peel Biochars

The proximate analysis was performed to confirm the amount of ash, fixed carbon, volatile material, and water content in banana peel biochar. Results reflected that BPB have 4.72% fixed carbon, 72.45% volatile material, 12.5% moisture, and 10.23% ash contents.

3.2. Characterization of Banana Peel Biochars

The spectrum 3407 cm^{-1} is assigned to OH stretching vibrations which may be due to the presence of moisture on BPB. The spectrum at 2920 and 2913 cm^{-1} was due to the CH-stretching vibration while the band at 2259 cm^{-1} showed the presence of C=C stretching vibration. The 1705 and 1726 cm^{-1} bands indicated C=O which showed the presence of carboxylic groups on BPB. The band at 1400 cm^{-1} reflected the CH bending vibration which showed stable binding and significant in adsorption process. There was a change in the CH stretching vibration at spectra 2130 cm^{-1} which reflected the presence of the methoxyl group as a result of the removal of lipids and lignin (Figure 1). The increase in the intensity of spectra after adsorption was due to the presence of C=C stretching vibration. Hydroxyl and carboxylic groups affected *RB5* adsorption [61]. The decrease in C-O-H stretching vibration peak in secondary cyclic alcohol was recorded after adsorption.

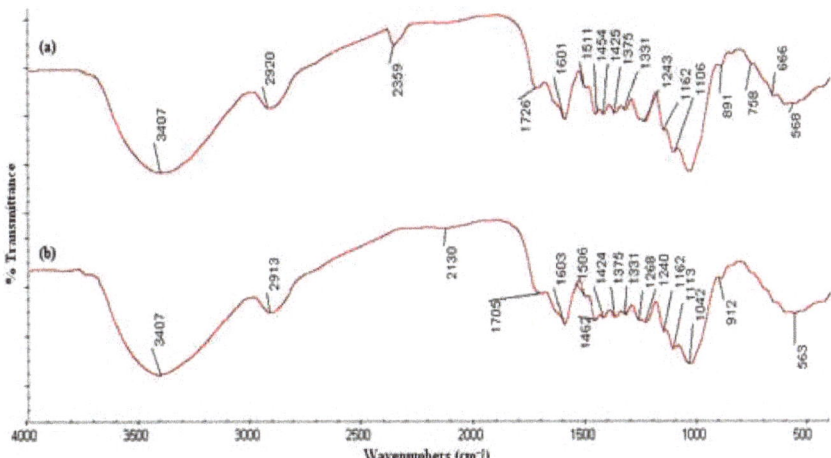

Figure 1. FTIR of banana peel biochars (**a**) before *RB5 dye* adsorption and (**b**) after *RB5 dye* adsorption.

The SEM micrograph of banana peel biochar before and after the adsorption process was recorded at the resolution of ×5000 magnification using 30 μm particle sizes. It was found that the morphology of the BPB before adsorption (Figure 2a) was different from banana peel biochars after adsorption of *RB5* (Figure 2b). In Figure 2a, the micrograph revealed the rough and porous surface of the banana peel biochar and it is due to the presence of lignin, pectin, and vicious compounds [61]. Banana peel biochar after adsorption with *RB5* can be observed with a rough and irregular surface because of the chemical alteration of the surface. Due to the *dye* uptake, lignin was oxidized and produced hydroxyl, carbonyl, and carboxyl groups which enhanced lignin solubility [62]. Functional groups present on the surface are responsible for improving the adsorption process through electrostatic interactions and chemisorption-based processes.

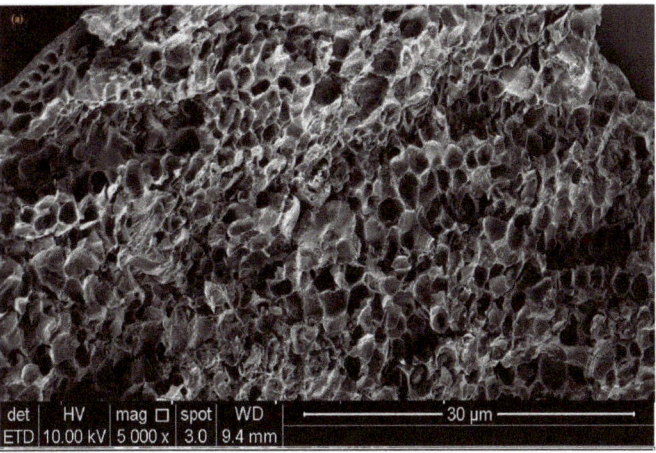

Figure 2. *Cont.*

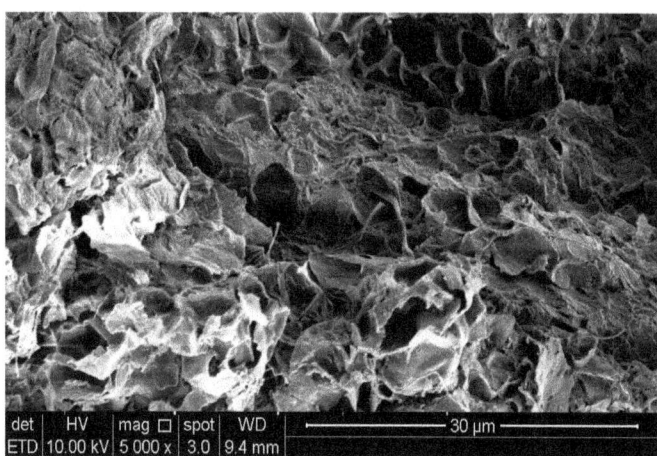

Figure 2. Scanning electron micrographs of banana peel biochar (**a**) before *RB5 dye* adsorption and (**b**) after *RB5 dye* adsorption.

3.3. Effect of Different Parameters on Reactive Black 5 Dye Adsorption by Banana Peel Biochar

3.3.1. pH

The solution pH plays a significant role in adsorption of *dye* on BPB. The degree of ionization, surface charge of adsorbent, and nature of *dye* solution were affected by the pH. The pH regulates electrostatic interactions between the functional groups available on the BPB surface and *dye* solution. The effect of pH on *RB5 dye* removal from an aqueous solution (25–150 mg/L *RB5* amount) was analyzed through the change in pH from 3 to 11 at 27 ± 2 °C. Maximum 96% removal of 75 mg/L *RB5 dye* was reported by BPB at pH 3 while 77, 56, and 34% *RB5* removal was observed at pH 7, 9, and 11, respectively (Figure 3a). The *RB5* removal was reported more at less pH due to the involvement of H^+. The surface of adsorbent was positively charged at low pH and attracted *RB5* which is anionic *dye*. At low pH, more *RB5* adsorption by BPB was due to electrostatic attraction [24].

3.3.2. Particle Size

Reactive Black 5 adsorption was assessed by three types of BPB particle sizes such as 0–170, 230–300, and 320–500 µm. As the BPB size reduces, adsorption of *RB5* molecules increases, because of the enlarged surface area of small particles, hence the surface area of BPB was directly proportional to *RB5* absorption (Figure 3a). For large-sized particles, diffusion resistance to mass transport is high and internal surface cannot be used for adsorption and due to this, less *dye* amount was adsorbed.

3.3.3. Contact Time

Exposure duration of interaction between BPB and *dye* is an important factor which plays a pivotal function in the kinetics of the adsorption process. The *dye* removal percentage was enhanced by increasing contact time (Figure 3b). The removal of *RB5 dye* was significant at the earlier stages in comparison with the last stage of the procedure which may be due to the availability of free sites on banana peel biochar. However, after 120 min there was no significant change in the adsorption efficiency, and it was considered as the equilibrium point for the adsorption process. The impact of exposure time for adsorption of *RB5 dye* was calculated to analyze equilibrium time. Reactive Black 5 *dye* removal of 77, 81, and 87% was recorded after 30, 60, and 90 min, respectively, for 75 mg/L *RB5 dye*. Two hours (120 min) was taken as equilibrium time in adsorption process as after 2 h, increase in *dye* adsorption was not reported.

3.3.4. Adsorbent Dose

The adsorbent amount can affect the adsorption adequacy. Reactive Black 5 *dye* removal was increased from 65 to 69% as BPB enhanced from 0.2 to 0.6 g. Highest 97% of *RB5 dye* removal was recorded with 0.8 g of BPB. More *dye* uptake rate was observed with high biochar amount due to the rise in active sites because of increased surface area and functional groups accessible for adsorption, these facilitate frequent binding of *RB5* on adsorption sites (Figure 3c).

3.3.5. *Dye* Concentration

The initial concentration of *dye* exhibits the significant effect on the adsorption capacity of the process. *Dye* concentration imparts energy for the regulation of mass transfer resistance of molecules between solid (adsorbent) and liquid (*dye* solution) stages [63]. Removal of *RB5 dye* by 0.8 g BPB was observed with different *dye* concentrations (25–150 mg/L). A significant amount of color was removed at a low concentration of *dye* whereas with high *RB5* concentration, the rate of *dye* removal was decreased as the adsorbent surface was completely infused (Figure 3d). High *RB5 dye* removal efficiency at a low concentration may be because of more interaction of *dye* molecules with the active sites available on the BPB. Reduction in adsorption efficiency by increasing *dye* concentration might be due to the saturation of active spaces of BPB or less vacancy of adsorbent sites or enhanced repulsive electrostatic force between surface of BPB and *dye* solution.

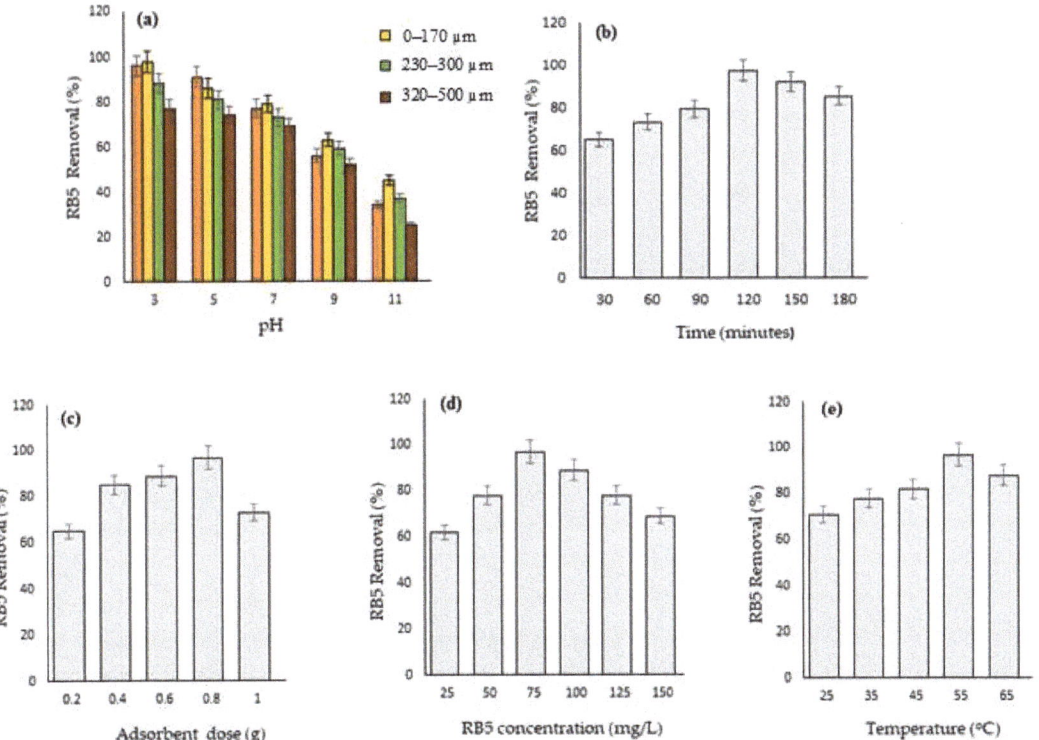

Figure 3. Impact of (**a**) pH and particle size for *RB5 dye* (75 mg/L), (**b**) contact period for *RB5 dye* (75 mg/L), (**c**) BPB dosage for *RB5 dye* (75 mg/L), (**d**) concentration of *dye*, (**e**) temperature on *RB5* removal (75 mg/L) by BPB.

3.3.6. Temperature

Reactive Black 5 *dye* adsorption on BPB was investigated under various temperature ranges from 25, 35, 45, 55, and 65 °C. *RB5 dye* exhibited 71 and 88% sorption at 25 and 65 °C respectively. Maximum 97% sorption of *RB5* was observed at 55 °C (Figure 3e). Temperature is an important parameter and it affects the transfer process and adsorption kinetics of *dyes*. A higher *RB5* sorption rate at high temperature was because of the increase in availability of sites on surface and more porosity and pore volume of adsorbent. Results reflected that the adsorption process was exothermic in nature.

3.4. Adsorption Isotherm

The adsorption isotherm indicated *RB5* molecules dissemination between liquid and solid stages under equilibrium at constant temperature. The isotherm model provides significant information on the mechanism of sorption, surface characteristics, and BPB ability. The isotherm results of *RB5 dye* sorption on BPB was analyzed by Langmuir and Freundlich models (Table 2; Figure 4). The Langmuir isotherm model is based on the assumption that monolayer adsorption occurs at homogeneous active sites on adsorbent structure, whereas the Freundlich model describes that adsorption occurs at non-uniform surfaces.

Table 2. Isotherm constants for *RB5 dye* adsorption by banana peel biochar.

Isotherm	Equation	Parameters	Value
Langmuir	$C_e/q_e = 1/q_e K_L + C_e/q_m$	q_m (mg/g) K_L (l/mg) R^2	7.58 0.0053 0.9489
Freundlich	$\ln q_e = \ln K_F + (1/n) \ln C_e$	$1/n$ K_F (mg/g) R^2	7.813 1.90294 0.4471

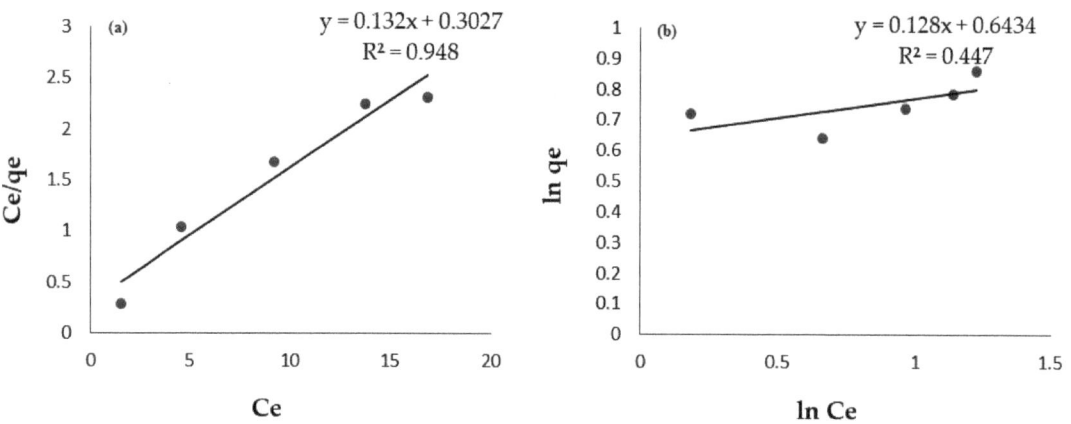

Figure 4. (**a**). Langmuir isotherm for *RB5 dye* adsorption by BPB (**b**). Freundlich isotherm for *RB5 dye* adsorption by BPB.

In this investigation, the Langmuir isotherm exhibited best fit model as it showed more correlation coefficient (R^2 = 0.9489) compared with Freundlich. It exhibited monolayer coverage of *RB5 dye* on banana peel biochar. After calculation, the values for Langmuir constants were q_m = 7.58 mg/g and k = 0.0053 mg^{-1} and Freundlich constants were K_F = 1.90294 and n = 7.813 and R^2 = 0.4471.

3.5. Adsorption Kinetic Models

A kinetic study provides information about adsorption efficiency and direction of reaction. Kinetic models were used to verify *RB5 dye* adsorption by BPB. The coefficient of determination (R^2) was 0.4111 and 0.9946 for pseudo-first and second-order models, respectively. Due to the high correlation coefficient value, the pseudo-second order kinetic model was followed (Table 3; Figure 5). Pseudo-second order kinetics exhibit chemisorption as the rate limiting step which was due to the physico-chemical interactions between the two phases. Data exhibited that the sorption procedure was controlled by uptake between the *RB5* molecules and BPB surface. The pseudo-second order model was found suitable in earlier findings such as *RB5 dye* adsorption by pumpkin seed husks [64], coffee waste [65], and macadamia seed husks [66].

Table 3. Kinetic variables for *RB5 dye* adsorption on banana peel biochars.

Model	Equation	Parameters	Value
Pseudo-first order	$\ln(q_e - q_t) = \ln q_e - k_1 t$	k_1 (min^{-1})	0.0096
		q_e (mg/g)	3.1852
		R^2	0.4111
Pseudo-second order	$t/q_t = 1/k_2 q_e + t/q_e$	K_2 (g/mg min)	0.0243
		q_e (mg/g)	8.9606
		R^2	0.9946

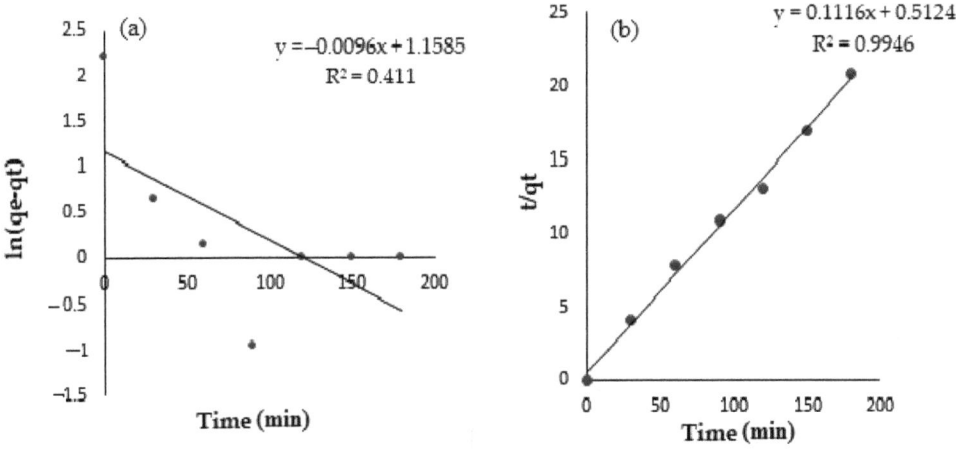

Figure 5. (a) Pseudo-first order model and (b) Pseudo-second order model for *RB5 dye* adsorption by BPB.

3.6. Thermodynamic Analysis

The change in free energy ($\Delta G°$), enthalpy ($\Delta H°$), and entropy ($\Delta S°$) were analyzed for RB 5 adsorption on banana peel biochar.

$$G° = -2.303RT \log K_d \text{ and } K_d = q_e/C_e \qquad (9)$$

$$\Delta G° = \Delta H° - T \Delta S° \qquad (10)$$

By repositioning the equation, we obtain $\log K_d = \Delta H^0/RT - (\Delta S^0)/R$, by applying the curve fitting method and $\Delta H°$ and $\Delta S°$ were calculated. Sorption experiments were conducted at 25, 35, 45, 55, and 65 °C (Table 4). The negative value of $\Delta G°$ at different temperature showed spontaneous nature of *RB5 dye* sorption on BPB [67].

Table 4. Thermodynamic variables for *RB5 dye* adsorption by banana peel biochar.

S.No.	Temperature (°C)	$\Delta G°$ (kJ/mol)	$\Delta H°$ (kJ mol^{-1})	$\Delta S°$ (J/K)
1.	25	−2363.79		
2.	35	−1506.13		
3.	45	−30.21	−11.223	30.457
4.	55	−3754.22		
5.	65	−596.03		

The negative value of $\Delta H°$ (−11.223 kJ mol^{-1}) established the exothermic nature of the process. The positive value of $\Delta S°$ (30.457 J/K) reflected an increase in the adsorbate content in solid state. Increased impermanence at solid-liquid confluence was recorded in sorption. The positive value of $\Delta S°$ reflects the randomness and stability of the adsorption procedure. Results revealed that *RB5 dye* adsorption on BPB was spontaneous and the exothermic procedure was consistent with our findings recorded in isotherm experiments.

3.7. Regeneration Analysis

Recyclability of BPB is an important parameter for the evaluation of total expenditure of the adsorption process which inhibits secondary pollution. Regeneration a is key indicator for the evaluation of the performance of an adsorbent. The regeneration process requires proper selection of eluent and it depends on the type of adsorbent and adsorption mechanism. Reactive Black 5 *dye* solution contains both positive and negative functional groups so both basic and acidic media are required to desorb *dyes* from the biochar surface. In an acidic medium, the solution consists of H$^+$ that attaches with the *dye* molecules with negative functional groups and desorbs from the adsorbent surface. Similarly, in basic medium, the *dye* molecules containing positive functional groups were removed [68]. Therefore, in the present analysis for obtaining maximum recovery, BPB was washed first with 1N HCl after that, the same biochar was washed with 1N NaOH. However, with the progression in the number of cycles, the efficacy of the *dye* removal reduced, which might be due to the blockage of the adsorption sites present in the micropores. For the regeneration analysis, 75 mg/L *RB5 dye* solution was used with 0.8 g banana peel biochar; as in batch experiments maximum 97% *RB5 dye* removal was observed with 75 mg/Ladsorbate concentration and 0.8 g BPB. Reproduced biochar exhibited 78, 62, 52, 43, and 38% *RB5 dye* adsorption efficiency from the first to fifth cycle (Figure 6). Hence, regenerated banana peel biochars can be reutilized for the uptake of *RB5 dye*.

3.8. Evaluation of Phytotoxicity

The phytotoxic effect of *RB5 dye* solution was analyzed prior and after the treatment with BPB by germination and morphological variables of tomato (*Solanum lycopersicum* L. variety Heera) seeds. Experiments were conducted in three sets: in the control set, distilled water was used for watering/irrigation of tomato seeds, whereas two other sets were arranged in which tomato seeds of second set were irrigated with *RB5 dye* solution (75 mg/L) and in the third set BPB-treated *RB5 dye* solution was used for treatment of tomato seeds. *RB5 dye* concentration (75 mg/L) was selected among the other *dye* concentrations for the evaluation of phytotoxic effects of *RB5 dye* on tomato seeds as maximum 97% *dye* removal was obtained with this concentration by 0.8 g of banana peel biochar. The seed *germination percentage* was calculated by counting the total number of tomato seeds germinated and after 8 days of seedling growth. Seedlings were used for the estimation of radicle and plumule length, *vigor index*, and analysis of biochemical components such as chlorophyll, sugar, and protein. Marked alterations were recorded among treatment for seed germination, seedling length, and *vigor index* of tomato seeds. In the control, 96% germination was noticed while tomato seeds treated with *RB5 dye* (75 mg/L) reflected

10% germination. Tomato seed germination was increased to 81% in *RB5* solution with BPB treatment. The length of radicle and plumule were 2.97 and 13.67 cm in the control but reduced to 0.19 and 1.93 cm with *RB5 dye*. Banana peel biochars-treated *dye* solution exhibited escalation in length of seedling and *vigor index* of tomato in comparison with *RB5 dye*. The tomato seeds *vigor index* showed the order: Control > *RB5 dye* solution treated with BPB > *RB5 dye* (Table 5). The biochemical components such as chlorophyll, sugar and protein were analyzed both in the control and treated tomato seedlings. The maximum amount of chlorophyll, sugar, and protein were reported in the control. Banana peel biochar-treated *RB5 dye* solution exhibited a significant reduction, 70, 71, and 76%, respectively, in total chlorophyll, sugar, and protein contents of tomato seedlings over the control. The reduction in biochemical parameters might be due to the adverse impact of *RB5 dye* on the physiological activities of tomato seeds. Similar results were reported by Kapoor and Sivamani [53].

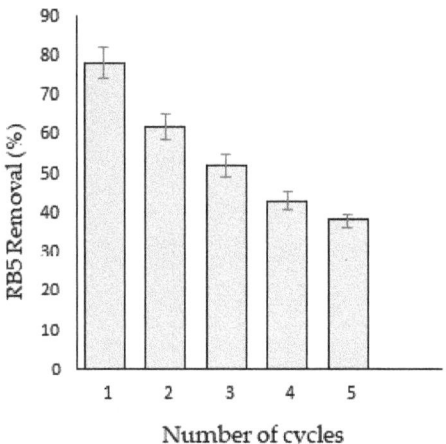

Figure 6. Reactive Black 5 adsorption by banana peel biochars up to five cycles.

Table 5. Phytotoxic effects of *RB5 dye* solution before and after treatment with banana peel biochar on germination, morphological and biochemical variables of *Solanum lycopersicum* L.

Treatment	Germination (%)	Length of Plumule (cms)	Length of Radicle (cms)	Vigor index	Total Chlorophyll (mg/g FW)	Sugar Content (mg/g DW)	Protein Content (mg/g FW)
Control	96 ± 1.41 [a]	13.67 ± 0.57 [a]	2.97 ± 0.29 [a]	15974.4 [a]	3.28 ± 0.36 [a]	3.74 ± 0.32 [a]	19.03 ± 0.57 [a]
RB5 dye solution (75 mg/L)	10 ± 0.71 [c]	1.93 ± 0.18 [d]	0.19 ± 0.01 [c]	212 [d]	0.99 ± 0.09 [c]	1.07 ± 0.06 [c]	4.53 ± 0.16 [d]
BPB-treated RB5 dye solution	81 ± 1.1 [b]	11.4 ± 0.74 [a]	2.4 ± 0.49 [a]	11178 [b]	2.33 ± 0.35 [b]	2.31 ± 0.32 [b]	15.57 ± 0.47 [b]

Values are mean ± standard error mean of 3 replicates from three independent experiments. Values showing different letters indicate significant difference among treatment at $p < 0.05$ significant level as per ANOVA.

3.9. Performance of Banana Peel Biochars

The prepared BPB capacity for *RB5 dye* adsorption was compared with similar studies as reported in Table 6. The development of adsorbent from biomass waste provides substitution of commercially available activated carbon and enhances cost effectiveness of the process [67]. Sorption ability (q_{max}) was utilized for comparison and it is in the line with previous findings, showing that *RB5 dye* can be easily adsorbed on BPB. The biochar prepared from different waste biomass resources showed a wide range of adsorption capacity for RB5 dye. It might be due to the difference in the surface area, pore size and availability of functional groups on the surface of biochar.

Table 6. Adsorption capacity (q_{max}) of RB5 dye with various adsorbents.

Biochars	Optimum Experimental Conditions	q_{max} (mg/g)	References
Coconut shell	pH = 2; exposure time = 60 min	0.82	Jozwiak et al. [69]
Pumpkin seed husk	pH = 3; exposure time = 60 min	1	Kowalkowska and Jozwiak [64]
Macadamia seed husk	pH = 3; exposure time = 510 min	1.21	Felista et al. [66]
Cotton fibers	pH = 3; exposure time = 240 min	2.74	Jozwiak et al. [70]
Potato peel	pH = 3; exposure time = 120 min	3.61	Samarghandy et al. [71]
Fly ash	pH = 5.64; exposure time = 60 min	7.94	Eren and Acar [72]
Pumpkin seed hulls	pH = 2; exposure time = 30 min	9.18	Celebi [46]
Eggshells	pH = 6; exposure time = 15 min	18.46	Celebi [46]
Coffee waste	pH = 7; exposure time = 50 min	77.52	Wong et al. [65]
Wood waste	Temperature = 25 °C; exposure time = 90 min	35.67	Figueiredo Do Nascimento [73]
Tobacco stalk biomass	pH = 2; exposure time = 120 min	92.84	Shah et al. [74]
Banana peel biochars	pH = 3; exposure time = 120 min	7.58	This study

Adsorption is considered as the most promising technology owing to its low cost, high selectivity and ease of operation. Banana peel biochar can be applied as an efficient adsorbent for removal of anionic *dyes* as observed in the findings of the present study. Further investigations are required for the utilization of the potential waste materials easily available at zero cost for its effective translation from laboratory scale treatment to real industrial effluent treatment at a large scale.

4. Conclusions

Azo *dye* treatment is an arduous task as these *dyes* are electron deficient xenobiotic compounds and recalcitrant to degradation. An FTIR analysis confirmed interactions between *RB5 dye* and functional groups available on the BPB surface. The Langmuir adsorption isotherm model best represented the experimental points and reflected highest 7.58 mg/g adsorption capacity. The negative value of ΔH^0 reflected that sorption was spontaneous and an exothermic process. Reproduced BPB showed positive results up to five successive cycles for removal of *RB5*. Hence, biochar prepared from renewable bio-waste i.e., banana peel, is a simple, inexpensive, sustainable, and efficient adsorbent for *RB5 dye* removal from contaminated water. Our findings represent promising alternatives for *RB5* removal from aqueous phases, but needs further research on a larger scale.

Author Contributions: R.T.K.: Methodology, Investigation, Writing—original draft; M.R.: Validation, Supervision Funding acquisition; M.R.S.: Writing—review and editing; M.A.K.: Writing—review and editing, Funding acquisition; M.S.: Writing—review and editing. All authors have read and agreed to the published version of the manuscript.

Funding: The authors acknowledge the financial support through Researchers Supporting Project number (RSP-2021/345), King Saud University, Riyadh, Saudi Arabia.

Institutional Review Board Statement: Not applicable.

Informed Consent Statement: Not applicable.

Data Availability Statement: Authors confirm to further provide the data related in this work if necessary.

Acknowledgments: Moonis Ali Khan acknowledges the financial support through Researchers Supporting Project number (RSP-2021/345), King Saud University, Riyadh, Saudi Arabia and the authors extend their appreciation to Amity University, Aarhus University, and Universiti Sains Malaysia for providing research facilities.

Conflicts of Interest: The authors declare no conflict of interest.

Abbreviations

The following abbreviations are used in this manuscript:

ANOVA	Analysis of Variance
BPB	Banana Peel Biochar
DMRT	Duncan's Multiple Range Test
FTIR	Fourier Transform Infrared Spectroscopy
RB5	Reactive Black 5
SEM	Scanning Electron Micrograph

References

1. Chowdhary, P.; Bharagava, R.N.; Mishra, S.; Khan, N. Role of industries in water scarcity and its adverse effects on environment and human health. In *Environmental Concerns and Sustainable Development: Air, Water and Energy Resources*; Shukla, V., Kumar, N., Eds.; Springer: Singapore, 2020; Volume 1, pp. 235–256.
2. Jawad, A.H.; Hum, N.N.M.F.; Farhan, A.M.; Mastuli, M.S. Biosorption of methylene blue dye by rice (*Oryza sativa* L.) straw: Adsorption and mechanism study. *Desalin. Water Treat.* **2020**, *190*, 322–330. [CrossRef]
3. Varjani, S.; Rakholiya, P.; Yong, N.H.; You, S.; Teixeira, J.A. Microbial degradation of dyes: An overview. *Bioresour. Technol.* **2020**, *314*, 123728. [CrossRef] [PubMed]
4. Sackey, E.A.; Song, Y.; Yu, Y.; Zhuang, H. Biochars derived from bamboo and rice straw for sorption of basic red dyes. *PLoS ONE* **2021**, *16*, e0254637. [CrossRef]
5. Shindhal, T.; Rakholiya, P.; Varjani, S.; Pandey, A.; Ngo, H.H.; Guo, W.; Ng, H.Y.; Taherzadeh, M.J. A critical review on advances in the practices and perspectives for the treatment of dye industry wastewater. *Bioengineered* **2021**, *12*, 70–87. [CrossRef] [PubMed]
6. Dassanayake, R.S.; Acharya, S.; Abidi, N. Recent advances in biopolymer-based dye removal technologies. *Molecules* **2021**, *26*, 4697. [CrossRef]
7. Ying, Z.; Chen, X.; Li, H.; Liu, X.; Zhang, C.; Zhang, J.; Yi, G. Efficient adsorption of methylene blue by porous biochar derived from soybean dreg using a one-pot synthesis method. *Molecules* **2021**, *26*, 661. [CrossRef] [PubMed]
8. Jamee, R.; Siddique, R. Biodegradation of synthetic dyes of textile effluent by microorganisms: An environmentally and economically sustainable approach. *Eur. J. Microbiol. Immunol.* **2019**, *9*, 114–118. [CrossRef] [PubMed]
9. Slama, H.B.; Chenari Bouket, A.; Pourhassan, Z.; Alenezi, F.N.; Silini, A.; Cherif-Silini, H.; Oszako, T.; Luptakova, L.; Golinska, P.; Belbahri, L. Diversity of synthetic dyes from textile industries, discharge impacts and treatment methods. *Appl. Sci.* **2021**, *11*, 6255. [CrossRef]
10. Desore, A.; Narula, S.A. An overview on corporate response towards sustainability issues in textile industry. *Environ. Dev. Sustain.* **2018**, *20*, 1439–1459. [CrossRef]
11. Mohamed, R.R.; Abu Elella, M.H.; Sabaa, M.W.; Saad, G.R. Synthesis of an efficient adsorbent hydrogel based on biodegradable polymers for removing crystal violet dye from aqueous solution. *Cellulose* **2018**, *25*, 6513–6529. [CrossRef]
12. Goswami, M.; Chaturvedi, P.; Sonwani, R.K.; Gupta, A.D.; Singhania, R.R.; Giri, B.S.; Rai, B.N.; Singh, H.; Yadav, S.; Singh, R.S. Application of Arjuna (*Terminalia arjuna*) seed biochar in hybrid treatment system for the bioremediation of Congo red dye. *Bioresour. Technol.* **2020**, *307*, 123203. [CrossRef]

13. Miyah, Y.; Lahrichi, A.; Idrissi, M.; Boujraf, S.; Taouda, H.; Zerrouq, F. Assessment of adsorption kinetics for removal potential of crystal violet dye from aqueous solutions using Moroccan pyrophyllite. *J. Assoc. Arab Univ. Basic Appl. Sci.* **2017**, *23*, 20–28. [CrossRef]
14. Ardila-Leal, L.D.; Poutou-Pinales, R.A.; Pedroza-Rodríguez, A.M.; Quevedo-Hidalgo, B.E. A brief history of colour, the environmental impact of synthetic dyes and removal by using laccases. *Molecules.* **2021**, *26*, 3813. [CrossRef]
15. Abd-Elhamid, A.I.; Emran, M.; El-Sadek, M.H.; El-Shanshory, A.A.; Soliman, H.M.A.; Akl, M.A.; Rashad, M. Enhanced removal of cationic dye by eco-friendly activated biochar derived from rice straw. *Appl. Water Sci.* **2020**, *10*, 45. [CrossRef]
16. Barathi, S.; Karthik, C.; Nadanasabapathi, S.; Padikasan, I.A. Biodegradation of textile dye Reactive Blue 160 by *Bacillus firmus* (Bacillaceae: Bacillales) and non-target toxicity screening of their degraded Products. *Toxicol. Rep.* **2020**, *7*, 16–22. [CrossRef] [PubMed]
17. Tang, L.; Yu, J.; Pang, Y.; Zeng, G.; Deng, Y.; Wang, J.; Ren, X.; Ye, S.; Peng, B.; Feng, H. Sustainable efficient adsorbent: Alkali-acid modified magnetic biochar derived from sewage sludge for aqueous organic contaminant removal. *Chem. Eng. J.* **2018**, *336*, 160–169. [CrossRef]
18. Wanyonyi, W.C.; Mulaa, F.J. Alkaliphilic enzymes and their application in novel leather processing technology for next-generation tanneries. *Adv. Biochem. Eng. Biotechnol.* **2020**, *172*, 195–220. [PubMed]
19. Rovira, J.; Domingo, J.L. Human health risks due to exposure to inorganic and organic chemicals from textiles: A review. *Environ. Res.* **2019**, *168*, 62–69. [CrossRef] [PubMed]
20. Hossen, M.Z.; Hussain, M.E.; Hakim, A.; Islam, K.; Uddin, M.N.; Azad, A.K. Biodegradation of reactive textile dye Novacron super black g by free cells of newly isolated *Alcaligenes faecalis* az26 and *Bacillus* spp. obtained from textile effluents. *Heliyon* **2019**, *5*, e02068. [CrossRef] [PubMed]
21. Jozwiak, T.; Filipkowska, U.; Szymczyk, P.; Zyśk, M. Effect of the form and deacetylation degree of chitosan sorbents on sorption effectiveness of Reactive Black 5 from aqueous solutions. *Int. J. Biol. Macromol.* **2017**, *95*, 1169–1178. [CrossRef] [PubMed]
22. Viana, D.F.; Salazar-Banda, G.R.; Leite, M.S. Electrochemical degradation of Reactive Black 5 with surface response and artificial neural networks optimization models. *Sep. Sci. Technol.* **2018**, *53*, 2647–2661. [CrossRef]
23. Parmar, A.; Shah, A. Cytogenotoxicity of azo dye Reactive Red 120 (RR120) on fish *Catla catla*. *Environ. Exp. Biol.* **2019**, *17*, 151–155.
24. Munagapati, V.S.; Wen, J.C.; Pan, C.L.; Gutha, Y.; Wen, J.H.; Reddy, G.M. Adsorptive removal of anionic dye (Reactive Black 5) from aqueous solution using chemically modified banana peel powder: Kinetic, isotherm, thermodynamic and reusability studies. *Int. J. Phytoremediation* **2020**, *22*, 267–278. [CrossRef] [PubMed]
25. Manimaran, D.; Sulthana, A.S.; Elangovan, N. Reactive black 5 induced developmental defects via potentiating apoptotic cell death in Zebrafish (*Danio rerio*) embryos. *Pharm. Pharmacol. Int.* **2018**, *6*, 449–452.
26. Gurses, A.; Açıkyıldız, M.; Gunes, K.; Gurses, M.S. Colorants in health and environmental aspects. In *Dyes and Pigments*; Springer: New York, NY, USA, 2016; pp. 69–83.
27. Mohanty, S.S.; Kumar, A. Biodegradation of Indanthrene Blue RS dye in immobilized continuous upflow packed bed bioreactor using corncob biochar. *Sci. Rep.* **2021**, *11*, 13390. [CrossRef] [PubMed]
28. Ahmad, A.; Mohd-Setapar, S.H.; Chuong, C.S.; Khatoon, A.; Wani, W.A.; Kumar, R.; Rafatullah, M. Recent advances in new generation dye removal technologies: Novel search for approaches to reprocess wastewater. *RSC Adv.* **2015**, *5*, 30801–30818. [CrossRef]
29. Park, J.H.; Wang, J.J.; Meng, Y.; Wei, Z.; DeLaune, R.D.; Seo, D.C. Adsorption/desorption behavior of cationic and anionic dyes by biochars prepared at normal and high pyrolysis temperatures. *Colloids Surf. A Physicochem. Eng. Asp.* **2019**, *572*, 274–282. [CrossRef]
30. Maqbool, Z.; Shahid, M.; Azeem, F.; Shahzad, T.; Mahmood, F.; Rehman, M.; Ahmed, T.; Imran, M.; Hussain, S. Application of a dye-decolorizing Pseudomonas aeruginosa strain ZM130 for remediation of textile wastewaters in aerobic/anaerobic sequential batch bioreactor and soil columns. *Water Air Soil Pollut.* **2020**, *231*, 386. [CrossRef]
31. Bhatti, H.N.; Safa, Y.; Yakout, S.M.; Shair, O.H.; Iqbal, M.; Nazir, A. Efficient removal of dyes using carboxymethyl cellulose/alginate/polyvinyl alcohol/rice husk composite: Adsorption/desorption, kinetics and recycling studies. *Int. J. Biol. Macromol.* **2020**, *150*, 861–870. [CrossRef]
32. Noreen, S.; Bhatti, H.N.; Iqbal, M.; Hussain, F.; Sarim, F.M. Chitosan, starch, polyaniline and polypyrrole biocomposite with sugarcane bagasse for the efficient removal of Acid Black dye. *Int. J. Biol. Macromol.* **2020**, *147*, 439–452. [CrossRef] [PubMed]
33. Wekoye, J.N.; Wanyonyi, W.C.; Wangila, P.T.; Tonui, M.K. Kinetic and equilibrium studies of Congo red dye adsorption on cabbage waste powder. *Environ. Chem. Ecotoxicol.* **2020**, *2*, 24–31. [CrossRef]
34. Vakili, M.; Rafatullah, M.; Salamatinia, B.; Abdullah, A.Z.; Ibrahim, M.H.; Tan, K.B.; Gholami, Z.; Amouzgar, P. Application of chitosan and its derivatives as adsorbents for dye removal from water and wastewater: A review. *Carbohydr. Polym.* **2014**, *113*, 115–130. [CrossRef] [PubMed]
35. Shaban, M.; Abukhadra, M.R.; Shahien, M.G.; Ibrahim, S.S. Novel bentonite/zeolite-NaP composite efficiently removes methylene blue and Congo red dyes. *Environ. Chem. Lett.* **2018**, *16*, 275–280. [CrossRef]
36. Degermenci, G.D.; Degermenci, N.; Ayvaoglu, V.; Durmaz, E.; Çakır, D.; Akan, E. Adsorption of reactive dyes on lignocellulosic waste; characterization, equilibrium, kinetic and thermodynamic studies. *J. Cleaner Prod.* **2019**, *225*, 1220–1229. [CrossRef]

37. Amin, M.T.; Alazba, A.A.; Shafiq, M. Comparative study for adsorption of methylene blue dye on biochar derived from orange peel and banana biomass in aqueous solutions. *Environ. Monit Assess.* **2019**, *191*, 735. [CrossRef] [PubMed]
38. Astuti, W.; Sulistyaningsih, T.; Kusumastuti, E.; Thomas, G.; Kusnadi, R.Y. Thermal conversion of pineapple crown leaf waste to magnetized activated carbon for dye removal. *Bioresour. Technol.* **2019**, *287*, 121426. [CrossRef] [PubMed]
39. Wong, S.; Tumari, H.H.; Ngadi, N.; Mohamed, N.B.; Hassan, O.; Mat, R.; Saidina Amin, N.A. Adsorption of anionic dyes on spent tea leaves modified with polyethyleneimine (PEI-STL). *J. Clean Prod.* **2019**, *206*, 394–406. [CrossRef]
40. Li, F.; Zimmerman, A.R.; Hu, X.; Yu, Z.; Huang, J.; Gao, B. One-pot synthesis and characterization of engineered hydrochar by hydrothermal carbonization of biomass with $ZnCl_2$. *Chemosphere.* **2020**, *254*, 126866. [CrossRef]
41. Masto, R.E.; Ansari, M.A.; George, J.; Selvi, V.A.; Ram, L.C. Co-application of biochar and lignite fly ash on soil nutrients and biological parameters at different crop growth stages of *Zea mays*. *Ecol. Eng.* **2013**, *58*, 314–322. [CrossRef]
42. Oyewo, O.A.; Onyango, M.S.; Wolkersdorfer, C. Application of banana peels nanosorbent for the removal of radioactive minerals from real mine water. *J. Environ. Radioact.* **2016**, *164*, 369–376. [CrossRef] [PubMed]
43. Vilardi, G.; Di Palma, L.; Verdone, N. Heavy metals adsorption by banana peels micro-powder: Equilibrium modeling by non-linear models. *Chin. J. Chem. Eng.* **2018**, *26*, 455–464. [CrossRef]
44. El-Zawahry, M.M.; Abdelghaffar, F.; Abdelghaffar, R.A.; Hassabo, A.G. Equilibrium and kinetic models on the adsorption of Reactive Black 5 from aqueous solution using *Eichhornia crassipes*/chitosan composite. *Carbohydr. Polym.* **2016**, *136*, 507–515. [CrossRef] [PubMed]
45. Rehman, K.; Shahzad, T.; Sahar, A.; Hussain, S.; Mahmood, F.; Siddique, M.H.; Siddique, M.A.; Rashid, M.I. Effect of Reactive Black 5 azo dye on soil processes related to C and N cycling. *PeerJ* **2018**, *6*, e4802. [CrossRef] [PubMed]
46. Celebi, H. The applicability of evaluable wastes for the adsorption of Reactive Black 5. *Int. J. Environ. Sci. Technol.* **2018**, *16*, 135–146. [CrossRef]
47. Giri, B.S.; Gun, S.; Pandey, S.; Trivedi, A.; Kapoor, R.T.; Singh, R.P.; Abdeldayem, O.M.; Rene, E.R.; Yadav, S.; Chaturvedi, P.; et al. Reusability of brilliant green dye contaminated wastewater using corncob biochar and Brevibacillus parabrevis: Hybrid treatment and kinetic studies. *Bioengineered* **2020**, *11*, 743–758. [CrossRef] [PubMed]
48. Cebrian, G.; Condon, S.; Manas, P. Physiology of the inactivation of vegetative bacteria by thermal treatments: Mode of action, influence of environmental factors and inactivation kinetics. *Foods* **2017**, *6*, 107. [CrossRef] [PubMed]
49. Bharathi, K.S.; Ramesh, S.P.T. Fixed-bed column studies on biosorption of crystal violet from aqueous solution by *Citrullus lanatus* rind and *Cyperus rotundus*. *Appl. Water Sci.* **2013**, *3*, 673–687. [CrossRef]
50. Ng, J.C.Y.; Cheung, W.H.; McKay, G. Equilibrium studies of the sorption of Cu(II) ions onto chitosan. *J. Colloid Interface Sci.* **2002**, *255*, 64–74.
51. Lagergren, S. About the theory of so-called adsorption of soluble substances. *Kungliga Svenska Vetenskapsakademiens Handlingar.* **1898**, *24*, 1–39.
52. Ho, Y.S.; Mckay, G. Pseudo-second-order model for sorption processes. *Process Biochem.* **1999**, *34*, 451–465. [CrossRef]
53. Kapoor, R.T.; Selvaraju Sivamani, S. Exploring the potential of *Eucalyptus citriodora* biochar against direct red 31 dye and its phytotoxicity assessment. *Biomass Convers. Biorefin.* **2021**, *24*, 1–12. [CrossRef]
54. Yaashikaaa, P.R.; Senthil Kumar, P.; Varjani, S.; Saravanan, A. A critical review on the biochar production techniques, characterization, stability and applications for circular bioeconomy. *Biotechnol. Rep.* **2020**, *28*, e00570. [CrossRef]
55. Ipeaiyeda, A.R.; Okeoghene Tesi, G. Sorption and desorption studies on toxic metals from brewery effluent using eggshell as adsorbent. *Adv. Nat. Sci.* **2014**, *7*, 15–24.
56. ISTA. *International Rules for Seed Testing*; International Seed Testing Association, ISTA Secretariat: Bassersdorf, Switzerland, 2008.
57. Abdul-Baki, A.A.; Anderson, J.D. Viability and leaching of sugars from germinating seeds by textile, leather and distillery industries. *Ind. J. Environ. Protec.* **1973**, *11*, 592–594.
58. Lichtenthaler, H.K. Chlorophyll and carotenoids: Pigments of photosynthetic biomembranes. In *Methods Enzymology*; Packer, L., Douce, R., Eds.; Academic Press: San Diego, CA, USA, 1987; pp. 350–382.
59. Hedge, J.E.; Hofreiter, B.T. Estimation of carbohydrate. In *Methods in Carbohydrate Chemistry*; Whistler, R.L., Be Miller, J.N., Eds.; Academic Press: New York, NY, USA, 1962; pp. 17–22.
60. Lowry, O.H.; Rosebrough, N.J.; Farr, A.L.; Randall, R.J. Protein measurement with the Folin phenol reagent. *J. Biol. Chem.* **1951**, *193*, 265–275. [CrossRef]
61. Oyekanmi, A.A.; Ahmad, A.; Hossain, K.; Rafatullah, M. Adsorption of Rhodamine B dye from aqueous solution onto acid treated banana peel: Response surface methodology, kinetics and isotherm studies. *PLoS ONE* **2019**, *14*, e0216878. [CrossRef] [PubMed]
62. Lee, S.L.; Park, J.H.; Kim, S.H.; Kang, S.W.; Cho, J.S.; Jeon, J.R.; Lee, Y.B.; Seo, D.C. Sorption behavior of malachite green onto pristine lignin to evaluate the possibility as a dye adsorbent by lignin. *Appl. Biol. Chem.* **2019**, *62*, 37. [CrossRef]
63. Regti, A.; Laamari, M.R.; Stiriba, S.E.; Haddad, M.E. Removal of Basic Blue 41 dyes using *Persea americana*-activated carbon prepared by phosphoric acid action. *Int. J. Ind. Chem.* **2017**, *8*, 187–195. [CrossRef]
64. Kowalkowska, A.; Jozwiak, T. Utilization of pumpkin (*Cucurbita pepo*) seed husks as a low-cost sorbent for removing anionic and cationic dyes from aqueous solutions. *Desalin. Water Treat.* **2019**, *171*, 397–407. [CrossRef]
65. Wong, S.; Ghafar, N.A.; Ngadi, N.; Razmi, F.A.; Inuwa, I.M.; Mat, R.; Amin, N.A.S. Effective removal of anionic textile dyes using adsorbent synthesized from coffee waste. *Sci. Rep.* **2020**, *10*, 2928. [CrossRef]

66. Felista, M.M.; Wanyonyi, W.C.; Ongera, G. Adsorption of anionic dye (Reactive black 5) using macadamia seed husks: Kinetics and equilibrium studies. *Sci. Afr.* **2020**, *7*, e00283. [CrossRef]
67. Karthick, K.; Namasivayam, C.; Pragasam, L.A. Removal of direct red 12B from aqueous medium by ZnCl2 activated Jatropha husk carbon: Adsorption dynamics and equilibrium studies. *Ind. J. Chem. Technol.* **2017**, *24*, 73–81.
68. Akter, M.; Rahman, F.B.A.; Abedin, M.Z.; Kabir, S.M.F. Adsorption characteristics of banana peel in the removal of dyes from textile effluent. *Textiles* **2021**, *1*, 361–375. [CrossRef]
69. Jozwiak, T.; Filipkowska, U.; Bugajska, P.; Kalkowski, T. The use of coconut shells for the removal of dyes from aqueous solutions. *J. Ecol. Eng.* **2018**, *19*, 129–135. [CrossRef]
70. Jozwiak, T.; Filipkowska, U.; Brym, S.; Kope'c, L. Use of aminated hulls of sunflower seeds for the removal of anionic dyes from aqueous solutions. *Int. J. Environ. Sci. Technol.* **2020**, *17*, 1211–1224. [CrossRef]
71. Samarghandy, M.R.; Hoseinzade, E.; Taghavi, M.; Hoseinzadeh, S. Biosorption of reactive black 5 from aqueous solution using acid-treated biomass from potato peel waste. *Bioresources* **2011**, *6*, 4840–4855.
72. Eren, Z.; Acar, F.N. Adsorption of Reactive Black 5 from an aqueous solution: Equilibrium and kinetic studies. *Desalination* **2006**, *194*, 1–10. [CrossRef]
73. Do Nascimento, B.F.; de Araujo, C.M.B.; do Nascimento, A.C.; da Costa, G.R.B.; Gomes, B.F.M.L.; da Silva, M.P.; da Motta Sobrinho, M.A. Adsorption of Reactive Black 5 and Basic Blue 12 using biochar from gasification residues: Batch tests and fixed-bed breakthrough predictions for wastewater treatment. *Bioresour. Technol. Rep.* **2021**, *15*, 100767. [CrossRef]
74. Shah, J.A.; Mirza, C.R.; Butt, T.A.; Khalifa, W.M.A.; Gasmi, H.H.; Haroon, H.; Khan, M.S.; Ali, M.A.; Zeb, I.; Shah, S.H.; et al. Tobacco stalk waste biomass holds multilayer and spontaneous adsorption capabilities for Reactive Black 5 dye: Equilibrium modelling and error function analysis. *Pol. J. Environ. Stud.* **2021**, *30*, 2301–2312. [CrossRef]

Article

Baseline Assessment of Heavy Metal Pollution during COVID-19 near River Mouth of Kerian River, Malaysia

Mohammad Nishat Akhtar [1], Mohd Talha Anees [2], Emaad Ansari [1], Jazmina Binti Ja'afar [1], Mohammed Danish [3] and Elmi Abu Bakar [1,*]

1. School of Aerospace Engineering, Universiti Sains Malaysia, Nibong Tebal 14300, Malaysia; nishat@usm.my (M.N.A.); emaadansari@student.usm.my (E.A.); s211133016@studentmail.unimap.edu.my (J.B.J.)
2. Department of Geology, Faculty of Science, University of Malaya, Kuala Lumpur 50603, Malaysia; talhaanees@um.edu.my
3. Bioresource Tech. Division, School of Industrial Technology, Universiti Sains Malaysia, George Town 11800, Malaysia; danish@usm.my
* Correspondence: meelmi@usm.my

Citation: Akhtar, M.N.; Anees, M.T.; Ansari, E.; Ja'afar, J.B.; Danish, M.; Bakar, E.A. Baseline Assessment of Heavy Metal Pollution during COVID-19 near River Mouth of Kerian River, Malaysia. *Sustainability* **2022**, *14*, 3976. https://doi.org/10.3390/su14073976

Academic Editors: Mohd Rafatullah and Masoom Raza Siddiqui

Received: 26 December 2021
Accepted: 17 March 2022
Published: 28 March 2022

Publisher's Note: MDPI stays neutral with regard to jurisdictional claims in published maps and institutional affiliations.

Copyright: © 2022 by the authors. Licensee MDPI, Basel, Switzerland. This article is an open access article distributed under the terms and conditions of the Creative Commons Attribution (CC BY) license (https://creativecommons.org/licenses/by/4.0/).

Abstract: River water quality is a serious concern among scientist and government agencies due to increasing anthropogenic activities and uncontrolled industrial discharge to rivers. The present study was conducted near the river mouth of the Kerian River to assess heavy metal pollution during COVID-19 pandemic-lockdown conditions and post-COVID-19 pandemic-unlock conditions. Twelve samples of shallow, middle, and bottom depths were collected at four locations along a 9.6 km reach. A concentration of eight heavy metals including Cadmium, Chromium, Copper, Iron, Manganese, Nickel, Lead, and Zinc were extracted through atomic absorption spectrometry. Total suspended solid was measured during laboratory experimentation. The results showed that, during the pandemic, concentrations of Nickel, Zinc, and Iron were high at shallow, middle, and bottom depths, respectively. Decreasing orders of heavy metal concentration are variable at different depths due to either their high sinking tendency with other existing components of water matrix or the anthropogenic source. However, almost all values of heavy metals are under the permissible limit of National Water Quality Standards of Malaysia and Food and Drug Administration. A possible reason for the lack of heavy metal pollution may be the restriction of anthropogenic activities during the COVID-19 pandemic. Additionally, no significant differences were observed in total suspended solid.

Keywords: heavy metals; Kerian River; pollution; anthropogenic activities; water quality

1. Introduction and Background

River water pollution is one the most critical issues in the world. Surface water quality, especially river water quality, is declining due to anthropogenic activities and uncontrolled discharge of anthropogenic sources [1,2]. Anthropogenic sources can be in the form of industrial waste, discharge from agricultural land, mining, and sewage. Fertilizers and pesticide used in agricultural land washed during precipitation and drain into river causes increment in nitrate and phosphate concentration [3,4]. Total suspended solid concentration also increases due to soil erosion from agricultural land. Uncontrolled discharge of industrial waste in river water containing pollutants such as zinc, cyanide, copper, lead, mercury, and cadmium causes fish death and an increment in toxic levels [5]. Pollution due to heavy metal is also a serious issue because it is non-degradable by natural processes and its existence in soil and sediment leads to rapid release as it sinks into watercourses [6]. A concentration of essential heavy metals under an acceptable limit is good for health; however, if it exceeds the acceptable limit, these heavy metals become harmful and extremely toxic for humans, animals, and aquatic ecosystem health [7,8].

In the last decade, several studies have analyzed heavy metal pollution in different parts of Malaysia. Ishadi et al. [9] examined water quality and habitat suitability of a hemipteran community upstream of the Kerian River. They used three heavy metals but did not compare with any water quality standards. Ibrahim et al. [10] compared the presence of heavy metals in river water and pumping-well water for a Riverbank filtration (RBF) system upstream of the Kerian River. They found that, out of 10 heavy metals, iron and arsenic exceed the standard values set by the Ministry of Health, Malaysia. The probable reason of this was due to the excess use of pesticides on the agricultural land by which the upstream of the Kerian River is mostly surrounded. Billah et al. [11] investigated metal contamination in the tropical Miri estuary of Sarawak, Malaysia, and found that iron was the highest contaminated metal. Their study was substantial to portray deterioration in water quality due to anthropogenic pollution, though it would have been interesting to differentiate the reading, had the data been collected during the COVID-19 lockdown period. Chowdhury et al. [12] assessed water quality effected due to anthropogenic pollution sources from the Sungai Selangor basin. They found that most sampling stations fall under Class 3 of the National Water Quality Standards of Malaysia (NWQSM), indicating that extensive treatment is required. Nonetheless, the differentiation could have been substantiated if the additional data were recorded during the COVID-19 lockdown. Ibrahim et al. [6] analyzed metal contamination at nine stations along the Sg. Sembilang due to anthropogenic and natural sources. They found that some heavy metals exceeded the NWQSM limit. Zanuri et al. [13] assessed the marine water quality of Penang Island to investigate the mass mortality of cultured fishes. They found that the concentration of cadmium, copper, iron, and nickel exceeded the permissible levels according to Malaysia Marine Water Quality Class 2. Due to this, the area may no longer be suitable for aquaculture or recreational purposes. It was also noted that their study was carried out during pre-COVID-19 times, whereby the data was collected from 2016 to 2017. Lee Goi [14] studied pre- and post-COVID-19 water qualities of Malaysian rivers using published papers and newspaper articles. They found that, in pre-COVID-19 conditions, 53% of the river's water quality in Malaysia was categorized as slightly polluted or polluted. While, in post COVID-19 condition, some polluted river became clearer than previous conditions. Their work provides an insight into the analysis of river water quality during pre- and post-COVID-19 periods. Razak et al. [15] studied heavy metal pollution in the Linggi River, Negeri Sembilan, Malaysia. They found that concentrations of heavy metals were under the permissible limits of NWQSM, but that the index showed low-level heavy metal contamination. Moreover, aluminum and zinc came under a medium potential risk, while arsenic and manganese came under low potential risk that impacted negatively on aquatic organisms and human health.

Overall, none of the studies have been reported near the river mouth of the Kerian River, where several industrial and agricultural activities have been increasing. However, this study was conducted during the COVID-19 pandemic, when there were restrictions on anthropogenic activities such as industrial lockdown near the vicinity of the Kerian river (limitation in industrial waste release) and after the COVID-19 pandemic, when the restrictions were removed. Therefore, the objective of this study is to assess heavy metal pollution near the river mouth of the Kerian River along a 9.6 km reach covering several industrial and agricultural waste drain areas. With regard to the industries, it is highlighted that the Kerian River is surrounded by industries such as semi-conductor manufacturing plants, paper, palm oil, rubber, and furniture factories. This study would be helpful in understanding the current status of pollution in the Kerian River. As none of the studies have previously reported in this area, the results of this study could be a reference for future heavy metal pollution studies.

2. Materials and Methods

2.1. Study Area and Data Collection

The study area was the downstream part of the Kerian River, situated in the state of Perak, the northern part of the Peninsular Malaysia. The area lies between latitudes 5°8′24″ and 5°10′41.94″ N and longitudes 101°24′12.89″ and 100°29′56.44″ E (Figure 1).

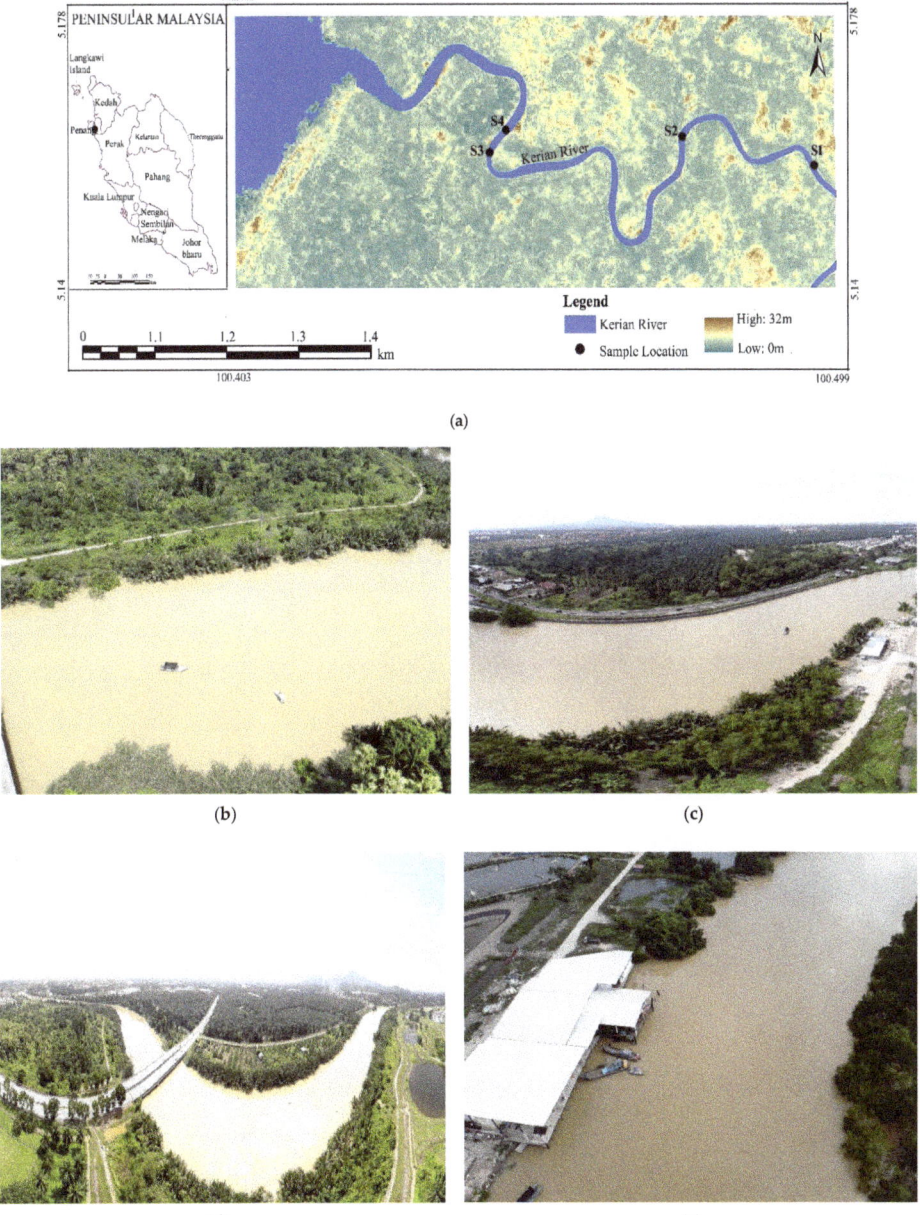

Figure 1. Study area (**a**) location of study area; water sampling sites (**b**) S1 (**c**) S2 (**d**) S3 (**e**) S4.

Total length of the Kerian River is 104 km, and it originates from the Bintang Range and flows south-westward towards the Malacca Straits near the town of Nibong Tebal [16]. In this study, a 9.6 km reach near the river mouth was selected due to the presence of settlement and industrial area. Average annual rainfall and temperature in the area is 2560.3 mm and 28 °C, respectively. As the area situated near the river mouth, the elevation ranged from 0 to 32 m.

Nibong Tebal is home to several manufacturing industries which are located across the vicinity of Kerian River. The domain of these industries comprises of metal machining, rubber industry, paper industry, and furniture industry [17]. In addition to this, the area also accommodates food-based industry, i.e., sugar, sauce, and biscuit factories. Nonetheless, overall, Nibong Tebal has 617 registered companies. These companies have an estimated turnover of RM 8.716 billion and employ a number of employees estimated at 17,025. The population of Nibong Tebal town, as of 2021, is 40,072 as per the GeoNames geographical database. From an agricultural point of view, the area around Kerian River is home to sugar cane, rice, and palm tree plantation. Land use such as forested areas, paddy fields, palm oil plantations, orchards, and areas of settlement are distributed along the Kerian River basin [18]. The existing environment of the Kerian River has been heavily developed into agricultural lands [19]. The area is divided into two categories: non-agricultural and agricultural lands. The major crops are rubber, rice (Kerian Rice Irrigation Schemes), and palm oil. Apart from the aforementioned positive parameters, there lies a threat of industrial wastewater pollution. In this regard, it is an urgent need to curb industrial waste pollution in Kebun Kuyung, near Nibong Tebal in the Seberang Perai Selatan district. Nonetheless, aquatic life in Sungai Kerian, especially fish and prawns, that are the main catch for small fishers here, will be more seriously affected and reduced, subsequently threatening their income.

In the proposed work, twelve water samples at four sites along the 9.6 km reach were collected to study the water quality condition of the Kerian River during and after the pandemic, whereby, in the manuscript, terms such as "during pandemic" are known as the lockdown period, and "after pandemic" or "post-pandemic" are known as the unlock period. Water samples were collected during rainy season due to the high possibility of pollution during high flows. During the pandemic, samples were collected in January 2021, while, in post-pandemic, samples were collected in January 2022. At each site, three samples were collected at different depths, such as shallow, middle, and bottom. Location and distance of different sites are given in Table 1.

Table 1. Locations of water samples collected for heavy metal pollution assessment.

Water Samples	Latitude	Longitude	Location	Distance (km)
S1	5°9′33″	100°26′37″	Taman Ilmu	0
S2	5°9′42″	100°28′28″	Kampung Chelsa	3
S3	5°9′30″	100°29′43″	Kampung Sungai Tok Tuntung	9
S4	5°9′47″	100°26′50″	Kampung Sungai Tok Tuntung	9.6

2.2. Laboratory Experiments and Data Processing

Three liters of each collected water sample were first concentrated in a sandy oven at 80 °C until the volume reached 50 mL. A total of 4 mL of concentrated sulfuric acid (Merck, Kenilworth, NJ, USA, 98%) was added to each sample and digested by Digesdahl apparatus for 3 min. Then, 10 mL hydrogen peroxide (Merck, 30%) was added and heated until oxidation was completed. After cooling, each sample filtered by filter (Whatman filter Merck, 0.45 m). The filtrate was diluted by deionized water for a final volume of 50 mL. The prepared samples were analyzed by a Graphite furnace atomic absorption spectrometry (GFAAS, Modal AAnalyst300) to determine the metals.

Vacuum filtration was considered to be a reliable approach to measure sediment weight in a sample. In the conventional method, potassium permanganate was usually added in order to allow the sediments to deposit at the base before filtration. However, in the present study, since the samples were to be used for atomic absorption spectroscopy (AAS) for determination of sediment composition, we avoided the addition of potassium permanganate to the samples. We passed water samples through the filter paper in the vacuum filtration apparatus (Figure 2). After water filtration, we kept the filter papers in the drying oven at a temperature of 104 °C for 24 h. Finally, the sediment load in each sample was measured by calculating the difference in weight of filter paper before and after the experiment. The analytical weight balance used in the present study had count of a least 0.01 mg.

Figure 2. Vacuum filtration setup.

2.3. Heavy Metal and Statistical Analysis

Basic statistics such as mean, standard deviation, minimum and maximum values of total suspended solid (TSS), and heavy metal concertation at different sites were compared to obtain their variations. Decreasing orders of heavy metals were also analyzed. Heavy metal values were compared with National Water Quality Standards of Malaysia, Food and Drug Administration, drinking water standards, irrigation water standards, aquatic life standards, and surface water standards [20].

3. Results and Discussion

3.1. Experimental Results of Total Suspended Solid and Heavy Metals

Average laboratory results of TSS and eight heavy metals such as Cadmium (Cd), Chromium (Cr), Copper (Cu), Iron (Fe), Manganese (Mn), Nickel (Ni), Lead (Pb), and Zinc (Zn) at shallow, middle, and bottom depths are presented in Table 2. Detailed results are given in Table A1, incorporated in Appendix A.

Table 2. Average laboratory results of TSS (mg/L) and eight heavy metals (ppm) at different depths.

Parameters	Time	Depths		
		Shallow	Middle	Bottom
TSS	During COVID	17.75	53.00	14.00
	Post COVID	24.25	64.00	8.25
Cd	During COVID	0.0014	0.0025	0.0015
	Post COVID	0.0049	0.0156	0.0076
Cr	During COVID	0.0002	0.0002	0.0002
	Post COVID	0.1441	0.1660	0.1179
Cu	During COVID	0.0012	0.0029	0.0032
	Post COVID	0.0890	0.0858	0.0858
Fe	During COVID	0.0011	0.0018	0.0036
	Post COVID	5.1748	5.4810	3.7255
Mn	During COVID	0.0014	0.0021	0.0006
	Post COVID	0.1746	0.1372	0.1599
Ni	During COVID	0.0047	0.0018	0.0017
	Post COVID	0.1674	0.1068	0.3098
Pb	During COVID	0.0009	0.0008	0.0012
	Post COVID	0.1466	0.2403	0.1508
Zn	During COVID	0.0021	0.0036	0.0023
	Post COVID	0.7301	0.2704	0.2104

SD = Shallow depth, MD = Middle depth, BD = bottom depth.

3.2. Average Concentration Order of TSS and Heavy Metals during and after the Pandemic

Industrial and agricultural waste discharge into river is one of the major concerns in developing countries. A high concentration of heavy metals causes water pollution that deteriorates water quality and affects human health. The average results of TSS and heavy metal concentration at shallow, middle, and bottom depths in the Kerian River are shown in Table 2. Based on the mean concentration of heavy metals during the COVID-19 pandemic-lockdown period, decreasing order at shallow depth in the Kerian River is Ni > Zn > Cd > Mn > Cu > Fe > Pb > Cr. At middle depth, the decreasing order is Zn > Cu > Cd > Mn > Ni > Fe > Pb > Cr, while, at bottom depth, the decreasing order is Fe > Cu > Zn > Ni > Cd > Pb > Mn > Cr. From these orders, it is clear that Cr and Pb are almost in same position at different depths. The concentration of Ni is high in shallow water but medium at other depths, which indicates that Ni settling tendency is lower in river water. However, concentration of Fe is lower at shallow and middle depths, but high in bottom that indicates Fe settling tendency is high in Kerian River. Similarly, the concentration of Cd and Mn is lower at bottom depth, which indicates lower settling tendency in the Kerian River. The settling of heavy metals may be due to the different binding capacities of the different metals with other existing components of the water matrix, such as micro particles or micro vegetation, which may be in suspended, colloidal, or dissolved form.

After the pandemic-unlock period, the decreasing order of heavy metals changed. At a shallow depth, the order was Fe > Zn > Mn > Ni > Pb > Cr > Cu > Cd, at middle depth was Fe > Zn > Pb > Cr > Mn > Ni > Cu > Cd, and at bottom depth, the order was Fe > Ni > Zn > Mn > Pb > Cr > Cu > Cd. This clearly shows that the Fe concentration is highest at different depths, indicating a high Fe source in industrial waste. Cu and Cd are lowest at all depths, indicating the lowest concentration in industrial waste.

Average results showed that TSS concentration is high at middle depth followed by shallow depth and bottom depth. Slight variation was observed during and after the pandemic, which are discussed in the following sections.

3.3. Variation in TSS Concentration during and after COVID-19 along the Kerian River

During the pandemic-lockdown period, TSS concentration at shallow depth varies from 12 to 24 mg/L, with a standard deviation (SD) of 4.9 mg/L. This concentration at middle depth varies from 41 to 63 mg/L, with an SD of 8.6 mg/L, and, at bottom depth, it varies from 3 to 39 mg/L, with an SD of 14.5 mg/L. This indicates more variation in bottom depth. After the pandemic, the concentration increases to 26.3% at shallow depth and 14.3% at middle depth, though declines at bottom depth (42.9%) (Figure 3).

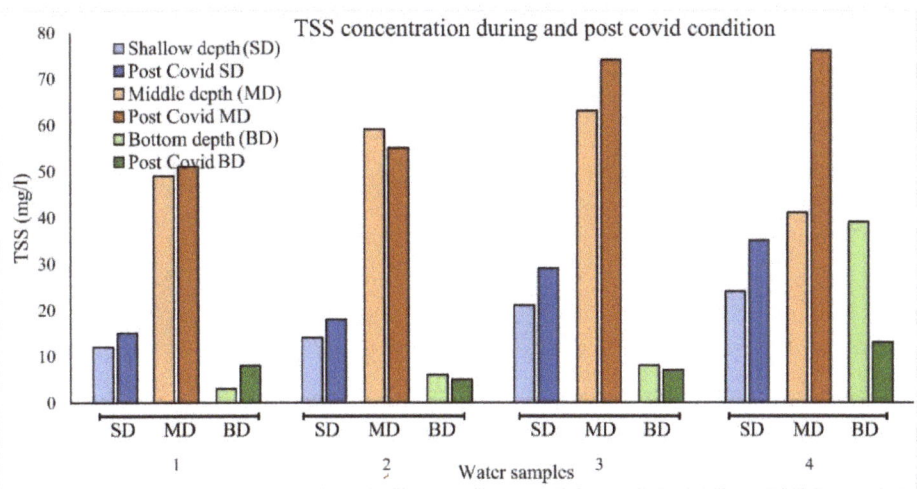

Figure 3. Concentration of TSS at shallow, middle, and bottom depths during and after the pandemic in the study area.

3.4. Variation in Heavy Metal Concentration during and after COVID-19 along the Kerian River

During the pandemic-lockdown period, Cd concentration at shallow depth varies from 0.0007 to 0.0022 ppm with a standard deviation (SD) of 0.00053 ppm. This concentration at middle depth varies from 0.0002 to 0.0075 ppm with an SD of 0.0029 ppm, and, at bottom depth, it varies from 0.0004 to 0.0024 with an SD of 0.0008 ppm. This indicates that more variation at middle depth is reported at site number 2. After the pandemic-unlock period, the concentration increased to 79.3% at shallow depth, 85.08% at middle depth, and 61.7% at bottom depth (Figure 4).

According to National Water Quality Standards of Malaysia (NWQSM), almost all TSS values are in natural condition during and after the pandemic except at middle depth (Class IIA/IIB). Results of middle depth indicate that conventional treatment is required for water supply and is sensitive to aquatic species.

According to National Water Quality Standards of Malaysia (NWQSM), almost all Cd values are under Class IIA/IIB during the pandemic, indicating that conventional treatment is required for water supply and is sensitive to aquatic species. According to Food and Drug Administration (FDA), Cd concentration for drinking water should not exceed 0.005 ppm. Compared to Cd concentration in the study area, most of the sites crossed the permissible limit. Fluctuation in Cd values at different sites indicates anthropogenic and industrial sources in the area. These sources are steel industry, fertilizers, and nuclear emission plants, metal plating and electroplating, plastic industry, and nickel–cadmium batteries [1]. After the pandemic-unlock period, half of the samples fall under Class V, indicating that they are not suitable for drinking and irrigation purposes.

During the pandemic-lockdown period, Cr concentration at shallow depth ranged from 0.0001 ppm to 0.0003 ppm with an SD of 0.0001 ppm. At middle depth, Cr concentration ranged from zero ppm to 0.0003 ppm with an SD of 0.0001 ppm, while, at the bottom depth, it varied from 0.0001 ppm to 0.0002 ppm, with an SD of 0.00004 ppm. More variation was observed at site numbers 2 and 4. After the pandemic-unlock period, the concentration increased to 99.8% at shallow depth, 99.9% at middle depth, and 99.8% at bottom depth (Figure 5).

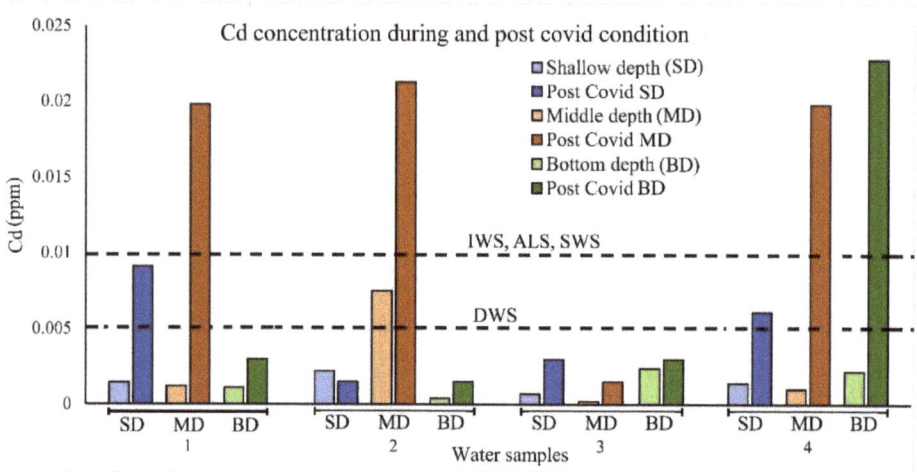

Figure 4. Concentration of Cadmium at shallow, middle, and bottom depths during and after the pandemic in the study area.

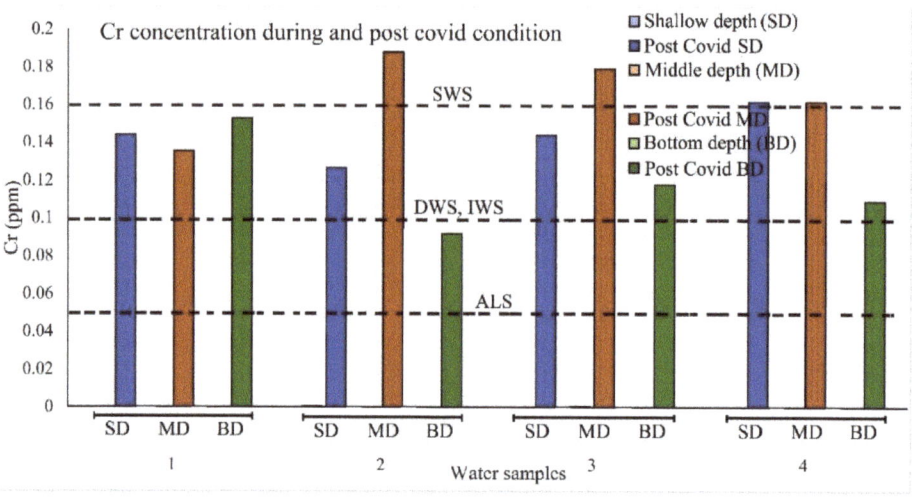

Figure 5. Concentration of Chromium at shallow, middle, and bottom depths during pandemic-lockdown and after the pandemic-unlock period in the study area.

According to NWQSM, all Cr values are in natural condition and indicate that no practical treatment is required for the water supply. According to FDA, the Cr concentration of the study area is under permissible limit (1 ppm). The lowest concentration of Cr in the study shows its source from natural deposits such as rocks and soil [1]. However, fluctuation in Cr values at different sites are from industrial waste discharge which contain very low Cr concentration. After the pandemic-unlock period, almost all samples come under Class V, thereby indicating that they are not suitable for drinking and irrigation purposes.

During the pandemic-lockdown period, Cu concentration at shallow depth varies from 0.001 ppm to 0.0013 ppm with an SD of 0.0001 ppm. At middle depth, it varies from 0.0012 ppm to 0.0047 ppm with an SD of 0.0013 ppm. Whereas, at the bottom depth, it varies from 0.0022 ppm to 0.004 ppm, with an SD of 0.0006 ppm. More variation was found at middle depth followed by bottom and shallower depth. High variation among different depths was found at site number 1. After the pandemic-unlock period, the concentration increases to 98.6% at shallow depth, 85.6% at middle depth, and 95.6% at bottom depth (Figure 6).

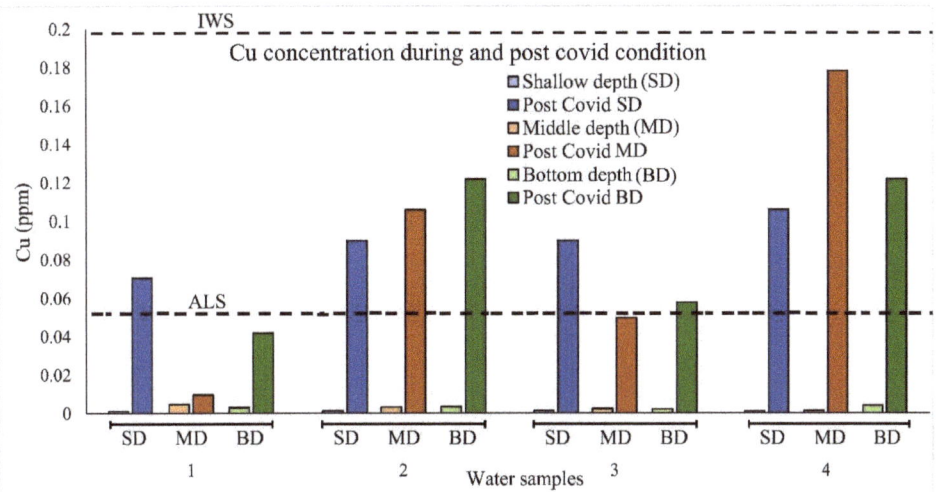

Figure 6. Concentration of Copper at shallow, middle, and bottom depths during pandemic-lockdown and after the pandemic-unlock period in the study area.

According to NWQSM, all Cu values are in natural condition, which indicates that no practical treatment is required for the water supply. According to FDA, Cu concentrations of the study area are under a permissible limit (1 ppm). Cu can be released from different sources such as chemical industry, mining, pesticide industry, and metal piping [1]. The second highest concentration at middle and bottom depth in the study showed its industrial source. After the pandemic-unlock period, all samples except a few come under Class V, indicating that they are not suitable for drinking and irrigation purposes.

During the pandemic-lockdown period, Fe concentration at shallow depth ranged from 0.0003 ppm to 0.002 ppm with an SD of 0.00061 ppm. At middle depth, it varied from 0.0011 ppm to 0.0029 ppm with an SD of 0.00068, while, at the bottom depth, it varied from 0.0002 ppm to 0.009 ppm with an SD of 0.0033 ppm. High variation was observed at bottom depth. After the pandemic-unlock period, the concentration increased to 99.8% at shallow depth, 99.9% at middle depth, and 99.1% at bottom depth (Figure 7).

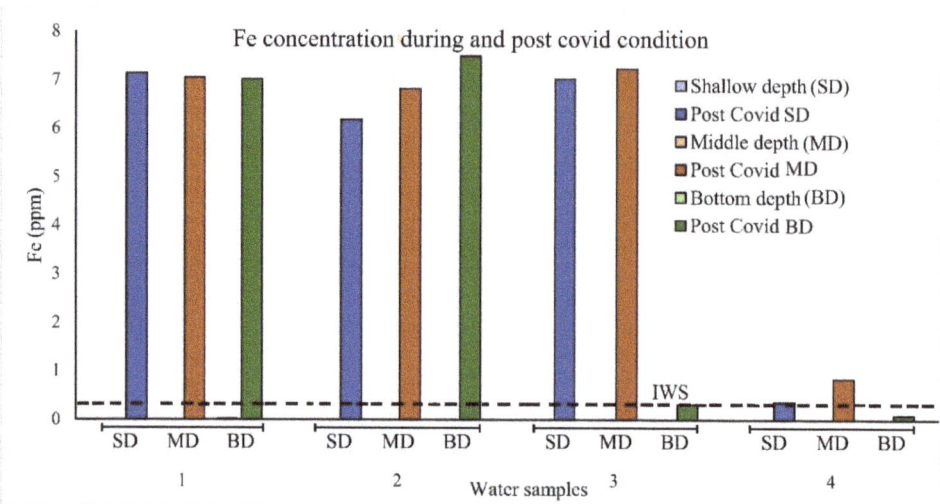

Figure 7. Concentration of Iron at shallow, middle, and bottom depths during pandemic-lockdown and after the pandemic-unlock period in the study area.

According to NWQSM, all Fe values are in natural condition, indicating that no practical treatment is required for the water supply. According to FDA, Fe concentrations of the study area are under the permissible limit (0.3 ppm). Generally, the source of Fe is from soil and rocks. A high difference in Fe values is reported only at site number 1, indicating an industrial or anthropogenic source. After the pandemic-unlock period, all samples except a few come under Class V, thereby indicating that they are not suitable for drinking and irrigation purposes

Mn concentration at shallow depth varies from 0.0006 ppm to 0.002 ppm with an SD of 0.0005 ppm. At middle depth, it varies from 0.0015 ppm to 0.0028 ppm with an SD of 0.0005 ppm, whereas, at the bottom depth, it varies from 0.0003 ppm to 0.0012 ppm, with an SD of 0.00035 ppm. Similar variation was observed at both shallow and middle depths. After the pandemic-unlock period, the concentration increases to 99.2% at shallow depth, 99.1% at middle depth, and 99.5% at bottom depth (Figure 8).

According to NWQSM, all Mn values are in natural condition, which indicates that no practical treatment is required for the water supply. According to FDA, Mn concentrations of the study area are under the permissible limit (0.05 ppm). As Mn values are under the permissible limit, its source must be natural, such as soil and rocks. It is interesting to observe a sudden drop of Mn values at site number 4 at both shallow and middle depths, and it slightly increases at the bottom depth. This may be due to adsorption or the ion exchange of Mn by riverbed material such as soil and sand. After the pandemic-unlock period, all samples fell under Class V, thereby indicating that they are not suitable for drinking and irrigation purposes.

During the pandemic-lockdown period, Ni concentration at shallow depth ranged from 0.0015 ppm to 0.0086 ppm with an SD of 0.0025 ppm. At middle depth, it ranged from zero ppm to 0.003 ppm with an SD of 0.0011 ppm. Whereas, at the bottom depth, it ranged from 0.0011 ppm to 0.0026 ppm, with an SD of 0.0006 ppm. High variation was observed at shallow depth followed by bottom and middle depths. After the pandemic-unlock period, the concentration increased to 96.4% at shallow depth, 94.9% at middle depth, and 99.2% at bottom depth (Figure 9).

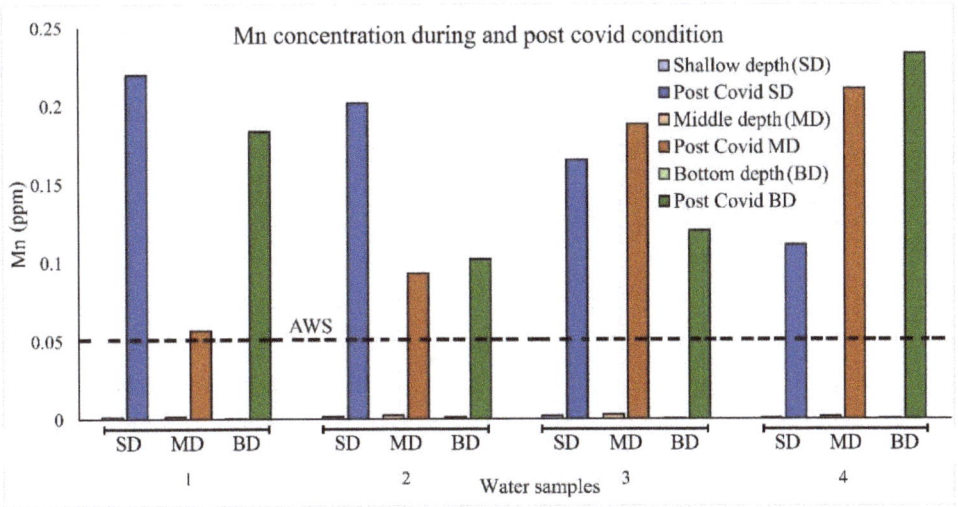

Figure 8. Concentration of Manganese at shallow, middle, and bottom depths during pandemic-lockdown and after the pandemic-unlock period in the study area.

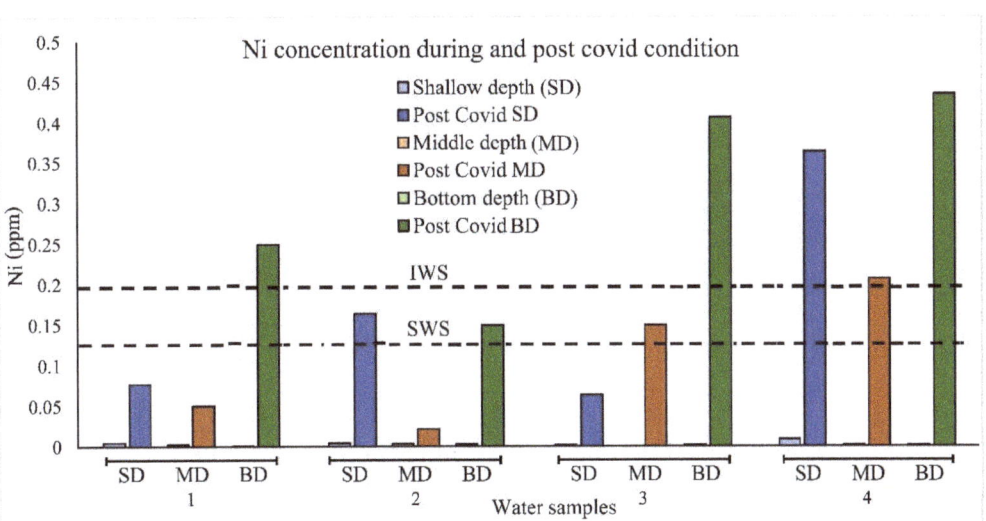

Figure 9. Concentration of Nickel at shallow, middle, and bottom depths during pandemic-lockdown and after the pandemic-unlock period in the study area.

According to NWQSM, all Ni values are in natural condition, indicating that no practical treatment is required for the water supply. According to FDA, Ni concentrations of the study area are under the permissible limit (0.1 ppm). As shown in Figure 8, at shallow depth, the sudden rise in Ni value at site number 4 may be due to industrial discharge or anthropogenic activity. After the pandemic-unlock period, all samples except a few fall under Class V, thereby indicating that they are not suitable for drinking and irrigation purposes.

During the pandemic-lockdown period, at shallow depth, Pb concentration ranged from 0.0004 ppm to 0.0013 ppm with an SD of 0.0003 ppm. Pb concentration at middle depth ranged from 0.0001 ppm to 0.0012 ppm with an SD of 0.00042 ppm. Whereas, at bottom depth, it ranged from 0.0004 ppm to 0.0022 ppm, with an SD of 0.0007 ppm. High variation was observed at bottom depth followed by middle and shallow depths. After the pandemic-unlock period, the concentration increased to 88.6% at shallow depth, 99.6% at middle depth, and 98.8% at bottom depth (Figure 10).

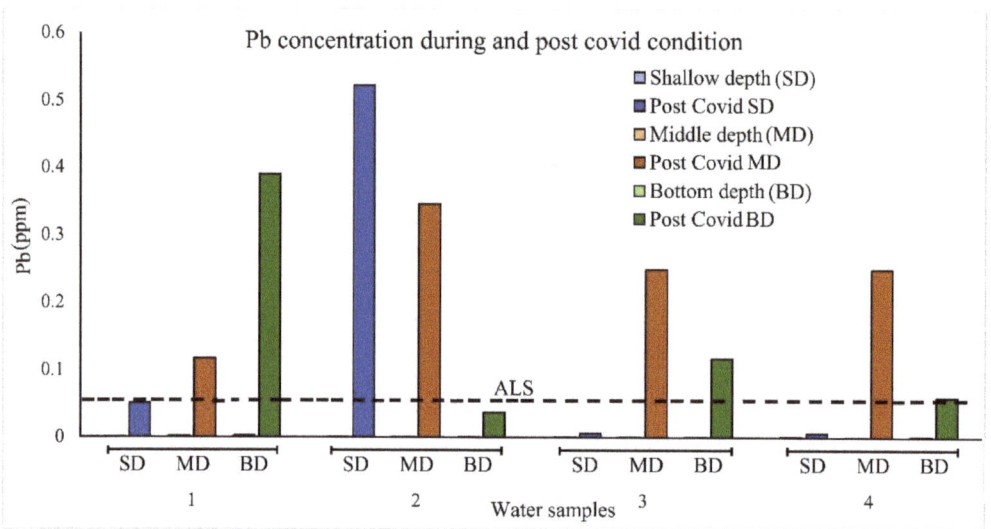

Figure 10. Concentration of Lead at shallow, middle, and bottom depths during pandemic-lockdown and after the pandemic-unlock period in the study area.

According to NWQSM, all Pb values are in natural condition, indicating that no practical treatment is required for the water supply. According to FDA, Pb concentrations of the study area are under the permissible limit (0.01 ppm). At site number 1, Pb concentration was high at the bottom depth; however, at site number 4, it was same the concentration at shallow depth. After the pandemic-unlock period, half of the samples except a few fall under Class V, thereby indicating that they are not suitable for drinking and irrigation purposes.

Zn concentration at shallow depth varies from 0.001 ppm to 0.0032 ppm with an SD of 0.0008 ppm. At middle depth, it varies from 0.0006 ppm to 0.006 ppm with an SD of 0.002 ppm. Whereas, at the bottom depth, it varies from 0.0014 ppm to 0.0034 ppm with an SD of 0.0008 ppm. High variation was observed at middle depth followed by shallow and bottom depths. After the pandemic-unlock period, the concentration increased to 96.6% at shallow depth, 60.7% at middle depth, and 92.7% at bottom depth (Figure 11).

According to NWQSM, all Zn values are in natural condition, indicating that no practical treatment is required for the water supply. According to FDA, Zn concentrations of the study area are under the permissible limit (5.0 ppm). Again, at site number 4, Zn values suddenly drop at both shallow and middle depths, while they increase at the bottom depth. This may be due to the adsorption and ion exchange of Zn by suspended sediments. After the pandemic-unlock period, half of the samples except a few come under Class V, thereby indicating that they are not suitable for drinking and irrigation purposes.

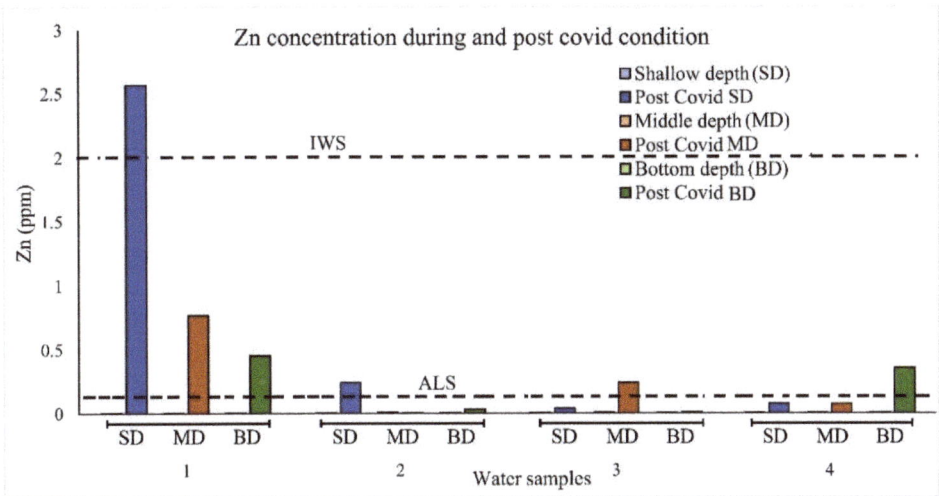

Figure 11. Concentration of Zinc at shallow, middle, and bottom depths during pandemic-lockdown and after the pandemic-unlock period in the study area.

3.5. Comparison Heavy Metals with Different Standards at Kerian River

Heavy metal values were compared with drinking water standards (DWS), irrigation water standards (IWS), aquatic life standards (ALS), and surface water standards (SWS) to understand the use of Kerian River water for different purposes. Overall, the results of the comparison showed that the Kerian River water is under permissible limits of drinking, irrigation, aquatic life, and surface water standards. Concentrations of heavy metals after the pandemic-unlock period crossed the standard limits, which is not required here to be compared. A summary of results is given in Table 3.

Table 3. Comparison of heavy metal concentration after the pandemic-unlock period at different depths with drinking water standards (DWS), irrigation water standards (IWS), aquatic life standards (ALS), and surface water standards (SWS).

Heavy Metals	Concentrations (mg/L) at Depths			DWS	IWS	ALS	SWS
	Shallow	Middle	Bottom				
Cd	0.0014	0.0023	0.0015	0.005	0.01	0.01	0.01
Cr	0.0002	0.0002	0.00018	0.1	0.1	0.05	0.16
Cu	0.0012	0.0029	0.0032	1.3	0.2	0.05	-
Fe	0.0011	0.0018	0.0035	0.3	-	-	-
Mn	0.0014	0.0021	0.0006	0.05	2	1	1
Ni	0.0047	0.0018	0.0018	-	0.2	-	0.144
Pb	0.0009	0.0008	0.0012	0	5	0.05	0.005
Zn	0.0021	0.0036	0.0022	5	2	<0.1	-

4. Conclusions

This study was conducted to analyze the baseline values of the total suspended solid and heavy metals during and post-COVID-19 pandemic. The study reports that concentrations of heavy metals are under permissible limits of National Water Quality Standards of Malaysia, Food and Drug Administration, and standards of drinking water, irrigation water, and aquatic life. However, there are variations in different depths at the same sampling site. The possible reason for those heavy metal values that suddenly drop from one site to another could be the high settling tendency in Kerian River due to binding capacities with other existing components of water matrix such as micro particles or micro vegetation. They may be in suspended, colloidal, or dissolved form. Whereas, for those heavy metal values that suddenly raised from one site to another, the possible reason could be an anthropogenic source. High concentrations of Ni, Zn, and Fe were reported at shallow, middle, and bottom depths, respectively. Overall, the possible reason for the lack of heavy metal pollution may be due to COVID-19 restrictions on anthropogenic activities. Whereas, before the post-pandemic period, heavy metal values increased from 60% to 100% from during pandemic conditions. This confirms that the increment is due to anthropogenic activities after releasing COVID-19 restrictions. Furthermore, no significant effect was observed on total suspended solid values in post-pandemic conditions.

Author Contributions: Conceptualization, M.N.A. and E.A.B.; Data curation, E.A. and J.B.J.; Formal analysis, M.N.A., M.T.A., J.B.J. and M.D.; Investigation, E.A. and M.D.; Methodology, M.N.A., M.T.A. and J.B.J.; Project administration, E.A.B.; Resources, M.D.; Supervision, E.A.B.; Writing—original draft, M.N.A.; Writing—review and editing, M.T.A. and E.A. All authors have read and agreed to the published version of the manuscript.

Funding: This study was funded by Research Creativity and Management Office, Universiti Sains Malaysia. The authors would like to acknowledge the grant RU Top-Down 1001/PAERO/870052.

Institutional Review Board Statement: Not applicable.

Informed Consent Statement: Not applicable.

Data Availability Statement: Not applicable.

Conflicts of Interest: The authors declare no conflict of interest.

Appendix A

Table A1. Laboratory results of eight heavy metals (ppm) with their standard deviations at different depths.

Parameters	Sample No. Depth	Water Sample 1			Water Sample 2			Water Sample 3			Water Sample 4		
		Shallow	Middle	Bottom	Shallow	Middle	Bottom	Shallow	Middle	Bottom	Shallow	Middle	Bottom
Cd	During pandemic	0.0014 ± 0.0002	0.0012 ± 0.0066	0.0011 ± 0.0060	0.0022 ± 0.0064	0.0075 ± 0.0066	0.0004 ± 0.0068	0.0007 ± 0.0076	0.0002 ± 0.0070	0.0024 ± 0.0056	0.0014 ± 0.0064	0.0010 ± 0.0064	0.0022 ± 0.0058
	Post pandemic	0.0091 ± 0.0000	0.0198 ± 0.0001	0.0030 ± 0.0001	0.0015 ± 0.0001	0.0213 ± 0.0000	0.0015 ± 0.0001	0.0030 ± 0.0000	0.0015 ± 0.0005	0.0030 ± 0.0001	0.0061 ± 0.0000	0.0198 ± 0.0001	0.0228 ± 0.0001
Cr	During pandemic	0.0002 ± 0.0002	0.0001 ± 0.0002	0.0002 ± 0.0002	0.0003 ± 0.0004	0.0002 ± 0.0000	0.0001 ± 0.0000	0.0002 ± 0.0002	0.0003 ± 0.0002	0.0002 ± 0.0004	0.0001 ± 0.0002	0.0000 ± 0.0002	0.0002 ± 0.0004
	Post pandemic	0.1441 ± 0.0001	0.1354 ± 0.0001	0.1528 ± 0.0000	0.1266 ± 0.0001	0.1878 ± 0.0000	0.0917 ± 0.0001	0.1441 ± 0.0000	0.1790 ± 0.0001	0.1179 ± 0.0000	0.1616 ± 0.0001	0.1616 ± 0.0001	0.1092 ± 0.0000
Cu	During pandemic	0.0012 ± 0.0003	0.0047 ± 0.0004	0.0031 ± 0.0004	0.0013 ± 0.0008	0.0032 ± 0.0001	0.0034 ± 0.0006	0.0012 ± 0.0008	0.0023 ± 0.0006	0.0022 ± 0.0004	0.0010 ± 0.0000	0.0012 ± 0.0002	0.0040 ± 0.0004
	Post pandemic	0.0705 ± 0.0001	0.0096 ± 0.0001	0.0417 ± 0.0001	0.0898 ± 0.0001	0.1058 ± 0.0002	0.1219 ± 0.0001	0.0898 ± 0.0000	0.0497 ± 0.0000	0.0577 ± 0.0002	0.1058 ± 0.0001	0.1780 ± 0.0000	0.1219 ± 0.0001

Table A1. Cont.

Parameters	Sample No. Depth	Water Sample 1			Water Sample 2			Water Sample 3			Water Sample 4		
		Shallow	Middle	Bottom	Shallow	Middle	Bottom	Shallow	Middle	Bottom	Shallow	Middle	Bottom
Fe	During pandemic	0.0011 ± 0.0010	0.0017 ± 0.0014	0.0029 ± 0.0040	0.0009 ± 0.0009	0.0014 ± 0.0004	0.0018 ± 0.0016	0.0003 ± 0.0006	0.0029 ± 0.004	0.0002 ± 0.0018	0.0020 ± 0.0016	0.0011 ± 0.0028	0.0032 ± 0.0022
	Post pandemic	7.1373 ± 0.0005	7.0440 ± 0.0002	7.0090 ± 0.0000	6.1808 ± 0.0009	6.8107 ± 0.0008	7.4872 ± 0.0001	7.0206 ± 0.0001	7.2306 ± 0.0001	0.3136 ± 0.0003	0.3603 ± 0.0004	0.8385 ± 0.0005	0.0920 ± 0.0001
Mn	During pandemic	0.0014 ± 0.0010	0.0015 ± 0.0014	0.0004 ± 0.0008	0.0016 ± 0.0008	0.0023 ± 0.0013	0.0012 ± 0.0005	0.0020 ± 0.0016	0.0028 ± 0.0004	0.0003 ± 0.0012	0.0006 ± 0.0023	0.0016 ± 0.0023	0.0005 ± 0.0007
	Post pandemic	0.2200 ± 0.0002	0.0567 ± 0.0004	0.1837 ± 0.0001	0.2018 ± 0.0002	0.0930 ± 0.0004	0.1020 ± 0.0001	0.1655 ± 0.0000	0.1882 ± 0.0010	0.1202 ± 0.0001	0.1111 ± 0.0001	0.2109 ± 0.0002	0.2336 ± 0.0003
Ni	During pandemic	0.0047 ± 0.0051	0.0025 ± 0.0051	0.0011 ± 0.0038	0.0040 ± 0.0034	0.0030 ± 0.0044	0.0026 ± 0.0062	0.0015 ± 0.0059	0.0000 ± 0.0078	0.0019 ± 0.0062	0.0086 ± 0.0088	0.0018 ± 0.0076	0.0013 ± 0.0050
	Post pandemic	0.0790 ± 0.0002	0.0505 ± 0.0001	0.2492 ± 0.0003	0.1636 ± 0.0001	0.0209 ± 0.0008	0.1493 ± 0.0001	0.0637 ± 0.0003	0.1493 ± 0.0004	0.4061 ± 0.0001	0.3633 ± 0.0001	0.2064 ± 0.0007	0.4347 ± 0.0005
Pb	During pandemic	0.0009 ± 0.0020	0.0012 ± 0.0020	0.0022 ± 0.0042	0.0004 ± 0.0014	0.0010 ± 0.0031	0.0004 ± 0.0022	0.0009 ± 0.0025	0.0001 ± 0.0025	0.0008 ± 0.0031	0.0013 ± 0.0026	0.0007 ± 0.0023	0.0013 ± 0.0020
	Post pandemic	0.0508 ± 0.0001	0.1169 ± 0.0004	0.3898 ± 0.0005	0.5220 ± 0.0000	0.3458 ± 0.0005	0.0373 ± 0.0002	0.0068 ± 0.0001	0.2492 ± 0.0000	0.1169 ± 0.0001	0.0068 ± 0.0006	0.2492 ± 0.0004	0.0593 ± 0.0007
Zn	During pandemic	0.0021 ± 0.0004	0.0028 ± 0.0005	0.0026 ± 0.0008	0.0022 ± 0.0002	0.0060 ± 0.0010	0.0016 ± 0.0008	0.0032 ± 0.0018	0.0050 ± 0.0001	0.0014 ± 0.0005	0.0010 ± 0.0004	0.0006 ± 0.0000	0.0034 ± 0.0015
	Post pandemic	2.5686 ± 0.0021	0.7675 ± 0.0006	0.4533 ± 0.0001	0.2416 ± 0.0001	0.0039 ± 0.0001	0.0298 ± 0.0001	0.0369 ± 0.0000	0.2380 ± 0.0000	0.0063 ± 0.0002	0.0733 ± 0.0001	0.0722 ± 0.0004	0.3522 ± 0.0001

References

1. Yunus, K.; Zuraidah, M.A.; John, A. A review on the accumulation of heavy metals in coastal sediment of Peninsular Malaysia. *Ecofeminism Clim. Chang.* **2020**, *1*, 21–35. [CrossRef]
2. Tahiru, A.A.; Doke, D.A.; Baatuuwie, B.N. Effect of land use and land cover changes on water quality in the Nawuni Catchment of the White Volta Basin, Northern Region, Ghana. *Appl. Water Sci.* **2020**, *10*, 198. [CrossRef]
3. Wang, Z.J.; Li, S.L.; Yue, F.J.; Qin, C.Q.; Buckerfield, S.; Zeng, J. Rainfall driven nitrate transport in agricultural karst surface river system: Insight from high resolution hydrochemistry and nitrate isotopes. *Agric. Ecosyst. Environ.* **2020**, *291*, 106787. [CrossRef]
4. Esteller, M.V.; Martínez-Valdés, H.; Garrido, S.; Uribe, Q. Nitrate and phosphate leaching in a Phaeozem soil treated with biosolids, composted biosolids and inorganic fertilizers. *Waste Manag.* **2009**, *29*, 1936–1944. [CrossRef] [PubMed]
5. Othman, F.; Chowdhury, M.S.U.; Wan Jaafar, W.Z.; Faresh, E.M.M.; Shirazi, S.M. Assessing risk and sources of heavy metals in a tropical river basin: A case study of the Selangor river, Malaysia. *Polish J. Environ. Stud.* **2018**, *27*, 1659–1672. [CrossRef]
6. Ibrahim, T.N.B.T.; Othman, F.; Mahmood, N.Z. Baseline Study of Heavy Metal Pollution in a Tropical River in a Developing Country. *Sains Malays.* **2020**, *49*, 729–742. [CrossRef]
7. Elmorsi, R.R.; Abou-El-Sherbini, K.S.; Abdel-Hafiz Mostafa, G.; Hamed, M.A. Distribution of essential heavy metals in the aquatic ecosystem of Lake Manzala, Egypt. *Heliyon* **2019**, *5*, e02276. [CrossRef] [PubMed]
8. Ali, H.; Khan, E. Bioaccumulation of non-essential hazardous heavy metals and metalloids in freshwater fish. *Risk Human Health Environ. Chem. Lett.* **2018**, *16*, 903–917. [CrossRef]
9. Ishadi, N.A.M.; Rawi, C.S.M.; Ahmad, A.H.; Abdul, N.H. The influence of heavy metals and water parameters on the composition and abundance of water bugs (Insecta: Hemiptera) in the kerian river basin, Perak, Malaysia. *Trop. Life Sci. Res.* **2014**, *25*, 61–79. [PubMed]
10. Ibrahim, N.; Aziz, H.A.; Yusoff, M.S. Heavy metals concentration in river and pumping well water for river bank filtration (RBF) system: Case study in Sungai Kerian. *J. Teknol.* **2015**, *74*, 59–67. [CrossRef]

11. Billah, M.M.; Mustafa Kamal, A.H.; Idris, M.H.; Ismail, J. Mangrove Macroalgae as Biomonitors of Heavy Metal Contamination in a Tropical Estuary, Malaysia. *Water Air Soil Pollut.* **2017**, *228*, 347. [CrossRef]
12. Chowdhury, M.S.U.; Othman, F.; Jaafar, W.Z.W.; Mood, N.C.; Adham, M.I. Assessment of pollution and improvement measure of water quality parameters using scenarios modeling for Sungai Selangor Basin. *Sains Malays.* **2018**, *47*, 457–469. [CrossRef]
13. Zanuri, N.B.M.; Abdullah, M.B.; Darif, N.A.M.; Nilamani, N.; Hwai, A.T.S. Case study of marine pollution in Teluk Bahang, Penang, Malaysia. *IOP Conf. Ser. Earth Environ. Sci.* **2020**, *414*, 012032. [CrossRef]
14. Lee Goi, C. The river water quality before and during the Movement Control Order (MCO) in Malaysia. *Case Stud. Chem. Environ. Eng.* **2020**, *2*, 100027. [CrossRef]
15. Razak, M.R.; Aris, A.Z.; Zakaria, N.A.C.; Wee, S.Y.; Ismail, N.A.H. Accumulation and risk assessment of heavy metals employing species sensitivity distributions in Linggi River, Negeri Sembilan, Malaysia. *Ecotoxicol. Environ. Saf.* **2021**, *211*, 111905. [CrossRef]
16. Nordin, N.F.M.; Mohamad, H.; Alarifi, H. Numerical modelling of seepage analysis using SEEP/W: A case study for the Kerian River Flood Mitigation Project (Phase 3) in Bandar Baharu, Kedah. *IOP Conf. Ser. Mater. Sci. Eng.* **2021**, *1101*, 12007. [CrossRef]
17. Ansari, E.; Akhtar, M.N.; Abdullah, M.N.; Othman, W.A.; Bakar, E.A.; Hawary, A.F.; Alhady, S.S. Image Processing of UAV Imagery for River Feature Recognition of Kerian River, Malaysia. *Sustainability* **2021**, *13*, 9568. [CrossRef]
18. Amelia, Z.S.; Che Salmah, M.R.; Abu Hassan, A. Diversity and distribution of dragonfly (Odonata: Insecta) in the Kerian River basin, Kedah-Perak, Malaysia. *USU Reposit.* **2006**, *14*. Available online: https://www.researchgate.net/publication/42320353_Diversity_and_Distribution_of_Dragonfly_OdonataInsecta_in_the_Kerian_River_Basin_Kedah-Malaysia (accessed on 19 November 2021).
19. Ansari, E.; Akhtar, M.N.; Bakar, E.A.; Uchiyama, N.; Kamaruddin, N.M.; Umar, S.N.H. *Investigation of Geomorphological Features of Kerian River Using Satellite Images BT—Intelligent Manufacturing and Mechatronics*; Bahari, M.S., Harun, A., Zainal Abidin, Z., Hamidon, R., Zakaria, S., Eds.; Springer: Singapore, 2021; pp. 91–101.
20. National Water Quality Standards for Malaysia. Available online: https://environment.com.my/wp-content/uploads/2016/05/River.pdf (accessed on 21 November 2021).

Article

Adsorption of Methylene Blue by Biosorption on Alkali-Treated *Solanum incanum*: Isotherms, Equilibrium and Mechanism

Hamza S. AL-Shehri [1,†], Hamdah S. Alanazi [2,†], Areej Mohammed Shaykhayn [2], Lina Saad ALharbi [2], Wedyan Saud Alnafaei [2], Ali Q. Alorabi [3], Ali S. Alkorbi [4] and Fahad A. Alharthi [2,*]

[1] Chemistry Division, King Khaled Military Academy, SANG, Riyadh 11495, Saudi Arabia; alshehrih@kkma.edu.sa
[2] Department of Chemistry, College of Science, King Saud University, Riyadh 11451, Saudi Arabia; hsenzi@ksu.edu.sa (H.S.A.); areej_shaykhayn@hotmail.com (A.M.S.); lenasaadx@outlook.com (L.S.A.); wedosmoon@gmail.com (W.S.A.)
[3] Chemistry Department, Faculty of Science, Al-Baha University, Al Baha 65731, Saudi Arabia; aalorabi@bu.edu.sa
[4] Empty Quarter Research Unit, Department of Chemistry, College of Science and Art in Sharurah, Najran University, Sharurah 68342, Saudi Arabia; assalem@nu.edu.sa
* Correspondence: fharthi@ksu.edu.sa
† These authors contributed equally to this work.

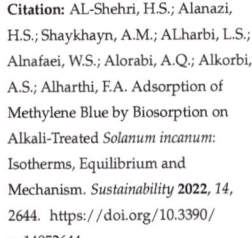

Citation: AL-Shehri, H.S.; Alanazi, H.S.; Shaykhayn, A.M.; ALharbi, L.S.; Alnafaei, W.S.; Alorabi, A.Q.; Alkorbi, A.S.; Alharthi, F.A. Adsorption of Methylene Blue by Biosorption on Alkali-Treated *Solanum incanum*: Isotherms, Equilibrium and Mechanism. *Sustainability* 2022, 14, 2644. https://doi.org/10.3390/su14052644

Academic Editors: Mohd Rafatullah and Masoom Raza Siddiqui

Received: 26 January 2022
Accepted: 16 February 2022
Published: 24 February 2022

Publisher's Note: MDPI stays neutral with regard to jurisdictional claims in published maps and institutional affiliations.

Copyright: © 2022 by the authors. Licensee MDPI, Basel, Switzerland. This article is an open access article distributed under the terms and conditions of the Creative Commons Attribution (CC BY) license (https://creativecommons.org/licenses/by/4.0/).

Abstract: In this study, a new bio-adsorbent (NASIF) was successfully prepared via chemical activation of *Solanum incanum* (SI) with hydrogen peroxide and sodium hydroxide reagents as an inexpensive and effective adsorbent for the removal of methylene blue (MB) from aqueous media. The morphology of the NASIF adsorbent surface and the nature of the potential MB interactions were examined by Fourier transform infrared spectroscopy (FTIR) and scanning electron microscopy (SEM) micrograph. FTIR results suggested that carboxyl, carbonyl, and hydroxyl groups were involved in MB adsorption on the NASIF surface. EDX analysis confirmed the successful increase of oxygen-containing functional groups during the chemical activation. The influence of important factors was studied using the batch method. The results revealed that the maximum removal efficiency was 98% at contact time: 120 min; pH: 6.5, adsorbent dose: 40 mg; and temperature-25 °C. Isothermal behavior was evaluated using three non-linear isotherm models, Langmuir, Freundlich, and D–R isotherm. MB adsorption onto NASIF adsorbent followed the Langmuir isotherm model with maximum monolayer capacity (mg/g) at 25 °C. Meanwhile, the PSO kinetics model was found to be better than PFO kinetic model for describing the adsorption process using kinetic models. Based on the D–R model, the free energy (E, kJ mol^{-1}) values were in the range of 0.090–0.1812 kJ mol^{-1}, which indicated that the MB adsorption onto NASIF may belong to physical adsorption. The adsorption mechanism of MB onto NASIF adsorbent mainly includes electrostatic attraction, π-π interaction, n-π interaction, and H-bonding. The thermodynamic parameters revealed that the adsorption process was a feasibility, spontaneous and exothermic process. Finally, the result of the present work could provide strong evidence of the potential of NASIF adsorbent for eliminating MB from aqueous media.

Keywords: *Solanum incanum*; isotherm; adsorption; mechanism; methylene blue; exothermic

1. Introduction

Synthetic dyes are colored organic compounds released in wastewater from different industries such as cosmetics, textiles, rubber, printing, leather, plastics, paper, printing, and pharmaceutical, which cause a series of serious hazards to our water resources [1–3]. Methylene blue (MB), Rhodamine B (RB), Malachite green (MG), crystal violet (CV), etc., are water-soluble dyes and have stable physical and chemical properties, and are difficult to degrade [4]. Methylene blue (MB), known as methyl thioninium chloride, is a cationic thiazine dye with the molecular formula [$C_{16}H_{18}N_3SCl$] [5]. It was widely used in many industries, as mentioned above. Excessive release of MB dye causes abnormal colorization of surface water, hinders sunlight penetration, and reduces oxygen dissolution, which affects the photosynthesis of aquatic plants [6]. Therefore, dye molecules cause serious threats to the environmental system and health of humans that include carcinogenicity, dysfunction of the brain, kidneys, and liver [7–9]. A large amount of MB (more than 7.0 mg kg^{-1}) can lead to mental disorders, high blood pressure, abdominal pain, and nausea [10]. Even low concentrations are very hazardous to the environment aquatic. Due to its harmfulness, it is important to remove these pollutants before their discharge into the environment.

Various approaches include photocatalytic degradation [11], ozone-based processes [12], membrane separation [13], free radical degradation [14], coagulation [15], chemical oxidation [16], and adsorption [17,18], etc. Among these approaches, adsorption has been considered as one of the most efficient methods for the removal of pollutants due to its low cost, high specific adsorption performance, economic method with efficient performance, high efficiency, and non-toxic adsorbents [19–21].

Various materials, either synthetic or natural, have been used for the elimination of MB dye from aqueous media. Materials with good adsorption performance, low cost, and natural affinity to pollutants are desired in the choice of adsorption materials, therefore, all these characteristics are found in bio-adsorbents. Additionally, biosorbents have numerous important characteristics such as eco-friendliness, biocompatibility, availability, and feasibility: hence their potential adsorption capacity to remove toxic pollutants from wastewater. Various biosorbents such as sugarcane Bagasse [22], *Terminalia catappa* (TC) shells [23], Guava leaves [24], *phoenix tree* leaf powder [25], *Platanus orientalis* leaves [26], *Rhus Coriaria* L [27], *Casuarina equisetifolia pine* [5], have been employed in the removal of methylene blue from aqueous solutions.

Solanum incanum is a perennial that grows naturally in southern Saudi Arabia and other countries (Tanzania, Kenya, Uganda, Australia, India, and Madagascar) [28]. The leaves and stems are yellowish-brown in color when ripe and they have small thorns, and the fruits are often 2–3 cm in diameter. *Solanum incanum* is used as medicine for managing hepatitis and reducing the risk of high blood pressure. *Solanum incanum* presents itself as a potential bio-adsorbent, owing to its wide availability in nature and with the objective of adding value to biomass in the polluted water treatment sector. To the best of our knowledge, there is no evidence from many studies on the potential use of activated *Solanum incanum* with H_2O_2 as a bio-adsorbent in the treatment of wastewater.

The aim of the present work was to activate *Solanum incanum* leaves and flowers by chemical activation with H_2O_2 and NaOH reagent to increase oxygen-containing functional groups on their surface and thus enhance the adsorptive properties. The activated adsorbents were characterized by SEM, EDX, elemental mapping, and FTIR. To the best of our knowledge, NASIF adsorbent has not been used yet for dye removal from aqueous media. The effects of contact time, adsorbent dose, initial MB concentration, and temperature on the removal of MB dye were achieved through the batch method. The kinetics and isotherm of MB adsorption on NASIF were investigated. The thermodynamic parameters have also been calculated. Additionally, the possible mechanisms for MB dye adsorption onto NASIF adsorbent were discussed.

2. Experiment

2.1. Chemicals and Instrumentation

All chemicals were used as received. Methylene blue (MB) and Rhodamine B (RB) dyes were obtained from Sigma Aldrich. Sodium hydroxide (NaOH), hydrochloric acid (HCl), and nitric acid (HNO_3) were obtained from BDH, England. Deionized water was used in this work. The analysis of functional groups of UTSIF, NASIF, and MB-saturated NASIF were performed using Fourier transform infrared spectroscopy (Nicolet iS50 FTIR, Thermo Scientific). The surface morphological features of UTSIF, NASIF, and MB-saturated NASIF were analyzed with scanning electron microscopy, and Energy-dispersive X-ray spectroscopy (Model JEM-2100F, JEOL, Japan). The MB and RB concentrations in the solutions were determined using a UV–Vis spectrophotometer (Perkin Elmer, 900T).

2.2. Chemical Treatment

For the present study, *Solanum incanum* leaves (SIL) and *Solanum incanum* flowers (SIF) were collected from Albaha region in Saudi Arabia. SIL and SIF materials were washed several times with deionized water to remove any adhering dirt. Then, the washed SIL and SIF were oven-dried at 100 °C for 24 h. Next, the dried SIL and SIF were crushed separately in the grinder to obtain fine powder. Chemical surface oxidation was carried out by immersing the SIF and SIL separately in 200 mL of 10% H_2O_2 under stirring for 24 h at room temperature, which developed surface functionalities and enhanced the adsorptive properties [29,30]. After that, SIF and SIL powders were washed several times with deionized water and then filtered. The SIL and SIF were activated separately with a 1 M NaOH (200 mL) solution for 24 h. They were then washed off with deionized water, filtered, and dried in an oven at 50 °C for 24 h to obtain NaOH-activated *Solanum incanum* leaves (NASIL) adsorbent and NaOH-activated *Solanum incanum* leaves (NASIF) adsorbent.

2.3. Batch Adsorption Studies

Initial evaluation of the *Solanum incanum* flower (SIF), *Solanum incanum* leaves (SIL), NaOH-activated *Solanum incanum* flower (NASIF), NaOH-activated *Solanum incanum* flower (NASIL) adsorbent for adsorption of MB and RB dyes was conducted via batch adsorption experiments. The MB dye adsorption experiment performed was achieved under different operating parameters such as concentration (50–300 mg/L), various temperatures (25, 35, and 45 °C), reaction time (5–210 min), and adsorbent dosage (5–100 mg). A stock MB solution for the adsorption tests was prepared by dissolving MB in certain deionized water and then diluted to the desired MB concentrations. The original pH of the MB dye stock solution was approximately 6.5. All experimental solutions in this study were performed at pH 6.5 without any adjustments. A 40 mg sample of NASIF was charged into a conical flask containing 50 mL of MB (50 mg/L) and then the suspension was allowed to equilibrate using a shaker at room temperature for 100 min. After that, the suspension was separated by centrifuging at 5000 rpm for 10 min. Finally, the concentrations of MB and RB were measured using UV–Vis spectrophotometer at 663 nm and 556 nm, respectively. The removal efficiency of dyes and adsorption capacity of adsorbents were calculated by Equations (1) and (2), respectively.

$$\% \text{ adsorption} = \frac{C_o - C_e}{C_o} \times 100 \quad (1)$$

$$q_e = (C_o - C_e)\frac{V}{m} \quad (2)$$

where C_o is the concentration of the initial dyes; C_e (mg/L) is the equilibrium dye concentration; m (g) is the adsorbent mass and V (L) is the dye solution volume.

3. Results and Discussion

3.1. Bio-Adsorbent Characterization

The FTIR technique is an effective technique to identify functional groups of adsorbents. The FTIR spectra of UTSIF, NASIF, and MB-saturated NASIF are presented in Figure 1. The FTIR spectra of UTSIF powder show distinct peaks at 3438, 3013, 2931, 2852, 1741, 1624, 1459, 1375, 1265, 1157, and 1105 cm^{-1}. In detail, the broad peak at 3438 cm^{-1} is due to –OH stretching vibration of carboxylic acids and phenols in lignin, cellulose, and hemicellulose, suggesting the existence of hydroxyl and carboxylic acid on the surface of UTSIF and NASIF. This band increased in intensity with a small shift to 3443 cm^{-1} after chemical activation [23,26,31]. The peaks at 3013, 2931, and 2852 cm^{-1} were attributed to the stretching vibration of =C-H aromatic, asymmetric, and symmetric –C-H in methylene groups, respectively [32]. The strong peaks at 1741 and 1624 cm^{-1} were due to the C=O (carbonyl, carboxyl) extending vibrations of keto-carbonyl gatherings and C=C bond [2,33,34]. After chemical modification (NASIF) the intensity of this peak increased. The peaks at 1454 cm^{-1} were due to bending vibrations of the aliphatic C–H bonds or carboxylate asymmetric. This peak sharply increased in intensity and shifted to 1462 cm^{-1}. In addition, a new peak appeared at 1520 cm^{-1} due to an increase in the oxygen-containing functionalities on the surface of NASIF. The peaks at 1375 cm^{-1}, 1265 cm^{-1}, and (1157–1105 cm^{-1}) were assigned to C-N, C-O, C-O-C bonds [33,35]. A new peak that was observed at 874 cm^{-1} and 812 cm^{-1} (after chemical activation) was due to the out-of-plane bending of aromatic C–H [36,37]. Characteristic changes observed in the FTIR spectrum of NASIF after MB adsorption indicated that some peaks increased/decreased in intensity and shifted or disappeared. In detail, the intensity of the –OH and C=O groups shifted to 3424 cm^{-1} and 1735 cm^{-1} due to bending of –OH and C=O groups on the surface of NASIF with $R(CH_3)_2N^+$ sites of MB by electrostatic interactions and H-bonding [35]. In addition, the intensity at 1658 cm^{-1} increased and shifted to 1606 cm^{-1}, which indicates π-π interactions between the aromatic ring molecule in the NASIF adsorbent and the benzene ring in the MB molecule. Also, a peak of 1520 cm^{-1} disappeared after MB adsorption onto NASIF surface adsorbent. Furthermore, the peak at 874 cm^{-1} was increased in intensity and shifted to 869 cm^{-1} due to H-bond formation between the N atoms of MB and H atoms of NASIF [38]. The observation supports interaction of MB with functional groups of NASIF via electrostatic, H-bonding, n-π interactions, and π-π interactions.

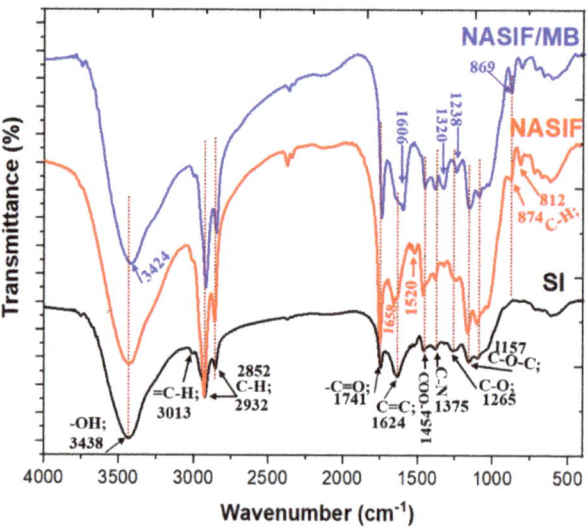

Figure 1. FTIR spectra of UTSIF, NASIF, and MB-saturated NASIF.

Figure 2 shows the morphology of UTSIF, NASIF, and MB-saturated NASIF. The surface of UTSIF shows a rough surface with pores and cracks (Figure 2a). After chemical activation, the porosity of NASIF surface increased (Figure 2b). After MB-saturated NASIF, the surface is covered and looks like a multilayered structure (Figure 2c). Thus, the SEM images of NASIF before and after MB-saturated NASIF display the differences in surface morphologies of these two adsorbents. The EDX of UTSIF adsorbent showed the existence of elements, mainly C (63.32%) and O (34.48%), with very trace elements of K and Ca (Table 1). The elemental mapping was carried out to identify the elements present in UTSIF, NASIF, and MB-saturated NASIF; therefore, the elemental mapping analysis in (Figure 3a) supports the presence of C, O, and other elements on the surface of the UTSIF adsorbent. After the chemical activation of UTSIF with H_2O_2 and NaOH (Table 1), it was observed that the weight% of oxygen increased to 44.08% while the weight% of carbon was reduced to 54.32%, indicating improved oxygen-containing functionalities on the surface of NASIF. The elemental mapping analysis confirms the distribution of elements after chemical activation; it was seen that the distribution of oxygen increased on the surface of the NASIF adsorbent (Figure 3b). After MB dye-saturated NASIF (Figure 4, a new peak of sulfide and nitrogen appeared with 0.81% and 0.02%, respectively (Table 1), indicating successful MB dye adsorption on NASIF adsorbent. The mapping results indicated the distribution of new elements (sulfide and nitrogen) over the NASIF surface (Figure 3c).

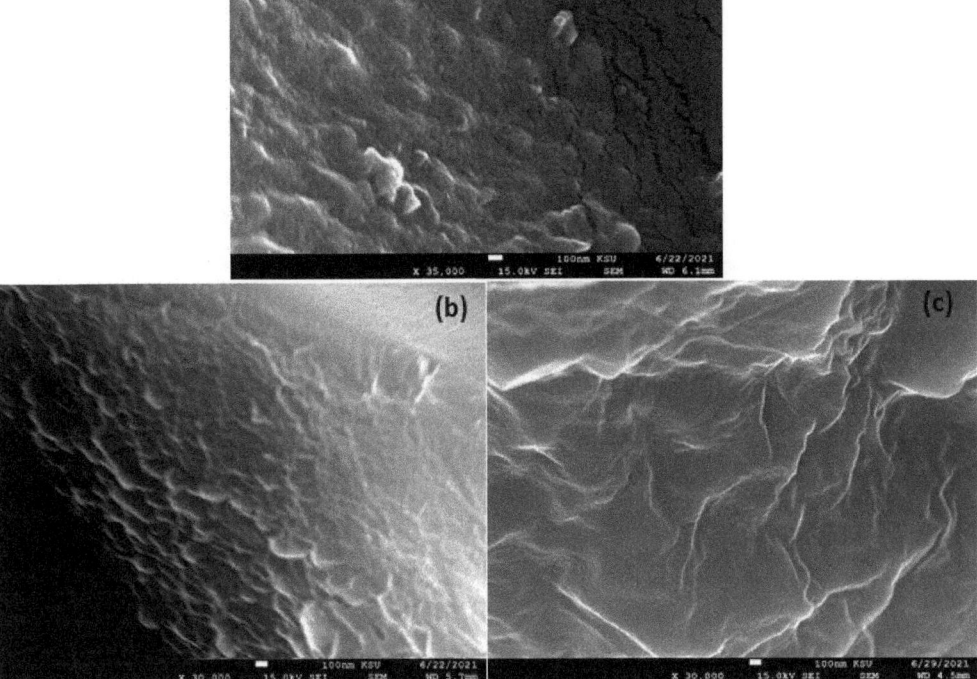

Figure 2. SEM images of (**a**) UTSIF, (**b**) NASIF, and (**c**) MB-saturated NASIF.

Table 1. Elemental analysis data.

Sample	Elemental Content (Wt.%)						
	C	O	Ca	K	Cl	S	N
UTSIF	63.14	34.48	0.76	1.36	0.25	-	-
NASIF	54.32	44.08	1.50	0.02	-	-	-
MB-saturated NASIF	66.97	32.14	-	-	0.07	0.81	0.02

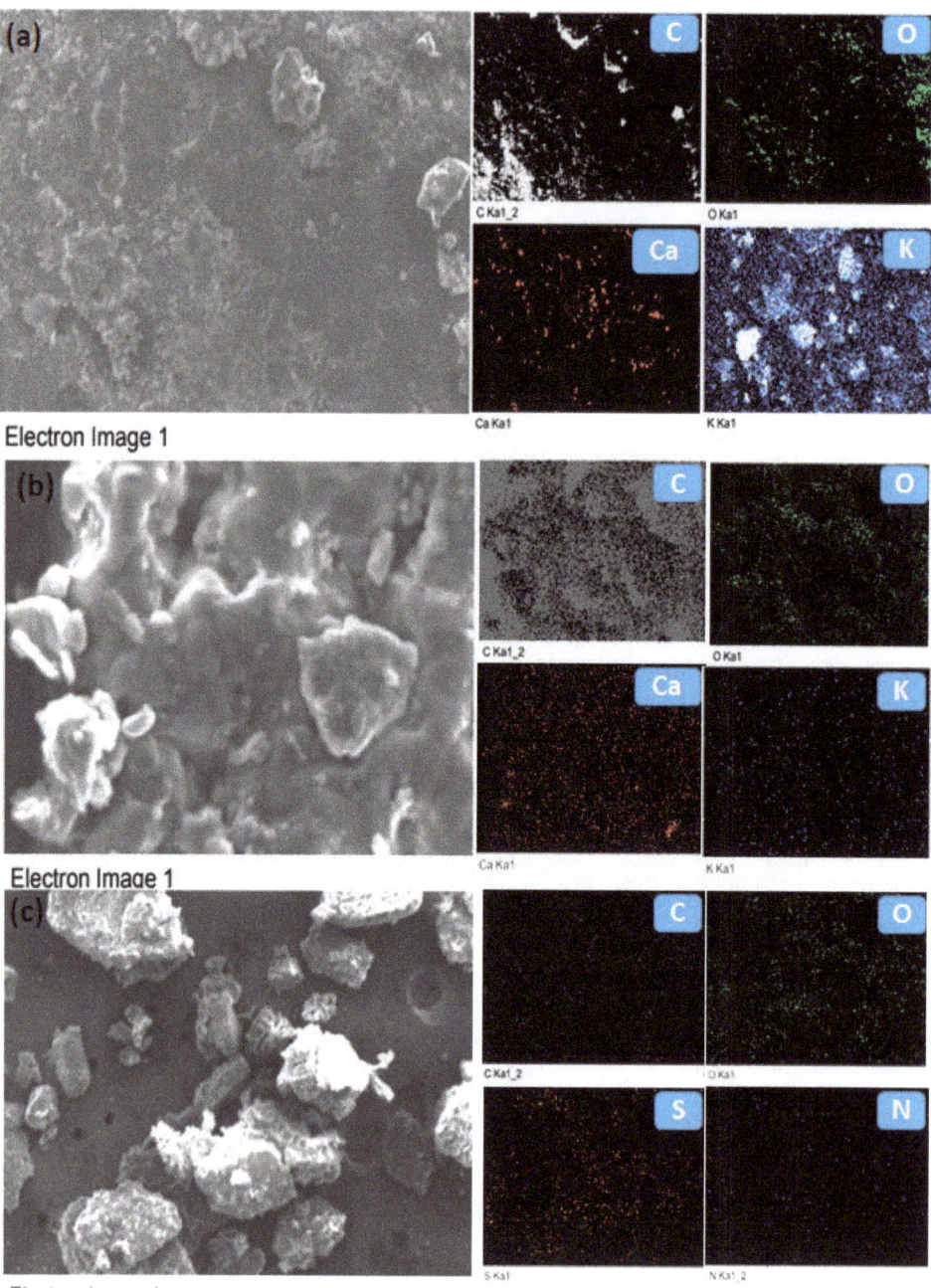

Figure 3. Elemental mapping analysis for (**a**) UTSIF, (**b**) NASIF, and (**c**) MB-saturated NASIF.

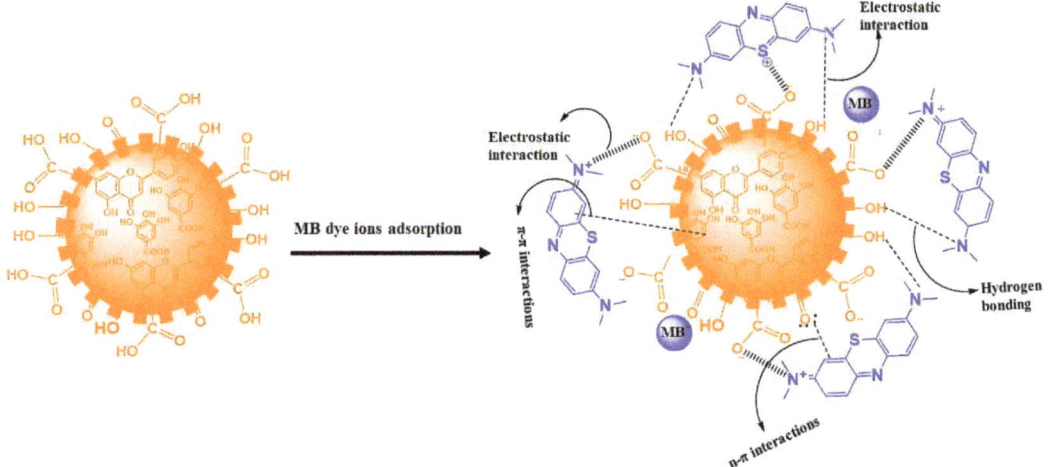

Figure 4. Mechanism of adsorption of MB using of NASIF adsorbent.

3.2. Adsorption Studies

3.2.1. Selectivity Studies of Adsorbent

Four adsorbents—UTSIF, UTSIL, NASIF, and NASIL—for the adsorption of cationic dyes (MB and RB dyes) from aqueous solutions were tested under specific condition parameters (dose: C_o: 50 mg L^{-1}; m: 40 mg; T: 25 °C; t: 24 h; pH: 6.5), as shown in Figure 5. The outcomes revealed that the four adsorbents show greater adsorption ability toward MB dye (95.6–99.0%) compared to RB dye (1.99–17.78%). In addition, the performance of NASIF and NASIL adsorbent was better than untreated UTSIF and UTSIL adsorbents toward the adsorption of MB dye. The high removal efficiencies of MB dye onto NASIF and NASIL adsorbents owes to the use of H_2O_2 and NaOH as activation agents [39,40].

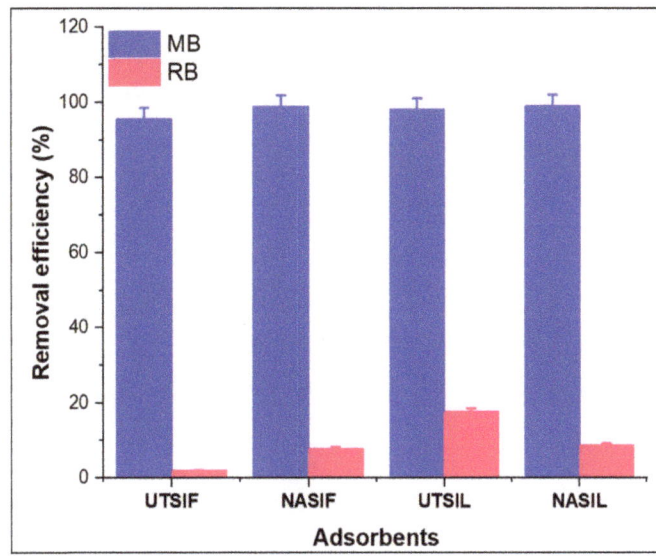

Figure 5. Selectivity studies of adsorbents toward MB and RB dyes.

3.2.2. Effect of Adsorbent Dose

The influence of adsorbent amount in the range from 0.05 mg to 0.1 mg on the adsorption process of MB using NASIF adsorbent has been examined as presented in Figure 6a. The result exhibited that the removal efficiency of MB dye increases significantly from 46.04% to 93.62% when the NASIF adsorbent mass rises from 0.01 g to 0.04 g. By improving the adsorbent amount, more adsorption active sites are available on the NASIF surface for interaction with MB molecules [9]. After this dose, no significant improvement in removal efficiency was observed. Oppositely, the adsorption capacity of NASIF adsorbent reduced significantly to 57.07 mg/g with the amount of NASIF adsorbent increasing to 0.04 g, due to the adsorption capacity being inversely proportional to the mass as per Equation (2) [41]. In addition, the rise in the amount of NASIF adsorbent may result in the aggregation of adsorbent and lead to some of the adsorption active sites overlapping, decreasing the adsorption capacity. These findings were similar to previous studies [2,42]. Thus, 40 mg of NASIF adsorbent was the optimum adsorbent dose in the subsequent studies.

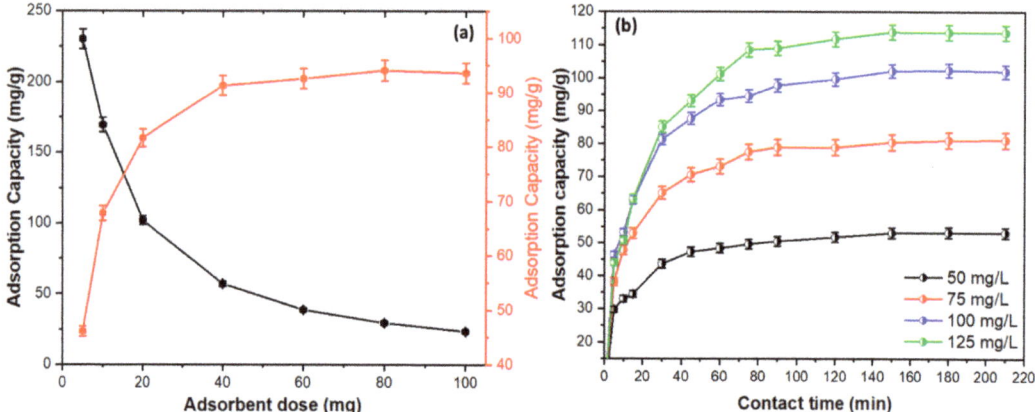

Figure 6. Effect of adsorbent dose (**a**), and contact time (**b**) on adsorption process.

3.2.3. Effect of Contact Time

The time-dependent adsorption capacity of NASIF was studied under different contact times (up to 210 min) and different MB concentrations (50, 75, 100, and 125 mg/L) as presented in Figure 6b (pH: 6.5; T: 25 °C; m: 0.04g). There is a rapid removal efficiency of MB dye (47.72%) within 5 min of the adsorption process due to the availability of abundant active sites on the surface of NASIF adsorbent. The adsorption capacities increased from 29.82 mg/g to 51.83 mg/g at 50 mg/L, 38.14 mg/g to 78.90 mg/g at 75 mg/L, 46.35 mg/g to 99.66 mg/g at 100 mg/L, and 43.92mg/g to 111.63 mg/g at 125 mg/L as the contact time increased from 5 min to 120 min, respectively. After 120 min, the adsorption capacity became almost constant. Therefore, an optimum time of 120 min for MB was selected in the subsequent studies. A similar result has been detected in the adsorption of MB dye by different bio-adsorption materials [43,44].

3.2.4. Effect of Initial MB Concentration

The influence of various concentrations of MB dye (50 to 300 mg/L) on the adsorption process at various temperatures (25 to 45 °C) were investigated (pH: 6.5; m: 0.04 g; t: 120 min), as indicated in Figure 7a. According to Figure 6a, with the initial MB concentration increasing from 50 to 300 mg/L, adsorption capacities of NASIF adsorbent improved from 55.38 mg/g to 127.46 mg/g at 25 °C, 51.76 mg/g to 121.30 mg/g at 35 °C, and 46.54 mg/g to 112.90 mg/g at 45 °C. The increase in adsorption capacity with increasing initial MB concentrations was attributed to the driving force increases [45].

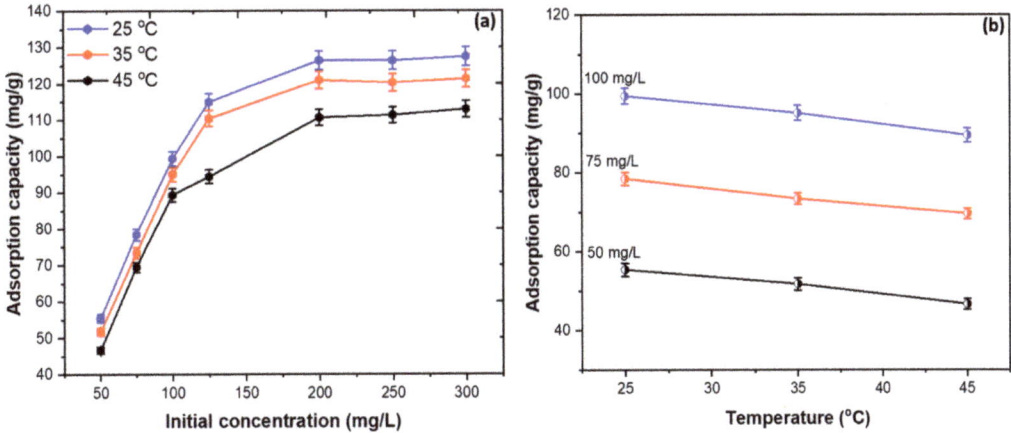

Figure 7. Effect of initial MB concentration (**a**), and temperature (**b**) on adsorption process.

3.2.5. Effect of Temperatures

The effect of various temperature (25, 35, 45 °C) on the adsorption process of MB using NASIF at various concentrations (50, 75, and 100 mg/L) under fixed condition parameters (pH: 6.5; T: 25 °C; m: 0.04 g; t: 120 min) adsorbent was studied as shown in Figure 7b. The results show that the adsorption ability of MB dye was reduced from 55.38 mg/g to 46.64 mg/g by increasing the test solution temperature from 25 °C to 45 °C at 50 mg/L, which shows that the adsorption of MB onto NASIF adsorbent is an exothermic process. This may be due to the destruction of binding sites between MB dye and NASIF adsorbent at 45 °C. This outcome is consistent with the literature reported in [41].

3.3. Adsorption Modeling

3.3.1. Isotherm Modeling

To study the adsorption behaviors of NASIF toward MB dye and mechanism adsorption, three nonlinear isotherm models (Langmuir Equation (3) [46], Freundlich Equation (4) [47], and D–R (Equations (5)–(7)) [48] isotherm models) were applied. The model parameters at different temperatures are presented in Table 2.

Table 2. Nonlinear isotherm models for MB adsorption on NASIF adsorbent.

Model	Temperature (K)		
	298	308	318
Langmuir			
q_m, mg/g	135.24	132.35	124.11
K_L (L/mg)	0.1262	0.0885	0.0266
R^2	0.98328	0.9484	0.94235
Freundlich			
K_f, (mg/g) (L/mg)$^{1/n}$	52.38	45.26136	36.03004
n	5.524	5.019	4.434
R^2	0.8073	0.7598	0.8016
Dubinin–R			
q_s, mg/g	120.14	117.34	109.47
K_{D-R} (mol^2 kJ^{-2})	15.235	33.422	61.739
E (kJ mol^{-1})	0.1812	0.1223	0.0900
R^2	0.8312	0.8988	0.96173

$$q_e = \frac{q_m K_L C_e}{1 + K_L C_e} \qquad (3)$$

$$q_e = K_F \, C_e^{1/n} \qquad (4)$$

$$q_e = q_s \, e^{-K_{D-R} \, \varepsilon^2} \qquad (5)$$

$$\varepsilon = RT\ln\left(1 + \frac{1}{C_e}\right) \qquad (6)$$

$$E = \frac{1}{\sqrt{2K_{D-R}}} \qquad (7)$$

where, q_e (mg/g); q_m (mg/g); C_e (mg/L) represent the equilibrium adsorption capacity, maximum adsorption capacity at monolayer adsorption, and MB molecule concentration at equilibrium, respectively. K_L (L/mg); K_F (mg/g) (L/mg)$^{1/n}$; K_{DR} (mol^2/kJ^2) represent the Langmuir constant, Freundlich constant, and D–R constant related to the adsorption energy, respectively. Variables n; ε; E(kJ/mol) represent the Freundlich isotherm intensity, Polanyi potential, and free energy, respectively. From Table 2 and Figure 8a–c and based on R^2, the adsorption data were observed to fit the Langmuir isotherm (R^2 = 0.98328) compared to the other models (D–R (R^2 = 0.8312), and Freundlich (R^2 = 0.8073)), indicating that the adsorption process can be described as monolayer adsorption and homogeneous surface with maximum monolayer capacity (135.24 mg/g). Based on Langmuir, the q_m values decreased with rising temperature, which emphasized that MB adsorption by NASIF is an exothermic reaction. Based on the D–R model, the adsorption free energy (E, kJ mol^{-1}) values were in the range of 0.090–0.1812 kJ mol^{-1}, which indicated that MB adsorption onto NASIF may belong to physical adsorption [49]. Based on the Freundlich isotherm, the n values were in the range at different temperatures, which indicates a favorable adsorption process [50].

3.3.2. Kinetics Modeling

To understand the mechanism of MB adsorption onto NASIF and estimate the efficiency of the NASIF, two nonlinear kinetics models (pseudo-first-order (PFO) Equation (8) [51], and pseudo-second-order (PSO) Equation (9)) were applied.

$$q_t = q_e \left(1 - e^{(-k_1 t)}\right) \qquad (8)$$

$$q_t = \frac{q_e^2 k_2 t}{1 + q_e \, k_2 \, t} \qquad (9)$$

where, k_1 (1/min) and k_2 (g/mg*min) represent the PFO and PSO constant, respectively. The parameters with coefficient of determination (R^2) values at various MB concentrations (50, 75, 100, and 125 mg/L) are summarized in Table 3. Based on R^2 (Table 3 & Figure 8d), the MB adsorption onto NASIF was found to be more suited to the PSO kinetic model compared to PFO kinetic model at different MB concentrations (50, 75, 100, and 125 mg/L), which suggests the effect of chemical interactions in the adsorption process. In addition, the value of experimental qe (qe$_{exp}$) for NASIF toward MB dye adsorption was 53.06 mg/g, which, being close to the calculated qe (qe$_{cal}$) = 53.74 mg/g at initial MB concentration (50 mg/L), confirmed the adsorption process followed by the PSO kinetics.

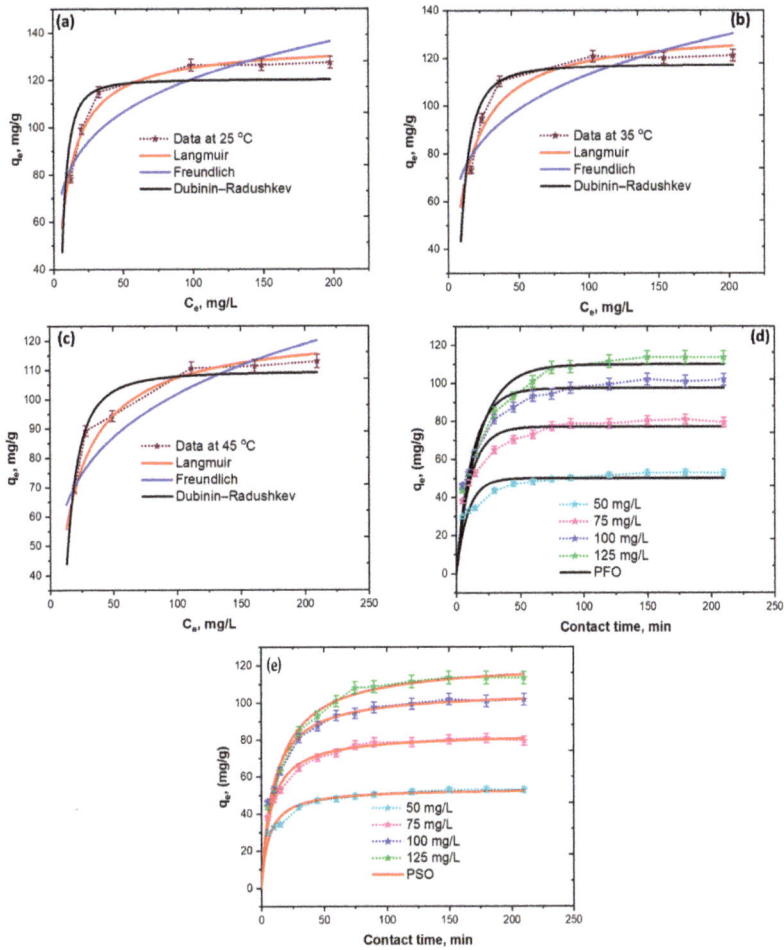

Figure 8. Nonlinear isotherm models at (**a**) 25 °C, (**b**) 35 °C, (**c**) 45 °C, and (**d**) Nonlinear kinetic models: pseudo-first-order and (**e**) pseudo-second-order for MB adsorption on NASIF.

Table 3. Nonlinear kinetics model for MB adsorption on NASIF adsorbent.

C_0 (mg/L)	$q_{e,exp}$ (mg/g)	Pseudo-First-Order			Pseudo-Second-Order		
		$q_{e1,cal}$ (mg/g)	K_1 (1/min)	R^2	$q_{e2,cal}$ (mg/g)	K_2 (g/mg-min)	R^2
50	53.06	50.1621	0.10976	0.93103	53.747	0.00319	0.9797
75	81.120	77.211	0.09167	0.95806	83.703	0.00161	0.9932
100	101.95	97.422	0.07849	0.95888	106.51	0.00106	0.9906
125	113.65	109.799	0.05836	0.96728	122.28	0.00065	0.9903

3.3.3. Thermodynamics Modeling

The thermodynamic analysis provides important insights into the adsorption nature and mechanism of adsorption. Therefore, the effect of solution temperature on MB adsorption onto NASIF was achieved through thermodynamic properties, as summarized in Table 4. Thermodynamic parameters including enthalpy change ($\Delta H°$, kJ/mol) and entropy change (ΔS°, kJ/mol·K) were calculated using van't Hoff equation Equation (11), while the free energy change (ΔG°, kJ/mol), was calculated using Equation (10) as follows:

$$\Delta G° = -RT \ln K_c \quad (10)$$

$$\ln K_c = -\frac{\Delta H°}{RT} + \frac{\Delta S°}{R} \text{ (Van't Hoff equation)} \quad (11)$$

where T is the solution temperature (K); R is the universal gas constant (8.314 J/mol·K); $K_c = q_e/c_e$ (L.g^{-1}) represents the thermodynamic equilibrium constant [52]. Figure 9 demonstrates the relevant van't Hoff plots. From Table 4, a negative ΔH° suggests that the interaction of MB dye adsorbed by NASIF is exothermic, which is supported by the reducing adsorption of MB onto NASIF with the temperature rise. The values of ΔH° were lower than 40 kJ mol^{-1}, further confirming that the MB adsorption onto NASIF belongs to the physisorption process [53]. The ΔS value was negative, which indicates the decrease in randomness at the solid-solution interface between NASIF and MB solution. Negative ΔG° under different temperatures (25, 35, 45 °C) indicates a spontaneous adsorption process.

Table 4. Thermodynamics parameters for the adsorption of MB on NASIF adsorbent.

Concentration of CV Dye (mg/L)	(−)ΔH° (kJ/mol)	(−)ΔS° (J/mol.K)	(−)ΔG° (kJ/mol)		
			298 K	308 K	318 K
50	38.34	109.68	5.63	4.59	3.44
75	22.97	61.80	4.59	3.86	3.37
100	17.13	44.30	3.910	3.52	3.02

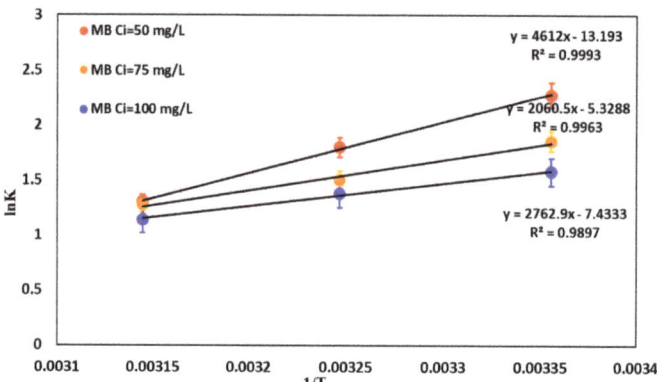

Figure 9. Adsorption thermodynamics.

3.4. Comparison with Other Adsorbents

The adsorption capacity of NASIF adsorbent toward the elimination of MB dye from aqueous solutions was compared to other bio-adsorbents, as presented in Table 5 [1,5,23,26,54–57]. The comparison indicates that the adsorption capacities of NASIF adsorbent were higher than those reported in many studies, due to increasing the oxygen-containing functionalities on the surface of NASIF by chemical activation with H_2O_2/NaOH reagents.

Table 5. Comparison of the maximum bio-adsorption capacities (q_m) of NASIF for MB adsorption from different biomaterials.

Adsorbent	q_m (mg/g)	Ref.
Salix babylonica	42.74	[1]
Casuarina equisetifolia pines	41.35	[5]
Terminalia catappa (TC) shells	88.62	[23]
Platanus orientalis leaves	114.94	[26]
Ficus carica bast	47.6	[54]
Carica papaya wood	32.25	[55]
Pinus durangensis	102	[56]
Casuarina equisetifolia pine/H_2SO_4	42.19	[57]
NASIF	135.24	This study

3.5. Adsorption Mechanism

The mechanism of MB dye adsorption onto NASIF was shown in Figure 1. To support the hypothesis given in Figure 1, FTIR analysis of the NASIF-loaded MB was conducted which suggest that the adsorption mechanism of MB on NASIF adsorbent mainly includes electrostatic attraction, n-π interaction, π-π interaction, and hydrogen bonding. There were distinctive changes observed in the FTIR spectrum of NASIF after MB adsorption indicated that some peaks increased/decreased in intensity and shifted or disappeared. In detail, the intensity of the –OH and C=O groups shifted from (3438 to 3424 cm^{-1}) and from (1741 to 1735 cm^{-1}), respectively. This shifting is due to the interactions of –OH and C=O groups on the surface of NASIF with $R(CH_3)_2N^+$ sites of MB by electrostatic interactions and H-bonding [35]. In addition, the intensity at 1658 cm^{-1} increased and shifted to 1606 cm^{-1}, which indicates π-π interactions between the aromatic ring molecule in the NASIF adsorbent and the benzene ring in the MB molecule. Further, the peak at 874 cm^{-1} increased in intensity and shifted to 869 cm^{-1} due to H-bond formation between the N atoms of MB and H atoms of NASIF [38]. Also, a peak at 1520 cm^{-1} disappeared after MB adsorption onto NASIF surface adsorbent, which suggests the formation of a n-π type interaction through donation of electron(s) from the oxygen in the carbonyl group present in the NASIF surface to the electron acceptors' aromatic rings in the MB dye [58,59]. These changes observed in the spectrum suggest interactions of MB dye with the functional groups of the NASIF in the adsorption process [21].

4. Conclusions

A new adsorbent, NaOH-activated *Solanum incanum* flower powder (NASIF), was successfully prepared by chemical activation method (H_2O_2/NaOH) as an inexpensive and effective adsorbent for the removal of methylene blue (MB) from aqueous media. The characterization of NASIF adsorbent was evaluated using FTIR and SEM, EDX, and elemental mapping. These technical results confirmed the adsorption of MB onto the NASIF adsorbent surface. The isothermal parameters indicated that the Langmuir model fit better than the Freundlich and D–R models, which confirmed the monolayer adsorption capacity of MB onto NASIF adsorbent. The MB adsorption onto NASIF adsorbent obeys the PSO kinetic model. The optimum conditions for NASIF adsorption of MB were contact time: 120 min; pH: 6.5, adsorbent dose: 40 mg; temperature-25 °C, with a maximum monolayer capacity of 135.24 mg/g. The D–R model revealed that the adsorption free energy (E, kJ mol^{-1}) values were in the range of 0.090–0.1812 kJ mol^{-1}, which indicated that MB adsorption onto NASIF may belong to physical adsorption. The adsorption of MB onto the NASIF adsorbent surface occurred by electrostatic attraction, π-π interaction, n-π interaction, and H-bonding. Negative ΔH and ΔG values indicate that the adsorption process was exothermic in nature and spontaneous.

Author Contributions: F.A.A. and A.Q.A. played an important role in the project design and execution, while A.M.S., L.S.A. and W.S.A. carried out the experimental work. H.S.A.-S., H.S.A. and A.S.A. played a crucial role in the characterization and calculations of the results. F.A.A. and A.Q.A. compiled the data and prepared the manuscript. All authors have read and agreed to the published version of the manuscript.

Funding: We thank the King Saud University, Deanship of Scientific Research for funding this work through the research group No. RG-1441-305. We gratefully acknowledge the financial support by Albaha University (Project No. 1441/3) and are grateful to the Scientific Research.

Conflicts of Interest: The authors declare no conflict of interest.

References

1. Kiwaan, H.A.; Mohamed, F.S.; El-Ghamaz, N.A.; Beshry, N.M.; El-Bindary, A.A. Experimental and electrical studies of zeolitic imidazolate framework-8 for the adsorption of different dyes. *J. Mol. Liq.* **2021**, *338*, 116670. [CrossRef]
2. Aldawsari, A.M.; Alsohaimi, I.H.; Al-Kahtani, A.A.; Alqadami, A.A.; Abdalla, Z.E.A.; Saleh, E.A.M. Adsorptive performance of aminoterephthalic acid modified oxidized activated carbon for malachite green dye: Mechanism, kinetic and thermodynamic studies. *Sep. Sci. Technol.* **2021**, *56*, 835–846. [CrossRef]
3. Alorabi, A.Q. Effective Removal of Malachite Green from Aqueous Solutions Using Magnetic Nanocomposite: Synthesis, Characterization, and Equilibrium Study. *Adsorpt. Sci. Technol.* **2021**, *2021*, 2359110. [CrossRef]
4. El-Bindary, M.A.; El-Desouky, M.G.; El-Bindary, A.A. Adsorption of industrial dye from aqueous solutions onto thermally treated green adsorbent: A complete batch system evaluation. *J. Mol. Liq.* **2022**, *346*, 117082. [CrossRef]
5. Chandarana, H.; Kumar, P.S.; Seenuvasan, M.; Kumar, M.A. Kinetics, equilibrium and thermodynamic investigations of methylene blue dye removal using *Casuarina equisetifolia* pines. *Chemosphere* **2021**, *285*, 131480. [CrossRef]
6. Fang, Y.; Liu, Q.; Zhu, S. Selective biosorption mechanism of methylene blue by a novel and reusable sugar beet pulp cellulose/sodium alginate/iron hydroxide composite hydrogel. *Int. J. Biol. Macromol.* **2021**, *188*, 993–1002. [CrossRef]
7. Alharthi, F.A.; Al-Zaqri, N.; el Marghany, A.; Alghamdi, A.A.; Alorabi, A.Q.; Baghdadi, N.; Al-Shehri, H.S.; Wahab, R.; Ahmad, N. Synthesis of nanocauliflower ZnO photocatalyst by potato waste and its photocatalytic efficiency against dye. *J. Mater. Sci. Mater. Electron.* **2020**, *31*, 11538–11547. [CrossRef]
8. Moghazy, R.M.; Labena, A.; Husien, S. Eco-friendly complementary biosorption process of methylene blue using micro-sized dried biosorbents of two macro-algal species (*Ulva fasciata* and *Sargassum dentifolium*): Full factorial design, equilibrium, and kinetic studies. *Int. J. Biol. Macromol.* **2019**, *134*, 330–343. [CrossRef]
9. Al-Wasidi, A.S.; AlZahrani, I.I.S.; Naglah, A.M.; El-Desouky, M.G.; Khalil, M.A.; El-Bindary, A.A.; El-Bindary, M.A. Effective Removal of Methylene Blue from Aqueous Solution Using Metal-Organic Framework; Modelling Analysis, Statistical Physics Treatment and DFT Calculations. *ChemistrySelect* **2021**, *6*, 11431–11447. [CrossRef]
10. Lyu, H.; Gao, B.; He, F.; Zimmerman, A.R.; Ding, C.; Tang, J.; Crittenden, J.C. Experimental and modeling investigations of ball-milled biochar for the removal of aqueous methylene blue. *Chem. Eng. J.* **2018**, *335*, 110–119. [CrossRef]
11. Soltani, N.; Saion, E.; Hussein, M.Z.; Erfani, M.; Abedini, A.; Bahmanrokh, G.; Navasery, M.; Vaziri, P. Visible light-induced degradation of methylene blue in the presence of photocatalytic ZnS and CdS nanoparticles. *Int. J. Mol. Sci.* **2012**, *13*, 12242–12258. [CrossRef] [PubMed]
12. Zhang, J.; Lee, K.-H.; Cui, L.; Jeong, T. Degradation of methylene blue in aqueous solution by ozone-based processes. *J. Ind. Eng. Chem.* **2009**, *15*, 185–189. [CrossRef]
13. Karisma, D.; Febrianto, G.; Mangindaan, D. Removal of dyes from textile wastewater by using nanofiltration polyetherimide membrane. In *IOP Conference Series: Earth and Environmental Science*; IOP Publishing: Bristol, UK, 2017; p. 12012.
14. Xiao, Z.; Li, Y.; Fan, L.; Wang, Y.; Li, L. Degradation of organic dyes by peroxymonosulfate activated with water-stable iron-based metal organic frameworks. *J. Colloid Interface Sci.* **2021**, *589*, 298–307. [CrossRef] [PubMed]
15. Guibal, E.; Roussy, J. Coagulation and flocculation of dye-containing solutions using a biopolymer (Chitosan). *React. Funct. Polym.* **2007**, *67*, 33–42. [CrossRef]
16. Dutta, K.; Mukhopadhyay, S.; Bhattacharjee, S.; Chaudhuri, B. Chemical oxidation of methylene blue using a Fenton-like reaction. *J. Hazard. Mater.* **2001**, *84*, 57–71. [CrossRef]
17. Alqadami, A.A.; Naushad, M.; Alothman, Z.A.; Ahamad, T. Adsorptive performance of MOF nanocomposite for methylene blue and malachite green dyes: Kinetics, isotherm and mechanism. *J. Environ. Manag.* **2018**, *223*, 29–36. [CrossRef]
18. Kiwaan, H.A.; Mohamed, F.S.; El-Bindary, A.A.; El-Ghamaz, N.A.; Abo-Yassin, H.R.; El-Bindary, M.A. Synthesis, identification and application of metal organic framework for removal of industrial cationic dyes. *J. Mol. Liq.* **2021**, *342*, 117435. [CrossRef]
19. Alorabi, A.Q.; Hassan, M.S.; Azizi, M.; Mohsen, Z.S.A.; Alzahrani, W.A.M.; Alzahrani, K.A.M.; Alghamdi, S.M.O.; Alharthi, F.F.M. Corrigendum to "Fe_3O_4-CuO-activated carbon composite as an efficient adsorbent for bromophenol blue dye removal from aqueous solutions". [Arab. J. Chem. 13(11) (2020) 8080–8091]. *Arab. J. Chem.* **2022**, *15*, 103508. [CrossRef]
20. Alorabi, A.Q.; Hassan, M.S.; Alam, M.M.; Zabin, S.A.; Alsenani, N.I.; Baghdadi, N.E. Natural Clay as a Low-Cost Adsorbent for Crystal Violet Dye Removal and Antimicrobial Activity. *Nanomaterials* **2021**, *11*, 2789. [CrossRef]

21. Alorabi, A.Q.; Alharthi, F.A.; Azizi, M.; Al-Zaqri, N.; El-Marghany, A.; Abdelshafeek, K.A. Removal of Lead(II) from Synthetic Wastewater by Lavandula pubescens Decne Biosorbent: Insight into Composition–Adsorption Relationship. *Appl. Sci.* **2020**, *10*, 7450. [CrossRef]
22. Bagotia, N.; Sharma, A.K.; Kumar, S. A review on modified sugarcane bagasse biosorbent for removal of dyes. *Chemosphere* **2021**, *268*, 129309. [CrossRef]
23. Hevira, L.; Zilfa; Rahmayeni; Ighalo, J.O.; Aziz, H.; Zein, R. Terminalia catappa shell as low-cost biosorbent for the removal of methylene blue from aqueous solutions. *J. Ind. Eng. Chem.* **2021**, *97*, 188–199. [CrossRef]
24. RGaikwad, W.; Kinldy, S.A.M. Studies on auramine dye adsorption on psidium guava leaves. *Korean J. Chem. Eng.* **2009**, *26*, 102–107. [CrossRef]
25. Han, R.; Wang, Y.; Zhao, X.; Wang, Y.; Xie, F.; Cheng, J.; Tang, M. Adsorption of methylene blue by phoenix tree leaf powder in a fixed-bed column: Experiments and prediction of breakthrough curves. *Desalination* **2009**, *245*, 284–297. [CrossRef]
26. Peydayesh, M.; Rahbar-Kelishami, A. Adsorption of methylene blue onto *Platanus orientalis* leaf powder: Kinetic, equilibrium and thermodynamic studies. *J. Ind. Eng. Chem.* **2015**, *21*, 1014–1019. [CrossRef]
27. Dülger, Ö.; Turak, F.; Turhan, K.; Özgür, M. Sumac Leaves as a Novel Low-Cost Adsorbent for Removal of Basic Dye from Aqueous Solution. *ISRN Anal. Chem.* **2013**, *2013*, 210470. [CrossRef]
28. Kaunda, J.S.; Zhang, Y.-J. Chemical constituents from the fruits of *Solanum incanum* L. *Biochem. Syst. Ecol.* **2020**, *90*, 104031. [CrossRef]
29. Xue, Y.; Gao, B.; Yao, Y.; Inyang, M.; Zhang, M.; Zimmerman, A.R.; Ro, K.S. Hydrogen peroxide modification enhances the ability of biochar (hydrochar) produced from hydrothermal carbonization of peanut hull to remove aqueous heavy metals: Batch and column tests. *Chem. Eng. J.* **2012**, *200–202*, 673–680. [CrossRef]
30. Huff, M.D.; Lee, J.W. Biochar-surface oxygenation with hydrogen peroxide. *J. Environ. Manag.* **2016**, *165*, 17–21. [CrossRef]
31. Alhumaimess, M.S.; Alsohaimi, I.H.; Alqadami, A.A.; Khan, M.A.; Kamel, M.M.; Aldosari, O.; Siddiqui, M.R.; Hamedelniel, A.E. Recyclable glutaraldehyde cross-linked polymeric tannin to sequester hexavalent uranium from aqueous solution. *J. Mol. Liq.* **2019**, *281*, 29–38. [CrossRef]
32. Naushad, M.; Alqadami, A.A.; Ahamad, T. Removal of Cd(II) ion from aqueous environment using triaminotriethoxysilane grafted oxidized activated carbon synthesized via activation and subsequent silanization. *Environ. Technol. Innov.* **2020**, *18*, 100686. [CrossRef]
33. Üner, O. Hydrogen storage capacity and methylene blue adsorption performance of activated carbon produced from Arundo donax. *Mater. Chem. Phys.* **2019**, *237*, 121858. [CrossRef]
34. Alshareef, S.A.; Otero, M.; Alanazi, H.S.; Siddiqui, M.R.; Khan, M.A.; Alothman, Z.A. Upcycling olive oil cake through wet torrefaction to produce hydrochar for water decontamination. *Chem. Eng. Res. Des.* **2021**, *170*, 13–22. [CrossRef]
35. Köseoğlu, E.; Akmil-Başar, C. Preparation, structural evaluation and adsorptive properties of activated carbon from agricultural waste biomass. *Adv. Powder Technol.* **2015**, *26*, 811–818. [CrossRef]
36. Gao, P.; Zhou, Y.; Meng, F.; Zhang, Y.; Liu, Z.; Zhang, W.; Xue, G. Preparation and characterization of hydrochar from waste eucalyptus bark by hydrothermal carbonization. *Energy* **2016**, *97*, 238–245. [CrossRef]
37. Kang, S.; Li, X.; Fan, J.; Chang, J. Characterization of hydrochars produced by hydrothermal carbonization of lignin, cellulose, D-xylose, and wood meal. *Ind. Eng. Chem. Res.* **2012**, *51*, 9023–9031. [CrossRef]
38. Tural, B.; Ertaş, E.; Enez, B.; Fincan, S.A.; Tural, S. Preparation and characterization of a novel magnetic biosorbent functionalized with biomass of Bacillus Subtilis: Kinetic and isotherm studies of biosorption processes in the removal of Methylene Blue. *J. Environ. Chem. Eng.* **2017**, *5*, 4795–4802. [CrossRef]
39. Guo, Y.; Yang, S.; Yu, K.; Zhao, J.; Wang, Z.; Xu, H. The preparation and mechanism studies of rice husk based porous carbon. *Mater. Chem. Phys.* **2002**, *74*, 320–323. [CrossRef]
40. le Van, K.; Thi, T.T.L. Activated carbon derived from rice husk by NaOH activation and its application in supercapacitor. *Prog. Nat. Sci. Mater. Int.* **2014**, *24*, 191–198. [CrossRef]
41. Do, T.H.; Nguyen, V.T.; Dung, N.Q.; Chu, M.N.; van Kiet, D.; Ngan, T.T.K.; van Tan, L. Study on methylene blue adsorption of activated carbon made from Moringa oleifera leaf. *Mater. Today Proc.* **2021**, *38*, 3405–3413. [CrossRef]
42. Alqadami, A.A.; Naushad, M.; Ahamad, T.; Algamdi, M.; Alshahrani, A.; Uslu, H.; Shukla, S.K. Removal of highly toxic Cd(II) metal ions from aqueous medium using magnetic nanocomposite: Adsorption kinetics, isotherm and thermodynamics. *Desalin. Water Treat.* **2020**, *181*, 355–361. [CrossRef]
43. Danish, M.; Ahmad, T.; Nadhari, W.; Ahmad, M.; Khanday, W.A.; Ziyang, L.; Pin, Z. Optimization of banana trunk-activated carbon production for methylene blue-contaminated water treatment. *Appl. Water Sci.* **2018**, *8*, 9. [CrossRef]
44. Rashid, J.; Tehreem, F.; Rehman, A.; Kumar, R. Synthesis using natural functionalization of activated carbon from pumpkin peels for decolourization of aqueous methylene blue. *Sci. Total Environ.* **2019**, *671*, 369–376. [CrossRef] [PubMed]
45. Tharaneedhar, V.; Kumar, P.S.; Saravanan, A.; Ravikumar, C.; Jaikumar, V. Prediction and interpretation of adsorption parameters for the sequestration of methylene blue dye from aqueous solution using microwave assisted corncob activated carbon. *Sustain. Mater. Technol.* **2017**, *11*, 1–11. [CrossRef]
46. Wallis, A.; Dollard, M.F. Local and global factors in work stress—The Australian dairy farming exemplar. *Scand. J. Work. Environ. Health* **2008**, *34*, 66–74.
47. Freundlich, H. Über die Adsorption in Lösungen. *Z. Phys. Chem.* **1907**, *57U*, 385. [CrossRef]

48. Dubinin, M.M. Equation of the characteristic curve of activated charcoal. *Chem. Zentr.* **1947**, *1*, 857.
49. Alqadami, A.A.; Naushad, M.; Othman, Z.A.A.L.; Alsuhybani, M.; Algamdi, M. Excellent adsorptive performance of a new nanocomposite for removal of toxic Pb(II) from aqueous environment: Adsorption mechanism and modeling analysis. *J. Hazard. Mater.* **2020**, *389*, 121896. [CrossRef]
50. Alqadami, A.A.; Khan, M.A.; Siddiqui, M.R.; Alothman, Z.A.; Sumbul, S. A facile approach to develop industrial waste encapsulated cryogenic alginate beads to sequester toxic bivalent heavy metals. *J. King Saud Univ. Sci.* **2020**, *32*, 1444–1450. [CrossRef]
51. Lagergren, S. About the theory of so-called adsorption of soluble substances. *Handlingar* **1898**, *24*, 1–39.
52. Lee, L.Y.; Gan, S.; Tan, M.S.Y.; Lim, S.S.; Lee, X.J.; Lam, Y.F. Effective removal of Acid Blue 113 dye using overripe *Cucumis sativus* peel as an eco-friendly biosorbent from agricultural residue. *J. Clean. Prod.* **2016**, *113*, 194–203. [CrossRef]
53. Khan, M.A.; Wabaidur, S.M.; Siddiqui, M.R.; Alqadami, A.A.; Khan, A.H. Silico-manganese fumes waste encapsulated cryogenic alginate beads for aqueous environment de-colorization. *J. Clean. Prod.* **2020**, *244*, 118767. [CrossRef]
54. Pathania, D.; Sharma, S.; Singh, P. Removal of methylene blue by adsorption onto activated carbon developed from *Ficus carica* bast. *Arab. J. Chem.* **2017**, *10*, S1445–S1451. [CrossRef]
55. Rangabhashiyam, S.; Lata, S.; Balasubramanian, P. Biosorption characteristics of methylene blue and malachite green from simulated wastewater onto *Carica papaya* wood biosorbent. *Surfaces Interfaces* **2018**, *10*, 197–215. [CrossRef]
56. Salazar-Rabago, J.J.; Leyva-Ramos, R.; Rivera-Utrilla, J.; Ocampo-Perez, R.; Cerino-Cordova, F.J. Biosorption mechanism of Methylene Blue from aqueous solution onto White Pine (*Pinus durangensis*) sawdust: Effect of operating conditions. *Sustain. Environ. Res.* **2017**, *27*, 32–40. [CrossRef]
57. Chandarana, H.; Suganya, S.; Madhava, A.K. Surface functionalized *Casuarina equisetifolia* pine powder for the removal of hetero-polyaromatic dye: Characteristics and adsorption. *Int. J. Environ. Anal. Chem.* **2020**, 1–15. [CrossRef]
58. Tang, Y.; Zhao, Y.; Lin, T.; Li, Y.; Zhou, R.; Peng, Y. Adsorption performance and mechanism of methylene blue by H_3PO_4-modified corn stalks. *J. Environ. Chem. Eng.* **2019**, *7*, 103398. [CrossRef]
59. Dinh, V.-P.; Huynh, T.-D.-T.; Le, H.M.; Nguyen, V.-D.; Dao, V.-A.; Hung, N.Q.; Tuyen, L.A.; Lee, S.; Yi, J.; Nguyen, T.D.; et al. Insight into the adsorption mechanisms of methylene blue and chromium(iii) from aqueous solution onto pomelo fruit peel. *RSC Adv.* **2019**, *9*, 25847–25860. [CrossRef]

Review

Degradation of Azo Dyes: Bacterial Potential for Bioremediation

Lucas Rafael Santana Pinheiro [1,2,*], Diana Gomes Gradíssimo [1,3], Luciana Pereira Xavier [1,2,3] and Agenor Valadares Santos [1,2,3,*]

[1] Laboratory of Biotechnology of Enzymes and Biotransformations, Institute of Biological Sciences, Federal University of Pará, Guamá, Belém 66075-110, PA, Brazil; dianagradissimo@gmail.com (D.G.G.); lpxavier@ufpa.br (L.P.X.)
[2] Faculty of Biotechnology, Institute of Biological Sciences, Federal University of Pará, Guamá, Belém 66075-110, PA, Brazil
[3] Post Graduation Program in Biotechnology, Institute of Biological Sciences, Federal University of Pará, Guamá, Belém 66075-110, PA, Brazil
* Correspondence: lucaspinheiro523@hotmail.com (L.R.S.P.); avsantos@ufpa.br (A.V.S.)

Abstract: The use of dyes dates to ancient times and has increased due to population and industrial growth, leading to the rise of synthetic dyes. These pollutants are of great environmental impact and azo dyes deserve special attention due their widespread use and challenging degradation. Among the biological solutions developed to mitigate this issue, bacteria are highlighted for being versatile organisms, which can be applied as single organism cultures, microbial consortia, in bioreactors, acting in the detoxification of azo dyes breakage by-products and have the potential to combine biodegradation with the production of products of economic interest. These characteristics go hand in hand with the ability of various strains to act under various chemical and physical parameters, such as a wide range of pH, salinity, and temperature, with good performance under industry, and environmental, relevant conditions. This review encompasses studies with promising results related to the use of bacteria in the bioremediation of environments contaminated with azo dyes in the most diverse techniques and parameters, both in environmental and laboratory samples, also addressing their mechanisms and the legislation involving these dyes around the world, showcasing the importance of bacterial bioremediation, specialty in a scenario in an ever-increasing pursuit for sustainable production.

Keywords: sustainability; effluent treatment; dyes; bioremediation; bacteria; wastewater; textile; consortium; BES; bioreactor

1. Introduction

The use of dyes for the aesthetic improvement of objects is an ancient practice, with historical records indicating that dyes of natural origin were already in use 3500 years BC. In the beginning, the coloring agents available (dyes and pigments) were only of natural origin, obtained from mineral sources, vegetables—such as those found in Mediterranean (*Rubia tinctorum*) and Brazilwood (*Paubrasilia echinata*)—which are mostly represented by chemical groups of naphthoquinones, anthraquinones and flavonoids, and those obtained from animals, such as those extracted from some insect species, like the cochineal (*Dactylopiidae coccus*). The coloring obtained with these dyes was applied to utensils, weapons, and dwellings, among others, having aesthetic and cultural importance [1,2].

From ancient times to the present moment of our history, dyeing technology has evolved with the discovery of new matrices and raw materials and the synthesis of new pigments and dyes. In 1856 a major discovery was accidentally made by William Henry Perkins, when he synthesized what came to be the first synthetic dye in history, mauvein [3]. Synthetic dyes have largely replaced natural dyes over the years due to their wide range of colors, cost-effectiveness, and resistance to fading by sunlight, water, perspiration, and different chemicals [4].

It is estimated that around 10,000 different dyes are currently being produced on an industrial scale, with an annual worldwide production volume of around 700,000 tons and about 10 to 15% of those are discarded into nature. This scenario generates serious consequences for the contaminated environment, such as interference with the entry of sunlight into the water, influencing photosynthetic organisms, causing damage to the oxygen level of the water, metabolic stress, neurosensorial damage, flora necrosis, death, and decreased growth of fauna, among others. Moreover, humans are also potential victims of these compounds, when discharged into nature without treatment, and can be quite toxic, either by oral or respiratory ingestion as well as mere skin contact [2,5,6]. The toxic effects of azo dyes, in particular their ability to promote mutations, are related both to the dyes themselves and to metabolites released upon their breakage or degradation, such as aromatic amines. The possibility of the dye breaking down and releasing these carcinogenic amines on contact with saliva or gastric juice is one of the factors evaluated in classifying the dyes as potentially hazardous to health. However, when ingested, the dye can also be reduced by the action of intestinal bacteria and, possibly, by the enzyme azoreductase present in the liver or intestinal wall, showing how complex the remediation of these toxins can be [7].

Therefore, it is necessary to understand the risks associated with the discarding of these dyes in the environment without prior treatment and how the use of microorganisms in the bioremediation of these contaminants is a viable alternative. The objective of this review is to evaluate the degradation of azo dyes specifically by bacteria, as well as the factors that influence these biological processes and the microbial mechanisms involved. In order to raise awareness about the importance of preventive measures in the discharge of untreated dyes, and some cases in which contamination by these pollutants was found in effluents are also presented, illustrating the importance, and urgency, of bacterial bioremediation for this sector.

2. Azo Dyes

The substances used to add color to many kinds of products are called colorants and include both the class of dyes and pigments. Pigments are mainly inorganic salts and oxides, insoluble in the substrate and commonly dispersed in crystal particles or powder form for application. Dyes, on the other hand, usually refer to organic substances that are soluble in the substrate and dispersed at the molecular level. Dyes promote more vivid colors than conventional pigments, and the characteristics of dyes depend primarily on their chemical structure, while for pigments the physical characteristics of their particles also influences the final color. Some examples of dyes are the azo, coumarin, and perylene groups [8].

Azo dyes are synthetic and organic chemical compounds applied in medicines, food, cosmetics, fabrics, among other products. These chemical classes are widely used in the dyeing industry and are present in 50–65% of commercial formulations, this is because of their stability and chemical versatility, having high fixation and resistance to light and moisture. These characteristics impact directly on their degradation in the environment, which has been a challenge [9,10].

The compounds of the azo group are chemically represented as (R-N = N-R'), with (-N = N-) being the chromophore group referred to as azo. According to the classification carried out by the International Union of Pure and Applied Chemistry (IUPAC), these compounds are derived from diazine HN = NH with substitution of hydrogens by hydrocarbyl, azobenzene, or diphenyldiazene groups, and may contain one to three azo bonds linking phenyl and/or naphthyl rings, which may be substituted by groups such as amino, chloro, nitro, and hydroxyl. They are characterized by their strong coloration and comprise approximately two-thirds of the synthetic dyes produced today, besides being the class of commercial organic dyes with the greatest structural diversity and the widest range of use [11,12].

However, the major problem related to the untreated disposal of these dyes in nature is that they and their byproducts, produced by the breakage of their azo bonds (aromatic amines), have been classified as highly carcinogenic compounds, representing a great risk to humans [13]. Thus, the search for ways to treat industrial effluents, especially those containing dyes of this class, has received much attention worldwide. Among the proposed solutions are chemical, physical, and biological processes. In this scenario bioremediation has received much attention for its lower costs compared to other mechanisms and high effectiveness in decoloring dyes in the affected environment, as well as generating less ecological impact. Among these biotechnological solutions we have the use of microorganisms, including bacteria [14].

Many bacteria have developed systems for decolorizing azo dyes contaminated medium, usually based on enzymatic mechanisms. However, the efficiency of these mechanisms depends on physicochemical parameters that may limit their activity, such as level of agitation, temperature, pH, dye concentration, structure of the dye, oxygen level, and carbon and nitrogen sources [4]. Different combinations of these parameters lead to different results in dye decolorization and degradation by influencing enzyme activity and microbial growth and maintenance in the medium. Understanding these relationships is crucial for improving the bioremediation process for a certain technique or bacterial strain, considering that these parameters in the environment can vary significantly in a few hours and that some dyes, as well as breakdown byproducts, can be toxic for the microbial population. In a laboratory environment it is important to consider that microbial metabolism can alter its degradation activity by nutrient depletion and variation in oxygen levels, for example, which do not necessarily represent real conditions. All of the factors mentioned above can prevent the complete breakdown of dyes and, consequently, the complete decolorization of the medium, since the mineralization efficiency of contaminating compounds relies on the capacity of the microorganism of performing a suitable metabolic response to degrade the contaminant under given environmental conditions, it is also necessary that these organisms are in sufficient concentration to proper handle with the entire amount of dye on the environment [15].

The bacterial degradation of azo dyes typically involves a two-step process, the first being a reductive cleavage of the azo bonds, leading to the formation of aromatic amines, which are potentially toxic and the second step is based on the degradation of these aromatic amines. The bacterial degradation process can occur in the presence or absence of oxygen; however, the biodegradation of these amines happens almost exclusively by aerobic processes. Considering this, the ideal method for the treatment of industrial waste contaminated by azo compounds is a combination, in the same process, of anaerobic and aerobic steps for the safest and most efficient removal of environmental and human risk factors related to these compounds [16].

Many the azo dyes and their breakdown molecules present toxicity, showing mutagenic and carcinogenic effects, affecting animals, plants, and humans alike, with harmful effects varying with the structure, reactivity, and substitution groups of the dyes. The toxic aromatic amines from dye breakdown are resistant to classical effluent treatment, persisting longer than the dyes and an increase in concentration of these substances in the medium may impair the dye degradation process because of their toxicity to bacterial life itself. Still, there are some microorganisms capable of producing enzymes that degrade these amines in an aerobic environment. For humans, inhalation and oral ingestion of the dyes and their by-products are the primary means of exposure that can lead to acute toxicity [17,18].

The toxicity assessment of the dyes and their by-products can be done using different techniques, which vary in sensitivity, resolution, and cost. Analyses using techniques such as Gas Chromatography–Mass Spectrometry (GC-MS) and Fourier-Transform Infrared Spectroscopy (FTIR) can be performed to confirm the degradation of the dye and/or the presence of aromatic amines, and then the medium containing the by-products of dye degradation is used in tests that evaluate possible negative biological effects being biotoxicity and phytotoxicity assays most often used for this purpose [19]. Another way to evaluate toxic effects is to expose the living organism to a solution containing different concentrations of dye, also evaluating chronic exposure to these contaminants [20], other toxicity tests that can be performed for these pollutants, including tests performed on human cells [21], with several studies proving the toxic, mutagenic, and carcinogenic effects of dyes and their by-products using plant and animal tests [19–24].

Legislation on the Use and Disposal of Dyes

Over the years the concern regarding the impact of human activities on the environment, especially industrial ones, has grown considerably. In view of this, standards on what is acceptable regarding not only the discarding of potentially toxic substances and waste in nature, but the industrial application of these dyes in some products have been regulated and even prohibited, and control agencies were created for monitoring and enforcing laws of this nature [25].

In the European Union there are some regulations on the use of azo dyes in consumer products, from clothing and toys, to cosmetics, and food, which obviously has stricter guidelines and norms. REACH (Registration, Evaluation, Authorization, and Restriction of Chemicals) lists 24 types of aromatic amines considered hazardous to humans and prohibits the use of azo dyes that produce 30 mg/kg or more of these amines in products that may have direct and prolonged contact with human skin. REACH also lists other colorants that are restricted for use in these same products when in concentrations above 0.1% of its weight. The EN-71 (European Norm 71) deals with the presence of this class of colorants in toys. The regulations (EC) 1223/2009 and (EU) 10/2011 deal, respectively, with the bane in the use of o-Dianisdine and Benzidine in cosmetic products and plastic products that come into contact with foodstuffs, which must keep the release of primary aromatic amines into the food within the limit of 0.01 mg/kg. There are a number of countries which in addition have their own legislation covering limitations related to azo dyes and the release of aromatic amines from them, such as Germany, the Czech Republic, and Switzerland [26].

In the United States there are no laws related specifically to azo dyes or restricting the use of them, however, there are restrictions on aromatic amines from these dyes. Due to the political system in the country, many states have their own laws restricting the presence of certain chemicals in certain products. Some examples that cover aromatic amines derived from certain azo dyes are California's Proposition 65, Washington's Children's Safe Product Act, and Vermont's Act 188 [27].

In Asia, the first example of a country to regulate chemicals, including azo dyes, was India, with a ban in 1997 on the handling of 112 dyes, including representatives of the azo group. In this country the list of banned aromatic amines is the same as in REACH Regulation 1907/2006. Other Asian countries have instituted their own regulations in which they include restrictions on azo dyes and/or aromatic amines, such as China (2005), South Korea (2010), Taiwan (2011), and Egypt (2012). In 2014 Japan joined this list when placed azo dyes as a hazardous substance and restricted the presence of 24 aromatic amines originated from these dyes in all textiles, leather, or fur products in a concentration of 30 mg/kg or more. This same concentration was used for the limitation of 22 aromatic amines in a regulation of Vietnam [28–31].

Other countries like Canada, France, Australia, Brazil, Pakistan, Malaysia, Turkey, and Morocco also have regulations related to industrial effluents that include in their guidelines specifications related to the color of these effluents which, consequently, also influences the concentration of dyes allowed in them [25].

Another relevant insight is that many-if not all-regulations of the cited countries also determine the methods of analysis to which the products will be subjected in order to determine whether they are in order with the regulations of that country. This is important, because different analytical methods have different limitations, ranges, sensibilities, and applications, and these must be considered when carrying out these controls. In addition, other parameters such as chemical and biological oxygen demand are also subject to regulation, and since the presence of azo dyes has an impact on these factors, the use of these dyes may also be indirectly restricted.

3. Bacteria in the Bioremediation of Azo Dyes

Biotechnology has been widely employed in the search for solutions to the degradation and elimination of dyes, mainly because biological solutions are effective and generate less negative impact on the environment [14]. When dealing with biological processes using bacteria, especially potentially pathogenic genera and species, the concern with a possible biological impact of them, when introduced into the environment for the bioremediation process, may arise. To attend to any unwanted negative effects, some strategies can be used, some of those include: (1) the use of isolated and purified enzymes or other bacterial products that act on the discoloration without needing the bacterial cell itself [32], (2) microbial bacteria/consortia isolated from the contaminated environment itself or similar environments, in order to increase the chance of integration of the bacteria with the environment and the existing microbiota [33], (3) application of genetic engineering techniques that can develop bacterial strains with programmed death, stopping bacterial metabolism in the absence of the target contaminant [34].

Some biological bioremediation systems also have the potential of generating more than one product, in addition of decolorization, following the example of bioelectrochemical systems (BES), which helps in mitigating the costs associated with biological processes [35]. Among these solutions we have bioremediation by heterotrophic bacteria, which have, more broadly, two mechanisms related to the degradation of dyes: biosorption and enzymatic action [36].

3.1. Bacterial Mechanisms of Azo Dye Degradation

3.1.1. Biosorption

The biosorption is related to both the adsorption and absorption processes, and bacteria capable of performing the removal of dyes by adsorption have already been described in the literature. Biosorption is directly correlated with the composition of lipids and heteropolysaccharides of the cell wall, in which different charged groups generate attractions between it and the azo dyes, therefore, dead and living cells, in this latter case called bioaccumulation, have the ability to perform biosorption. Taking into account the range of charged groups existing in the cell walls of microorganisms and the variety of structures of the dyes, a microorganism X that adsorbs/absorbs dye A may not adsorb/absorb dye B, which is processed by a microorganism Y. Pretreatment can promote changes in the biosorption capacity of cells, optimizing the process and achieving a better fit to a certain need, among the substances capable of performing these changes are acids, formaldehyde, bases, among others [36,37].

To be used as biosorbents, dead cells are more advantageous than living cells because they do not require nutrients, can be stored for a longer time, and can be regenerated by the application of organic solvents or surfactants. However, biosorption is not the most suitable mechanism for dye treatment, since the treatment of large volumes of contaminated material would lead to the generation of large amounts of biomass containing dyes and possibly other toxic products that should have a proper disposal, i.e., it does not completely solve the problem, since it often does not destroy the dye, only seizes it in a matrix: the biomass [37,38].

3.1.2. Enzymatic Degradation

The initial step for the decolorization of solutions with azo dyes, being it waste, industrial effluents, or environmental samples, is the reduction of the azo bond (-N = N-) in the chromophore group, this step can occur intra- or extracellularly and involves the transfer of four electrons in two steps, where in each step two electrons are transferred from the dye to its final electron acceptor, leading to its decolorization. Some groups of enzymes already identified as capable of performing this reduction are azoreductases and laccases. These two are the most addressed groups in the literature regarding these decolorization reactions [39,40]. The Figure 1 presents the general action mechanisms of these two enzymatic groups plus the peroxidase group which also acts on the azo chromophore group [41–43].

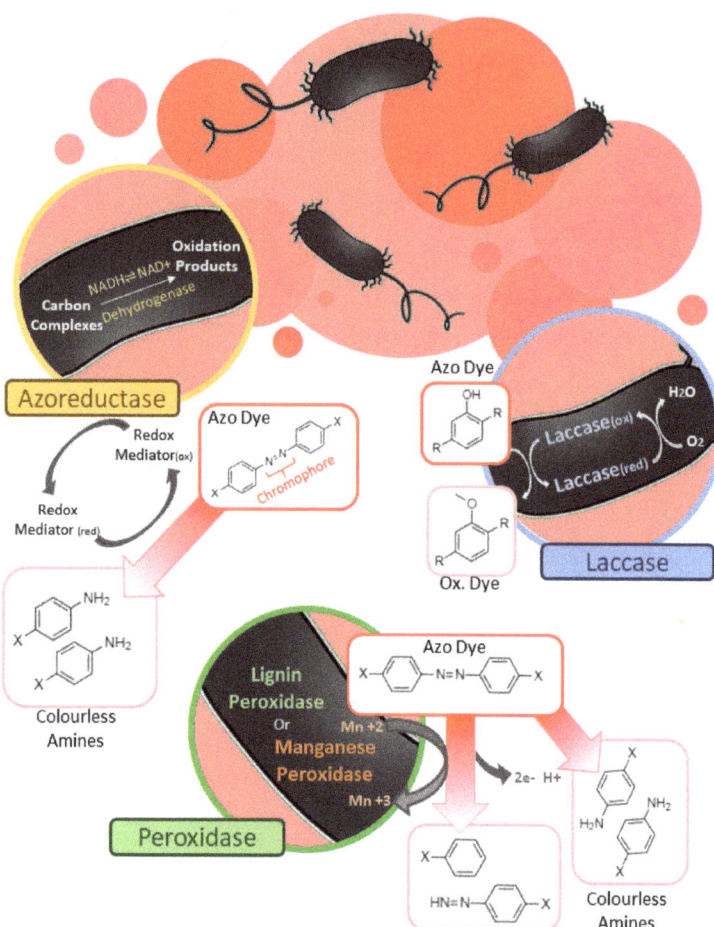

Figure 1. Schematic representation of three general bacterial enzymatic degradation mechanisms of azo chromophore group. Firstly, showing the enzymatic degradation by the action of azoreductases—yellow—in this example using NADH as an essential reducing agent for the cleavage of azo bonds, generating aromatic amines and thus discoloring the medium. Then—clockwise—we have the catalytic reaction cycle mediated by laccase—blue—with generation of oxidized substrate instead of potentially toxic amines, in addition to not requiring cofactors. Finally, peroxidase enzymes—green—such as lignin peroxidase and manganese peroxidase, two enzymes most commonly used for dye degradation, illustrating some possible products according to the cleavage of their bonds, which can be symmetric or asymmetric.

3.1.3. Enzymatic Degradation by Azoreductases

The azoreductases (e.g., EC 1.7.1.6 and EC 1.7.1.17) are the largest group of enzymes active in the biodegradation of azo dyes. They have the specific activity of reductive cleavage of azo bonds, resulting in aromatic amines, but to promote this reaction, azoreductases require reducing agents, such as $FADH_2$, NADPH, and NADH. They are more related to the anaerobic degradation of dyes because the presence of oxygen impairs this azo bond reduction step by competing for the reducing agents needed as electron acceptors for the azo bonds, which are also used by aerobic respiration. These enzymes are classified based on function into flavin-dependent and flavin-independent. The former class is subdivided into those enzymes that use as coenzymes: (1) NADH only; (2) NADPH only; and (3) both NADH and NADPH [40,44].

This group is quite varied and, depending on the source in which it is found, i.e., which organism, and even species, it is obtained from, it is possible to observe differences, such as in its catalytic activity, cofactor requirement and biophysical characteristics. Because of it, there is specificity between substrate and the types of azoreductases described so far, which varies in the requirement for cofactors and reducing agents and in the ability to resist oxygen [45].

Most azo dyes have a high molecular mass and are unlikely to cross the plasma membrane of cells. Therefore, microorganisms have a reduction mechanism related to the electron transport of azo dyes in the extracellular medium, so that there is a need for connection between the intracellular electron transport system and the dye molecules for degradation to occur. However, the action of azoreductases in the intracellular medium has also been identified and enzymes of this group have been found in bacteria, including in halophilic and halotolerant microorganisms [37,40,44].

3.1.4. Enzymatic Degradation by Laccases

Laccases (EC 1.10.3.2) are oxidases that have multiple structurally attached copper ions and are of great industrial interest due to their ability to utilize different substrates. They are able to non-specifically catalyze the degradation of azo dyes by acting on the phenolic group of the dye using a free radical mechanism that forms phenolic compounds generating fewer toxic aromatic amines. Moreover, these enzymes do not need other cofactors for their activation [36,44,46]. Although laccases do not need other cofactors to carry out their activity, they benefit from their presence in the medium. The presence of redox mediators can extend the range of dyes that this enzyme can degrade and significantly improve the degradation of dyes already covered in its range of action [35]. Bacterial laccases have great potential as biocatalysts due to their properties of high thermal stability, activity over a wide pH range, and resistance to denaturation by detergents, being already used to remove textile dyes and treat industrial effluents [46].

3.1.5. Enzymatic Degradation by Peroxidase

Peroxidases (EC 1.11.1) are hemoproteins that catalyze reactions in the presence of hydrogen peroxide and are mostly present in fungi, but also occur in some bacteria [47]. They possess the ability to degrade a wide range of dyes, as cited by Paszczynski and co-workers [48], where lignin peroxidase (EC 1.11.1.14) and manganese peroxidase (1.11.1.13) were indicated as directly involved in the degradation of dyes and xenobiotic compounds [44]. Another class, versatile peroxidase (1.11.1.16), was pointed out by Đurđić and co-workers [49] as having the ability to perform structure breakdown of azo dyes.

3.2. Bacterial Degradation of Commercial Colorants

The occurrence of bacteria in different environments and physicochemical conditions makes them an interesting focus of prospection (Table 1). In the case of dyes degradation, a wide range of variables has already been explored and it was identified that this group of microorganisms can degrade azo dyes under aerobic, microaerophilic, and anaerobic conditions, as isolated cultures or as microbial consortia, in the presence of various sources of carbon and nitrogen and in wide ranges of pH, temperature, salinity and other physical-chemical parameters. In addition, bioreactors have been used in several works in an attempt to increase the efficiency of the degradation process, especially by immobilization of microorganism or redox mediators [50].

Table 1. Dye degradation by bacterial strains—pure cultures—under various medium conditions.

Species	Dye	Optimum Values of Phisicochemical Parameters for Bacterial Decolorization	Degradation Mechanism	Local of Bacterial Isolation	Maximum Degradation	Reference
Shewanella marisflavi	Xylidine Ponceau 2R	20–30% of salinity	Flocculation and Enzymatic	China	≈100% (30% of salinity, anaerobic conditions and 22h incubation)	[51]
Pseudomonas extremorientalis	Congo Red	50 mg/L of dye concentration, 2.5–5% of salinity and 0.6 U/mL enzyme concentration	Enzymatic-Laccase	Tunísia	79.8 ± 2.1% (50 mg/L of dye concentration, 2.5–5% of salinity, 24h incubation and 0.6 U/mL enzyme concentration)	[52]
Aliiglaciecola lipolytica	Congo Red	35 °C, <100 mg/L of dye concentration, 0–1% of salinity, pH 6–7, >4 g/L of glucose.	Adsorption and Enzymatic-Laccase and Azoreductase	-	>90% (35 °C, 25 mg/L of dye concentration, 1% of salinity, pH 6 and 4 g/L of glucose)	[53]
Enterococcus faecalis, Shewanella indica, Oceanimonas smirnovii and Clostridium bufermentans	8 different dyes	Varied depending of bacteria strain and dye	Enzymatic-Azoreductase and phenol oxidases	China	96.5% (E. faecalis strain and C. bufermentans with Dye Acid Orange 7 when pH ranged from 5 to 8, respectively)	[54]
Bacillus sp.	7 different dyes	50–100 mg/L of dye concentration, pH 10, 30 °C, with glucose and yeast extract supplementation.	Enzymatic	Ethiopia	100% (pH 10, 30 °C, anoxic and anaerobic conditions)	[55]
Aeromonas hydrophila	Reactive Red 198 e Reactive Black 5	pH 5.5–10.0, temperature were and 20–35 °C under anoxic culture	Adsorption and Enzymatic	Taiwan	>90% (pH 5.5–10.0, temperature were and 20–35 °C under anoxic culture)	[56]
Comamonas sp.	Direct Red 5B	pH 6.5, 40 °C, static incubation conditions and 300–1100 mg/L of dye concentration.	Enzymatic-Laccase and Lignin Peroxidase	India	100% (pH 6.5, 40 °C and static incubation conditions)	[57]
Halomnas sp.	Remazol Black B	Varied depending of bacteria strain.	-	Iran	≈100% (40 °C)	[58]
Aeromonas sp.	Reactive Black	Microaerophilic conditions	-	India	≈100% (Microaerophilic conditions)	[59]
Oerskovia paurometabola	Acid Red 14	Anaerobic conditions	Enzymatic	Portugal	91% (anaerobic conditions)	[60]
Aeromonas hydrophila, Lysinibacillus sphaericus	Reactive Red 195	-	Enzymatic-Laccase and Azoreductase	India	91.96% (pH 8, 37 °C, 100 mg/L of dye concentration and sequential aerobic-microaerophilic conditions)	[61]
Bacillus sp.	4 different dyes	-	Enzymatic-Azoreductase	-	-	[62]
Bacillus sp.	5 different dyes	-	Enzymatic-Azoreductase	-	-	[63]
Aeromonas hydrophila, Lysinibacillus sphaericus	5 different dyes	-	Enzymatic-Azoreductase and Laccase	India	90.4% (pH 8, 37 °C, 100 mg/L of dye concentration and sequential aerobic-microaerophilic conditions)	[64]
Lysinibacillus fusiformis	Methyl Red	pH 7.5–8, 30 °C, 100 mg/L of dye concentration and 10–20% (v/v) of inoculum size	Enzymatic-Laccase, Azoreductase and Lignin Peroxidase	-	96% (aerobic condition, pH 7.5, 30 ± 2 °C, dye concentration of 100 mg/L and 10% (v/v) inoculum size)	[65]
Pseudomonas stutzeri	Acid Blue 113	-	Enzymatic-Azoreductase and Laccase	India	86.2% (static conditions, 37 °C and 300 ppm of dye)	[66]

Table 1. Cont.

Species	Dye	Optimum Values of Phisicochemical Parameters for Bacterial Decolorization	Degradation Mechanism	Local of Bacterial Isolation	Maximum Degradation	Reference
Aeromonas sp.	Methyl Orange	pH 6, 5–45 °C, 100–200 mg/L of dye concentration	Enzymatic-laccase, NADH-DCIP reductase, and azo reductase	China	≈100% (100–200 mg/L of dye concentration; with carbon and nitrogen supplementation; pH 6; 5–45 °C)	[67]
Proteus mirabilis	Reactive Blue 13	pH 7, 35 °C and anoxic conditions.	Enzymatic-Laccase, azoreductase and veratryl alcohol oxidase	Nigeria	≈90% (pH 7)	[68]
Pseudomonas putida, *Bacillus subtilis*	18 different dyes	-	Enzymatic-Azoreductase and Laccase	-	≈100%	[69]
Bacillus sp.	Red HE7B	-	Enzymatic-Azoreductase and Laccase	-	89% (30 °C, 50 mg/L of dye concentration, 24h incubation and under agitation conditions)	[70]

3.2.1. Pure Bacterial Colonies
Degradation in the Presence of Salts

As mentioned, bacteria are able to degrade dyes in different salt concentrations, a desirable attribute since inorganic salts are used in dyeing processes as mordants, helping in color fixation process, especially sodium chloride (NaCl) and sodium sulfate (Na_2SO_4). The bacterial degradation process has proven effective for dyeing at salt concentrations ranging from 0.25% to 10%. Many countries have no regulations on the amount of salt that can be discharged into the environment, or when they do, many allow the presence of these salts in high concentrations [25]. Taking this into account, the ability of bacteria to degrade dyes in saline environments becomes an important differential for the selection of microorganisms with real biotechnological potential of application in dyes bioremediation processes.

The work of Xu and co-workers [51] used a strain of *Shewanella marisflavi*, an exoelectrogenic bacterium-bacteria that has extracellular electron transfer pathways-isolated from marine sediments in China. Tests performed with the dye Xylidine Ponceau 2R showed that this strain is able to decolorize the medium at concentrations ranging from 0% to 20% NaCl. Higher salt concentrations also influence the solubility of the dye by increasing the ionic strength of the solution, leading to floc formation, and flocculation was also observed at lower NaCl concentrations when in the presence of *Shewanella marisflavi*. The researchers concluded that the tested bacteria had two decolorization mechanisms, performing only dye degradation up to 6% NaCl, then degrading and flocculating from 8% to 10% NaCl in the medium, and above this percentage, only underwent flocculation process.

Laccases are widely studied in fungi and plants whose ability to degrade phenolic compounds is well known. These enzymes have also been found in bacteria, showing ability to act in environments with high salt concentration and alkaline pH, presenting an advantage over fungi and plant laccases [52]. Neifar and co-workers [52] worked with a laccase-producing *Pseudomonas extremorientalis* strain isolated from oil-contaminated sediments in Tunisia, evaluating the degradation of Congo red dye. The enzyme produced by this strain showed great resistance to alkalinity, maintaining its activity in pH 7 to 10 and also resisted to salinity, maintaining almost 90% of its activity in the presence of sodium chloride at 17.5%.

Wang and co-workers [53] demonstrated the ability of a strain of the marine bacterium *Aliiglaciecola lipolytica* to decolorize medium containing Congo red dye at salt concentrations as high as 4%. However, the strain did not tolerate pH increase and has glucose as the best carbon source for its dye degradation activity, increasing the costs of the process. The study also pointed out that the bacteria use a degradation mechanism that involves the adsorption of the dye in their cells by means of extracellular polymeric substances (EPS), where part of it is degraded by a process involving the co-metabolism with glucose and the mediation of the enzymes azoreductase and intracellular laccase, in this process the non-degraded dye is encapsulated in the bacterial cell.

In a study by Zhuang and co-workers [54] the degradation of azo dyes used four bacteria isolated from coastal region in China, *Enterococcus faecalis*, *Shewanella indica*, *Oceanimonas smirnovii*, and *Clostridium bufermentans*. The isolated strains were able to degrade eight different dyes, achieving removal percentages above 70% for most of them. *E. faecalis* and *C. bufermentans* maintained dye decolorization rates above 80% at salt concentrations up to 7%. The four also demonstrated to have great resistance of their decolorization activity in the presence of ions, being little affected by the great majority of those tested; however, Cadmium (Cd^{2+}) and Copper (Cu^{2+}) were the ions that most interfered in the decolorization results, which is a problem to be considered, since they are metals frequently found in industrial effluents.

PH Range: Degradation in Alkaline Medium

Due to the use of sodium hydroxide and other basic components in textile dyeing processes and with the textile industry being one of the three sectors with the highest release of dye contaminated effluents into nature, it is important for this scenario that we use bacteria resistant to alkaline environments for the treatment of these effluents which can have pH up to 11.5. In this sense, the work of Guadie and co-workers [55] tested the decolorization capacity of a *Bacillus* sp. strain isolated from lakes with an alkaline characteristic. This strain proved to be able to decolorize medium with pH from 9 to 11 with efficiency above 90% and was also effective against the seven different dyes tested with complete decolorization been achieved in anoxic and anaerobic environments, and in the presence of oxygen, there was almost no decolorization of the medium.

Several studies have isolated and identified bacterial strains capable of decolorizing dyes in alkaline solutions. Chen et al. [56] worked with a strain of *Aeromonas hydrophila* capable of degrading the dyes Reactive Red 198 and Reactive Black 5 over a pH range of 5.5 to 10. The research of Jadhav and co-workers [57] isolated a strain of *Comamonas* sp. from contaminated soil of industrial environment able to degrade the dye Direct Red 5B in a pH range of 6–12, but having its best activity in neutral pH range, different from that was found by of Asad group [58] who worked with three different strains of bacteria belonging to genus *Halomonas* sp. isolated from textile effluents and observed that all of them performed the best decolorization in alkaline pH, with the highest activity achieved at the highest pH tested: 11. These three studies also showed, again, little or no decolorization activity in an aerobic environment.

Bioremediation in Aerobic and Anaerobic Environments

One of the major problems faced in the degradation of azo dyes in anaerobic environments is the formation of toxic products, notably mutagenic and/or carcinogenic aromatic amines (e.g., benzidine and 4-biphenylamine) that are only degraded in aerobic environments [7,59]. The work of Shah [59] evaluated the degradation of the dye Reactive Black using a strain of *Aeromonas* sp. in a microaerophilic environment until the color was no longer noticeable in the medium and then promoted the aeration of it to stimulate the oxidation of the aromatic amines-formed by the breakdown of the dye into non-toxic products, obtaining a discolored medium with a low degree of toxicity. This work proves that a dye bioremediation process performed outside anaerobic conditions can also be successful.

Franca and co-workers [60] used a different approach, where they performed the decolorization tests in an anaerobic environment and, after complete removal of the color, they used an aerobic environment to evaluate the ability of a *Oerskovia paurometabola* strain to metabolize the toxic products generated by the previous step. The result obtained was positive, with decolorization above 90% and removal of toxic products above 63%.

The research by Srinivasan and Sadasivam [61] also worked with the aerobic-microaerophilic approach for the degradation of azo dye, in this case Reactive Red 195 using strains of *Aeromonas hydrophila* and *Lysinibacillus sphaericus*. This work also used the molecular docking tool, in addition to decolorization tests, in order to understand the interaction between the amino acid residues of the enzymes—laccase and azoreductase—and the dye. The study pointed out the high efficiency of these strains in degrading the dye used and a positive correlation between the score of the docking studies and the percentage results of dye degradation. In silico approaches for azo dye degradation studies have been increasingly employed, also as a strategy to better use resources and research efforts [62–64].

Although the mechanism of azo dye decolorization commonly occurs under anaerobic conditions, the work of Sari and Simarani [65] indicates that there are bacteria capable of performing azo dye degradation in an aerobic environment. The study identified a strain of *Lysinibacillus fusiformis* capable of achieving a 96% decolorization rate of Methyl Red dye in 2 hours under neutral pH and temperature of 30 ± 2 °C under aerobic conditions. The study further pointed out that the bacteria's oxidoreduction mechanism involved the enzymes laccase, azoreductase and lignin peroxidase. As seen, azoreductase and laccase are highlighted in several studies related to the bioremediation of azo dyes by bacteria, such as the action of *Pseudomonas stutzeri* in the degradation of the dye Acid Blue 113 [66]. This strain obtained good decolorization results and a high resistance to elevated concentrations of the dye. Genomic studies indicated the presence of both enzymes in the decolorization system of the microorganism. Several other works pointed out the presence of these two enzymes in the bioremediation activity of azo dyes [58,67–70].

3.2.2. Microbial Consortia

The application of microbial consortia instead of isolated organisms for the treatment of effluents contaminated with dyes presents advantages (Table 2), especially in the fact that each strain can act on different targets of the organic molecule and/or consume different products generated by the breakdown of the dyes, thus generating a synergistic effects action where the enzymatic activity of a bacterial strain is positively influenced by the presence of other microorganisms, thus, increasing the degradation rate [71].

An example of this is recorded in the work of Masarbo and co-workers [72] in which the researchers evaluated the decolorization activity of pure bacterial strains and in consortia. The result shows that two of the consortia tested were able to achieve higher percentages of decolorization and in shorter time, when compared to the results of the strains tested individually. Furthermore, at the highest dye concentration tested, the percentage of degradation was at least 10% higher than the best result achieved by a single bacterial species.

Physicochemical Parameters in Microbial Consortia: Salinity, PH and Oxygenation

Other studies have identified consortia capable of decolorizing environments with a combination of parameters, such as high salinity and pH, closer to those found in real remediation scenarios. Guo and co-workers [73] assembled a consortium consisting mostly of genus *Halomonas*, *Marinobacter*, and *Clostridiisalibacter*, which was able to decolorize over 90% of the dye at 40 °C, withstand 10% salinity, and to achieve a decolorization rate above 70% at pH 8–12. However, the different strain's ability to endure increasing concentrations of the dye was low. Guo's group [74] also studied a consortium consisting mostly of the genus *Bacillus* and *Piscibacillus*. This consortium showed best activity at 50 °C, pH 10, and between 1% and 10% salinity. However, it poorly withstood higher pH values as well as the increase in dye concentration. Another factor worth mentioning is that, in both studies, the supplementation of the medium with yeast extract was essential for higher decolorization results.

Table 2. Dye degradation by bacterial consortium under various medium conditions.

Main Consorcium Species	Dye	Best Phisicochemical Parameters for Bacterial Decolorization	Degradation Mechanism	Local of Bacterial Isolation	Maximum Degradation	Reference
Pseudoarthrobacter sp. and *Gordonia* sp., (consortium PsGo); *Stenotrophomonas* sp., and *Sphingomona* sp. (consortium StSp)	Reactive Black 5	pH 11 (for PsGo) and Glucose as carbon source (for StSp)	-	Iran	85% (for PsGo) 75% (for StSp)	[71]
Bacillus sp., *Lysinibacillus* sp. and *Kerstersia* sp. at different combinations	Ponceau 4R	200 mg/L of dye	Enzymatic-Azoreductase	India	100% (with 200 mg/L of dye)	[72]
Halomonas sp., *Marinobacter* sp. and *Clostridiisalibacter* sp.	Methanil Yellow G	pH 8, salinity 5–10%, 40 °C, 100 mg/l of dye concentration and presence of yeast extract	Enzymatic	China	93.3% (with yeast extract and 100 mg/L of dye)	[73]
Piscibacillus sp. and *Bacillus* sp.	Methanil Yellow G	pH 10, 50 °C, yeast extract, 1% salinity and 100mg/L of dye concentration	Enzymatic-Laccase, Lignin peroxidase, Nicotinamide adenine dinucleotide-dichlorophenol indophenol reductase and Azoreductase	China	94.26% (with yeast extract)	[74]
Zobellella sp., *Rheinheimera* sp. and *Marinobacterium* sp.	Methanil Yellow	Microaerophilic conditions, pH 6, 400 mg/L of dye concentration and 3% salinity	Enzymatic-Azoreductase, Laccase and Lignin Peroxidase	-	≈100% (with 400 mg/L of dye concentration)	[75]
Staphylococcus sp. and *Bacillus* sp.	Remazole Brilliant Violet 5R	-	Enzymatic-Azoreductase, Laccase and Lignin Peroxidase	India	100% (with pH 6.5 and 37 °C)	[76]
Pseudomonas sp., *Bacillus* sp., *Bacillus* sp. and *Ochrobactrum* sp.	15 different dyes	pH 8, 40 °C, 100 mg/L of dye concentration and no salt presence	Enzymatic	India	100% (with pH 8)	[77]
Bacillus cereus, Pseudomonas fluorescenc, Staphylococcus aureus, Escherichia coli and *Lactobacillus* sp.	Remazol Yellow	-	-	India	97.84%	[78]
Enterococcus faecalis and *Klebsiella variicola*	Reactive Red 198	10–25 mg/L of dye concentration, pH 8 and 37 °C	Enzymatic	Iran	99.26% (with pH 8)	[79]
-	5 different dyes	pH 7–8, 35 ± 2 °C, static conditions, 0.5–1% salinity and 200 mg/L of dye concentration	-	India	98.8%	[80]

As we have seen before, another factor that affects the decolorization process is the presence of oxygen. In the work of Guo and co-workers [75], a consortium consisting mainly of *Zobellella*, *Rheinheimera*, and *Marinobacterium* was used for the degradation of the dye Methanil Yellow in a saline and microaerophilic environment. In a short time, assay, five hours, the consortium achieved above 80% decolorization at up to 5% salt concentration in the medium and optimum pH of 6. However, it showed great resistance to increasing dye concentration, decolorizing almost 100% at 400 mg/L concentration within those five hours. Laccase, azoreductase and lignin peroxidase were, once more, the enzymes identified in the degradation mechanism of this consortium.

The group of Shah [76] studied the behavior of a bacterial consortium consisting of five bacterial strains in microaerophilic and aerobic environments based on the reduction of the azo dye Remazole Brilliant Violet 5R. A difference was observed in the efficiency of the consortium regarding the degradation of the dye between these two environments, where, in a microaerophilic environment, the degradation reached 100% in less than 24 hours, this result was not seen in an aerobic environment. Another factor pointed out in the study as influencing the efficiency of degradation was the structure of the dye, where 29 different dyes were tested, and the efficiency of the consortium varied from less than 20% of degradation to more than 80% in some cases.

Khan and co-workers [77] also compared the efficiency of a bacterial consortium in microaerophilic and aerobic environments and with respect to several structurally different dyes. In the microaerophilic environment the decolorization was 100% and in the aerobic environment it was less than 10%. The consortium was found to withstand the presence of salt and temperature up to 45 °C, as well as a slightly alkaline pH. For the different dyes tested, the decolorization ranged from values above 92% to less than 25%.

Microbial Bioprospecting in Textile Effluents

The search for bacteria capable of degrading dyes can take advantage of the natural, or not so natural, selection of these organisms through bioprospecting in contaminated environments, selecting bacteria that underwent great environmental pressure, thanks to human action. Kannan and co-workers [78] evaluated the results of microbial consortia assembled with different combinations of five strains isolated from textile effluent. The best result obtained was from a consortium composed of *Bacillus cereus* and *Pseudomonas fluorescence* that achieved nearly 100% decolorization of Remazol Yellow dye, with nine of the ten combinations tested achieving decolorization above 78%. Another work that used microorganisms isolated from textile effluents was that of Eslami's group [79], which evaluated the action of a consortium of *Enterococcus faecalis* and *Klebsiella variicola* which obtained results of almost 100% removal of the dye Reactive Red 198.

However, it is worth noting that these two studies achieved good results with their consortia at mild parameters of temperature, pH, and salinity, which is often not the case with effluents contaminated by dyes, in a real case scenario. In addition, the consortia evaluated in both studies required more than two days for almost complete decolorization to occur.

Eskandari and co-workers [71] tested in their study two consortia with bacterial strains isolated from textile effluent and soil from a typically cold region, the Zagros Mountains in Iran. Their ability to degrade the dye Reactive Black 5 was evaluated, achieving the best results at mild temperatures, but in an alkaline range with pH between 9 and 11, which already represents an advantage over previous works, since it is common for industrial effluents containing dyes to have alkaline pH due to the use of substances of basic character in industrial processes.

The azo dyes have some representatives called pre-metallized dyes, which have metals in their structure previous to their application in the staining process [81]. Eleven consortia formed by bacteria isolated from areas contaminated with pre-metallized dyes in India were tested in the degradation of eighteen of these dyes. There was a wide variation in the percentage of degradation between consortia and dyes, with values ranging from 7.4% to 98.8% [80].

3.2.3. Bioreactors and Their Potential in the Bioremediation of Azo Dyes by Bacteria

Bioelectrochemical systems (BES) are represented by microbial fuel cells, microbial desalination cells, and microbial electrolysis cells. In these systems microorganisms perform the oxidation of compounds and the electrons generated in this process can be used in the production of energy and other compounds of interest [82]. BES have proven to be better compared to conventional anaerobic and electric or electrochemical processes by performing well in a shorter time, in a more cost-effective manner, and causing less negative environmental impact [83].

Microbial fuel cells have already been explored to generate energy allied to the treatment of effluents containing dyes. This type of treatment presents advantages such as: potential for energy production instead of its consumption; low sludge formation; operation at mild temperatures and atmospheric pressure and offers the possibility of performing oxidation (anode) and reduction (cathode) of the dyes [84]. In line with this, several works explore this possibility in the process of dye decolorization (Table 3), even adding steps for the detoxification of other harmful compounds.

Table 3. Bacterial dye degradation in bioreactor systems.

Main Bacteria	Dye	Reaction System	Parallel Study	Best Parallel Study Results	Maximum Degradation	Reference
Enterobacter, Desulfovibrio and Enterococcus	Alizarin Yellow R	Single chamber up-flow bioelectrochemical system (UBES)	-	-	87.74 ± 3.52%	[83]
Shewanella oneidensis	Acid Orange 7	Microbial Fuel Cell	COD * Reduction and Electricity production	Power density of 50 ± 4 mW m^{-2} and COD reduction 80.4 ± 1.2%	80%	[84]
Proteobacteria phyla	Congo red	Single chamber air-breathing cathode Microbial Fuel Cell	Sulfide removal and Electricity production	Maximum power density of 23.50 mW m^{-2} and 98% of Sulfide removal	81.5%	[85]
-	Acid Orange 7	Bioelectrochemical system (BES) combined with a membrane biofilm reactor	COD Reduction and Degradation of Sulfanilic acid	52.6 ± 3.2% of COD reduction and 64.7 ± 2.7% of Sulfanilic acid reduction	96.5 ± 0.6%	[86]
Unclassified genus	Reactive Brilliant Red X-3B	Biofilm electrode reactors (BERs)	COD Reduction	75.65% of COD removal and 21.13 mA m^{-2} of current density	75.27%	[87]
-	Reactive Brilliant Red X-3B	Microbial fuel cell (MFC)-biofilm electrode reactor (BER) coupled system	COD Reduction and Electricity production	Power density of 0.257 W m^{-3} and 88.62% of COD removal	97.77%	[88]
Geobacter sulfurreducens and Beta proteobacteria	Reactive Brilliant Red X-3B	Microbial fuel cell coupled constructed wetlands (CW-MFCs)	Electricity production	0.256 W m^{-3}	92.7%	[89]
Enterobacter and Enterococcus		Up-flow anaerobic sludge blanket (UASB) reactor	VFA production and COD removal efficiency	≈100% of COD removal efficiency	95.84 ±2.60	[90]
-	Direct Black 22	Sequencing batch reactors	COD Reduction and Ecotoxicity	81.4% of COD removal	81.4%	[91]
-	C.I. Basic Red 46	Anaerobic-aerobic sequencing batch reactor (SBR)	COD Reduction	>90% of COD removal	98%	[92]
Citrobacter sp., Enterococcus sp. and Enterobacter i	Remazol Black B	Upflow packed-bed reactor for continuous sequential microaerophilic–aerobic batch operations	-	-	95.87%	[93]
Serratia marcescens and Klebsiella oxytoca	Nylosan Yellow E2RL SGR	Sequencing batch reactor system, followed by ultrafiltration	COD Reduction	94% of COD removal	97%	[94]
Paludibacter, Trichococcus and Methanosarcina	Reactive Red 2	Anaerobic sequencing batch reactor	Ammonium removal and The effect of Fe3O4 on anaerobic treatment of azo dye	≈100% of Ammonium removal	≈100%	[95]
-	Yellow Dye	Aerobic Ganular Sludge (AGS)	Ammonium removal	≈100% of Ammonium removal	≈100%	[96]

* COD = Chemical Oxygen Demand.

Sulfide is considered to be toxic, corrosive and a threat to human health. It is commonly present in textile effluents containing dyes, formed due to the addition of sodium sulfide for reduction processes of azo compounds or conversion of other sulfur-containing substances. With this in mind, a single-chamber air-cathode microbial fuel cell was used by Dai and co-workers [85] for the simultaneous degradation of Congo red dye, bioelectricity generation, and sulfide oxidation. The results showed 98% sulfide removal and 88% decolorization, accompanied by the formation of maximum power of 23.50 mW m^{-2}. It was also evaluated that the sulfide concentration affects the sulfide oxidation rate as well as the dye degradation [85].

Mani and co-workers [84] studied the difference between the decolorization efficiency, electricity production and the decrease in chemical oxygen demand (COD) between feeding the fuel cell via the anode chamber containing the electrochemically active bacterium *Shewanella oneidensis* or via the cathode chamber containing the enzyme laccase. The conclusion of this work was that degradation with the laccase at the cathode is a more advantageous process, as it generates more stable and chemically simpler products, as well as lower COD (80.4% versus 69% of the anode) and higher power generation efficiency (50 mW m^{-2} versus 42.5 mW m^{-2} of the anode).

Other BES have already been studied in relation to their potential use for industrial effluent treatment, for example, the evaluation of their use in the removal and recovery of nutrients, especially with regard to nitrogen present in industrial effluents [82], as well as in the decrease of (COD) allied with the degradation of azo dyes, and the effects of their coupling with a continuous stirred reactor system and the increase of modules in a stacked BES has also been studied. In these with the following advantages were pointed out: (1) the use of a three-module system improved decolorization by 15% and 33% compared to systems with only two or one modules, respectively, achieving up to 80% removal [86]; (2) coupling with a continuous stirring reactor achieved 97% of color removal from the medium in just seven hours, being superior to the results obtained with these techniques alone, where approximately 54.9% and 91.4% decolorization were achieved [83]; (3) there was 75.6% reduction in (COD) and voltage has shown to affect the decolorization efficiency [87].

Other studies have performed couplings of systems aiming to improve the decolorization process and for the evaluation of energy generation. A microbial fuel cell system was coupled to a biofilm electrode reactor and the results indicated an increase of almost 30% in color removal efficiency compared to the process performed with these mechanisms decoupled [88]. In the system assembled with microbial fuel cells in combination with a constructed wetland it was observed that the higher the substrate biomass, the higher the decolorization and the lower the power generation, with these varying among the groups tested from 76.3% to 92.7% and 0.117W m^{-3} to 0.256W m^{-3}, respectively [89].

The use of bioreactors can also be achieved with various combinations and parameters variations such as concentrations of salts, presence of oxygen, and feed rate. Regarding the presence of oxygen, several possibilities have already been studied, namely tests in an anaerobic environment [90]; with continuous micro aeration; intermittent and without aeration [91]; anaerobic starting followed by aerobic [92]; and microaerophilic environments followed by aerobic [93].

In tests performed in a reactor with a fully anaerobic environment, the decolorization rate ranged from 62.98% to 95.84%, depending on the loading rate of dye in the reactor, which decreased the decolorization with its increase [90]. In the study by Menezes and co-workers [91] the reactor without aeration had the highest decolorization, followed by the intermittent aeration reactor and lastly the continuous aeration reactor, the rates being 81.4%, 76.8%, and 74.5%, respectively. However, the reactor without aeration produced waste with toxic substances, which was not observed in other conditions. Assadi and co-workers [92] pointed out better dye removal in an anaerobic environment, reaching almost 98%, and indicate that increasing the concentration of the dye negatively affects the decrease in COD. This same work also showed that the decolorization is also negatively affected by increasing the concentration of salt and nitrate ions in the medium. The assays without an anaerobic environment obtained maximum decolorization of 95.87% in the micro-aeration stage and removal of 23 mg/L of aromatic amines in the aerobic stage compared to the previous environment with micro-aeration [93].

Bioremediation can also be employed in other treatment processes; an example is the study by Korenak and his group [94] who combined the treatment of contaminated effluent performed in bioreactors with bacteria to the subsequent process of ultrafiltration. This combination of methods improved both the dye removal rate, 85% before ultrafiltration and 97% after, and the decrease in COD, 91% before and 94% after.

Studies also indicate the possibility of improving the anaerobic treatment with the use of Fe_3O_4, which generated a decrease in the lag phase of microbial growth, improved the decolorization rate, increased microbial resistance to increasing dye concentration, among other factors [95]. Moreover, ammonium removal has already been achieved with 92% to

100% rates, coupled with 89% to 100% dye removal in microbial treatment in bioreactors using aerobic granular sludge [96].

As for the inoculation of bacteria into the reactors, it can occur in multiple ways, depending on the operation system proposed. For example, sludge obtained from a wastewater treatment location can be added into the bioreactor to serve as a inoculum source, being composed, in theory, of a myriad of organisms adapted to the contaminated environment and in balanced association [85], or a specific pure culture bacteria can be grown in order to be inoculated directly in a anode of a microbial fuel cell [84]. A packed-bed column reactor can be inoculated by the circulation of the bacteria culture in the packed-bed column [93]. As demonstrated, there is more than one form of inoculating a bioreactor for bioremediation use, and the initial bacteria acclimatation/growth also varies with the bacteria strain and bioreactor operation. Because of the various possibilities when it comes to bacterial inoculum, this step is susceptible to improvement, being an interesting hub for microbial prospection and cultivation optimizations, including here the pre-treatment of the sludge for better results [97–99].

3.3. Degradation of Environmental and Industrial Samples

The use of bacteria to treat contamination caused by azo dyes can aim at both the treatment of effluents before their release into nature and the bioremediation of already contaminated natural environments. This topic deals with research conducted on the treatment of samples taken from contaminated environments and industrial effluents (Table 4) to show how efficient bacteria can be applied in remediating real samples in real cases of contamination.

Table 4. Bacterial degradation of azo dyes contaminated industrial effluents.

Main Bacteria	Wastewater Source	Degradation Mechanism	Country	Maximum Degradation and Experiment Conditions	Reference
Micrococcus luteus	Dyehouse	Adsorption and Enzymatic	Japan	Laboratory	[99]
Pseudomonas aeruginosa	-	Enzymatic-Azoreductase	India	62%-Laboratory	[100]
Pseudomonas sp.	Textile Industries	Enzymatic-Laccase	India	90%-Laboratory	[101]
Pseudomonas sp. and *Bacillus* sp.	Mill effluent outlet	-	India	*Pseudomonas* 95% *Bacillus* 97%-Laboratory	[102]
Pseudomonas aeruginosa, *Pseudomonas putida* and *Bacillus cereus*	Textile Factory	-	Egypt	92%-Laboratory	[103]
-	Dye Wastewater Plant	-	Korea	75%-Real production facility	[104]

3.3.1. Industrial Effluents

To illustrate how diverse the sources of bacteria capable of degrading azo dyes can be, the study of Ito and co-workers [99] isolated bacteria from the microbiota of human hands and classified them into two groups: azo dye decolorizers and anthraquinone dye decolorizers. The two strains chosen for further work, one from each group, were able to perform decolorization of the industrial effluent sample collected from a dyeing plant. Other works used *Pseudomonas* sp. strains in the decolorization of industrial effluent samples. The decolorization rates were different, of 62% when supplemented with nutrients and in the time of 7 days [100], up to 90% in sixty hours [101] and ranging between 87–95% in 48 hours [102]. The latter work further tested a *Bacillus* sp. strain and obtained decolorization of the textile effluent samples in the range of 92–97% in 48 hours. The work of Bayoumi and co-workers [103] also focused on bacteria from the genera *Bacillus* sp. and *Pseudomonas* sp. in the decolorization of textile effluent samples from an industrial city in Egypt, obtaining results between 84% and 92% of decolorization in a period of 48 hours.

Kalathil and co-workers [104] worked with microbial fuel cells to treat wastewater containing dyes in Daegu, South Korea, the tests were conducted with a retention time of forty-eight hours and the system was operated in open loop and in closed loop. The closed system had the highest color removal-almost 80%-while the open system showed only 62% decolorization. In addition, the closed loop also presented higher toxicity removal and a decrease in COD.

3.3.2. Environmental Samples

Still in the efforts to find bacterial strains capable of degrading azo dyes under conditions relevant to real-world application, different studies have evaluated the ability of these organisms to degrade dyes in environmental samples (Table 5). The work of Tara and co-workers [105] used a pilot scale floating wetland system coupled with dye degrading bacteria to treat wastewater from a textile industry in Pakistan. These macrocosms were installed using separate or combined plants and bacteria, and the symbiose between them improved the removal of organic and inorganic pollutants by decreasing chemical and biochemical oxygen demand by 92% and 91%, respectively, staining was reduced by 86%, and trace metals by 87%. Furthermore, the combination of these organisms also resulted in improved detoxification of the effluent, where no fish kills were observed after exposure to the treated textile effluent. When treated without the combination of bacteria and plants, the effluent still caused the death of some fish, highlighting the benefits of this synergy for bioremediation efforts.

Table 5. Bacterial degradation of azo dyes contaminated environmental samples.

Remediation site	Parallel Study	Degradation System	Country	Maximum Degradation	Reference
On-site Textile Industrial Wastewater	COD *, BOD ** and Trace metals removal	Floating Wetlands	Pakistan	86%	[105]
On-site Textile Industrial Wastewater	COD, BOD, heavy metals, nitrogen, phosphorous and total dissolved solids decrease.	Vertical flow constructed wetlands (VFCW) with bacterial endophytes	Pakistan	74%	[106]
On-site Textile Industrial Wastewater	-	Aerated wetland	Italy	82%	[107]
Dye Contaminated Soil	-	Continuous flow reactor	India	98%	[108]
Dye ontaminated soil	COD, BOD, TOC ***, heavy metals, nitrogen reduction and nitrogen, phosphorous and potassium increase.	Microbial consortia	India	98.87%(after 30 days) 99.25% (after 60 days)	[109]
Dye spiked soil	Optimum pH and temperature decolorization and effect of carbon and nitrogen sources addition	Bacterial consortia	India	97%	[110]
Yabagawa river sediments	Aromatic amines persistance	Natural river microbiota	Japan	-	[111]

* COD = Chemical Oxygen Demand. ** BOD = Biological Oxygen Demand. *** TOC = Total Organic Carbon.

Hussain and co-workers [106] also focused on a wetlands system to treat wastewater from a Pakistani industry but using pilot-scale vertical flow constructed wetlands (VFCW) system augmented with bacterial endophytes, which were selected based on their capabilities to improve plant growth and degrade dyes. The combination proposed was able to decrease chemical oxygen demand (81%), biochemical oxygen demand (72%), total dissolved solids (32%), color (74%), nitrogen (84%), phosphorous (79%), and heavy metals (Cr (97%), Fe (89%), Ni (88%), Cd (72%)). In addition, the treated wastewater was found to cause no harm based on a fish toxicity assay. Another study based on wetlands system (but with aeration) was carried out by Masi and co-workers [107]. The group assessed color removal based on three different wavelengths and different influent concentrations, achieving results that varied from a negative color removal (-58%) to a positive decolorization of 82%.

In India a continuous flow reactor was tested for the bioremediation of contaminated soil using a consortium of bacteria, achieving 85% color removal on the first day of operation, 90% on the second, and a steady 98% removal rate from the 13th day of operation [108]. Other groups have also worked with soil bioremediation. Vipul and co-workers [109] treated soil samples, collected from industrial area, with a bacterial community previously isolated from sludge samples of six sites contaminated with different organic compounds containing bacteria and fungi organisms. The microbial community was able to achieve a decolorization of 98.87%, 82.88% of COD removal and 89.82% of BOD removal after 30 days. As for the Tandon and co-workers' group [110], they treated dye contaminated soil with a bacterial consortium, achieving 97% and 96.25% of decolorization for two different dyes.

The Yabagawa River in Japan was studied over three years to assess the natural degradation of dyes and their breakdown products, the aromatic amines, by bacteria. This river had been suffering from the dumping of dyes and industrial effluents by a dye factory for more than 50 years, and with the closure of the factory in 2012, Ito and co-workers [111] were able to evaluate the natural recovery of this environment.

This work pointed out the persistence of dyes and their aromatic amines in the river sediments even years after the end of the discharge of industrial effluent in that environment and even without the presence of visible coloration in the water. It was also observed that the degradation rate of the dye varies with its concentration in the medium, i.e., the less dye the lower the rate of degradation, and the opposite occurred with the degradation of aromatic amines, which increased over time, reaching its highest rate one year after the end of the effluent discharge on the river. The variety of bacteria itself changed over time, going from an abundant variety of genera related to the degradation of azo dyes to a decrease in these and an increase in groups related to aniline degradation [111].

4. Conclusions

Azo dyes can be harmful to the environment and human health when disposed of without prior treatment, and the search for sustainable and less harmful production processes requires the development of new alternatives for effluent treatment that are efficient, cost-effective and of low environmental impact. Thus, bacterial bioremediation is a good alternative, given the versatility of this phylum that offers a range of possibilities, either with pure cultures or in consortia, tolerating different physicochemical parameters, in order to better adapt this process to various industrial wastes.

The application of these organisms in BES also brings the possibility of generating more than one salable product or service, making this process more attractive in terms of cost, an important bottleneck to be overcome in the implementation of biological systems. The application of bacteria to environmental samples also attests to this viability, being able to degrade dyes and their toxic by-products in environmentally relevant concentrations. Through the critical reading of the literature presented, scientific advances in this area can be evaluated, as well as the efforts to remedy the still deficient points, showing bacterial bioremediation to be an increasingly feasible process.

For the widespread application of bacterial bioremediation, several factors have to be considered, depending on the technique used, the characteristics of the environment to be remediated and of the bacteria strain, in this sense, the following points are relevant bottlenecks for large-scale application: (1) Bioreactor implementation and maintenance costs, (2) physicochemical parameters—which may vary over time, (3) space available for use of, e.g., wetlands or bioreactors, (4) availability of nutrients in the environment or in the textile effluent to be decontaminated, (5) presence/generation of suitable redox mediators for the enzymatic action of azo bond breaking, (6) engineering optimization in the transition from laboratory/pilot to industrial scale, (7) stricter local legislation forcing companies to treat their effluents properly, (8) co-relation between dye and bacteria/bacterial consortia or the presence of mixed dyes that can affect the bleaching given the bacterial suitability to each dye, (9) the use of industrial chemicals not considered in the laboratory tests, (10) changes in industrial dyeing techniques that modify the characteristics of its effluent and require adaptation of the bioremediation technique used, and (11) generation of toxic by-products that bacteria are not able to degrade, among other factors more specific to the numerous systems under study.

Despite the challenges in this sector, which this review tried to address under a critical approach, the good results obtained in laboratory, pilot scale and in some specific cases applied to real situations, together with the urgency for new sustainable solutions in large-scale industries make the search for biotechnological solutions a possible path in collective efforts in the search for cleaner and more responsible forms of production.

Author Contributions: Conceptualization, L.R.S.P., A.V.S. and D.G.G.; Writing—original draft preparation, L.R.S.P.; Writing—review and editing, A.V.S., D.G.G. and L.P.X.; Funding acquisition, A.V.S. All authors have read and agreed to the published version of the manuscript.

Funding: This study was financed in part by Coordenação de Aperfeiçoamento de Pessoal de Nível Superior-Brasil (CAPES)-Finance Code 001.

Institutional Review Board Statement: Not applicable.

Informed Consent Statement: Not applicable.

Data Availability Statement: Not applicable.

Acknowledgments: The authors would like to thank Pró-Reitoria de Pesquisa e Pós-Graduação da Universidade Federal do Pará (PROPESP/UFPA).

Conflicts of Interest: The authors declare no conflict of interest.

References

1. De Araújo, M.E.M. Corantes naturais para têxteis—Da antiguidade aos tempos modernos. *Conserv. Patrim.* **2006**, *3–4*, 39–51.
2. Kant, R. Textile Dyeing Industry an Environmental Hazard. *Nat. Sci.* **2012**, *4*, 22–26. [CrossRef]
3. Abel, A. The history of dyes and pigments: From natural dyes to high performance pigments. In *Colour Design: Theories and Applications*; Woodhead Publishing: Sawston, UK, 2012; pp. 557–587. [CrossRef]
4. Ajaz, M.; Shakeel, S.; Rehman, A. Microbial Use for Azo Dye Degradation—A Strategy for Dye Bioremediation. *Int. Microbiol.* **2020**, *23*, 149–159. [CrossRef] [PubMed]
5. Guaratini, C.C.I.; Zanoni, M.V.B. Corantes Têxteis. *Química Nova* **2000**, *23*, 71–78. [CrossRef]
6. Saxena, A.; Gupta, S. Bioefficacies of Microbes for Mitigation of Azo Dyes in Textile Industry Effluent: A Review. *BioResources* **2020**, *15*, 9858–9881. [CrossRef]
7. Chequer, F.M.D.; Lizier, T.M.; de Felício, R.; Zanoni, M.V.B.; Debonsi, H.M.; Lopes, N.P.; Marcos, R.; de Oliveira, D.P. Analyses of the Genotoxic and Mutagenic Potential of the Products Formed after the Biotransformation of the Azo Dye Disperse Red 1. *Toxicol. In Vitr.* **2011**, *25*, 2054–2063. [CrossRef] [PubMed]
8. Gürses, A.; Açıkyıldız, M.; Güneş, K.; Gürses, M.S. Dyes and Pigments: Their Structure and Properties. In *Dyes and Pigments*; Springer: Cham, Switzerland, 2016; pp. 13–29. ISBN 9783319338903.
9. Oliveira, D.P.D. Corantes como Importante Classe de Contaminantes Ambientais—Um Estudo de Caso. Ph.D. Thesis, Universidade de São Paulo, São Paulo, Brasil, 1 March 2005.
10. Gomes, L.M. Estudo da otimização do processo Fenton para o descoramento de corantes azo. Master's Thesis, Universidade de São Paulo, São Paulo, Brasil, 2009.
11. Bell, J.; Plumb, J.J.; Buckley, C.A.; Stuckey, D.C. Treatment and Decolorization of Dyes in an Anaerobic Baffled Reactor. *J. Environ. Eng.* **2000**, *126*, 1026–1032. [CrossRef]
12. Chung, K.-T. Azo Dyes and Human Health: A Review. *J. Environ. Sci. Health Part C* **2016**, *34*, 233–261. [CrossRef] [PubMed]
13. De Amorim, C.C.; Leão, M.M.D.; Moreira, R.d.F.P.M. Comparação Entre Diferentes Processos Oxidativos Avançados Para Degradação de Corante Azo. *Eng. Sanit. Ambient.* **2009**, *14*, 543–550. [CrossRef]
14. Singh, A.L.; Chaudhary, S.; Kayastha, A.M.; Yadav, A. Decolorization and degradation of textile effluent with the help of Enterobacter asburiae. *Indian J. Biotechnol.* **2015**, *14*, 101–106.
15. Kumar, A.; Kumar, A.; Singh, R.; Singh, R.; Pandey, S.; Rai, A.; Singh, V.K.; Rahul, B. Genetically Engineered Bacteria for the Degradation of Dye and Other Organic Compounds. In *Abatement of Environmental Pollutants*; Elsevier: Amsterdam, The Netherlands, 2020; pp. 331–350. ISBN 9780128180952.
16. van der Zee, F.P.; Villaverde, S. Combined Anaerobic–Aerobic Treatment of Azo Dyes—A Short Review of Bioreactor Studies. *Water Res.* **2005**, *39*, 1425–1440. [CrossRef] [PubMed]
17. Mahmood, S.; Khalid, A.; Arshad, M.; Mahmood, T.; Crowley, D.E. Detoxification of Azo Dyes by Bacterial Oxidoreductase Enzymes. *Crit. Rev. Biotechnol.* **2016**, *36*, 639–651. [CrossRef]
18. Lellis, B.; Fávaro-Polonio, C.Z.; Pamphile, J.A.; Polonio, J.C. Effects of Textile Dyes on Health and the Environment and Bioremediation Potential of Living Organisms. *Biotechnol. Res. Innov.* **2019**, *3*, 275–290. [CrossRef]

19. Rawat, D.; Sharma, R.S.; Karmakar, S.; Arora, L.S.; Mishra, V. Ecotoxic Potential of a Presumably Non-Toxic Azo Dye. *Ecotoxicol. Environ. Saf.* **2018**, *148*, 528–537. [CrossRef]
20. Parrott, J.L.; Bartlett, A.J.; Balakrishnan, V.K. Chronic Toxicity of Azo and Anthracenedione Dyes to Embryo-Larval Fathead Minnow. *Environ. Pollut.* **2016**, *210*, 40–47. [CrossRef] [PubMed]
21. Barathi, S.; Karthik, C.; Nadanasabapathi, S.; Padikasan, I.A. Biodegradation of Textile Dye Reactive Blue 160 by *Bacillus firmus* (*Bacillaceae*: Bacillales) and Non-Target Toxicity Screening of Their Degraded Products. *Toxicol. Rep.* **2020**, *7*, 16–22. [CrossRef]
22. Croce, R.; Cinà, F.; Lombardo, A.; Crispeyn, G.; Cappelli, C.I.; Vian, M.; Maiorana, S.; Benfenati, E.; Baderna, D. Aquatic Toxicity of Several Textile Dye Formulations: Acute and Chronic Assays with *Daphnia magna* and *Raphidocelis subcapitata*. *Ecotoxicol. Environ. Saf.* **2017**, *144*, 79–87. [CrossRef]
23. Duarte Baumer, J.; Valério, A.; de Souza, S.M.A.G.U.; Erzinger, G.S.; Furigo, A.; de Souza, A.A.U. Toxicity of Enzymatically Decolored Textile Dyes Solution by Horseradish Peroxidase. *J. Hazard. Mater.* **2018**, *360*, 82–88. [CrossRef]
24. North, A. Impact of Chronic Exposure of Selected Azo Dyes on the Model Organism *Caenorhabditis elegans*. Ph.D. Thesis, Texas Southern University, Houston, TX, USA, 2021.
25. Hessel, C.; Allegre, C.; Maisseu, M.; Charbit, F.; Moulin, P. Guidelines and Legislation for Dye House Effluents. *J. Environ. Manag.* **2007**, *83*, 171–180. [CrossRef]
26. Compliance Gate. Are Azo Dyes Banned in the European Union? Available online: https://www.compliancegate.com/azo-dye-regulations-european-union/#Are_azo_dyes_banned_in_the_European_Union (accessed on 6 April 2021).
27. Compliance Gate. Azo Dye Regulations in the United States: An Overview. Available online: https://www.compliancegate.com/azo-dye-regulations-united-states/ (accessed on 6 April 2021).
28. SGS. Textile and Clothing Chemical Safety Criteria in Asian Countries—When East Meets West. Available online: https://www.sgs.com/en/news/2012/04/textile-and-clothing-chemical-safety-criteria-in-asian-countries-when-east-meets-west (accessed on 6 April 2021).
29. Qima. Japan Regulation Restricting Azo Dyes in Household Products Comes into Force. Available online: https://www.qima.com/regulation/03-16/japan-azo (accessed on 6 April 2021).
30. Hktdc Hong Kong Means Business. India—Azo Requirement on Imported Apparel and Textile Products. Available online: https://hkmb.hktdc.com/en/1X09YJMT/hktdc-research/India-Azo-Requirement-on-Imported-Apparel-and-Textile-Products (accessed on 6 April 2021).
31. Tuvsud. Vietnam: New Regulation on Azo Dyes and Formaldehyde in Textile Products Published. Available online: https://www.tuvsud.com/en/e-ssentials-newsletter/consumer-products-and-retail-essentials/e-ssentials-1-2018/new-regulation-on-azo-dyes-and-formaldehyde-in-textile-products-published (accessed on 6 April 2021).
32. Buthelezi, S.; Olaniran, A.; Pillay, B. Textile Dye Removal from Wastewater Effluents Using Bioflocculants Produced by Indigenous Bacterial Isolates. *Molecules* **2012**, *17*, 14260–14274. [CrossRef]
33. Sriram, N.; Reetha, D.; Saranraj, P. Biological Degradation of Reactive Dyes by Using Bacteria Isolated from Dye Effluent Contaminated Soil. *Middle–East J. Sci. Res.* **2013**, *12*, 1695–1700. [CrossRef]
34. Li, Q.; Wu, Y.-J. A Fluorescent, Genetically Engineered Microorganism That Degrades Organophosphates and Commits Suicide When Required. *Appl. Microbiol. Biotechnol.* **2009**, *82*, 749–756. [CrossRef] [PubMed]
35. Pan, Y.; Zhu, T.; He, Z. Enhanced Removal of Azo Dye by a Bioelectrochemical System Integrated with a Membrane Biofilm Reactor. *Ind. Eng. Chem. Res.* **2018**, *57*, 16433–16441. [CrossRef]
36. Kuhad, R.C.; Sood, N.; Tripathi, K.K.; Singh, A.; Ward, O.P. Developments in Microbial Methods for the Treatment of Dye Effluents. In *Advances in Applied Microbiology*; Elsevier: Amsterdam, The Netherlands, 2004; Volume 56, pp. 185–213. ISBN 9780120026586.
37. Solís, M.; Solís, A.; Pérez, H.I.; Manjarrez, N.; Flores, M. Microbial Decolouration of Azo Dyes: A Review. *Process Biochem.* **2012**, *47*, 1723–1748. [CrossRef]
38. Khan, R.; Bhawana, P.; Fulekar, M.H. Microbial Decolorization and Degradation of Synthetic Dyes: A Review. *Rev. Environ. Sci. Biotechnol.* **2013**, *12*, 75–97. [CrossRef]
39. Saratale, R.G.; Saratale, G.D.; Chang, J.S.; Govindwar, S.P. Bacterial Decolorization and Degradation of Azo Dyes: A Review. *J. Taiwan Inst. Chem. Eng.* **2011**, *42*, 138–157. [CrossRef]
40. Singh, R.L.; Singh, P.K.; Singh, R.P. Enzymatic Decolorization and Degradation of Azo Dyes—A Review. *Int. Biodeterior. Biodegrad.* **2015**, *104*, 21–31. [CrossRef]
41. Keck, A.; Klein, J.; Kudlich, M.; Stolz, A.; Knackmuss, H.J.; Mattes, R. Reduction of Azo Dyes by Redox Mediators Originating in the Naphthalenesulfonic Acid Degradation Pathway of Sphingomonas Sp. Strain BN6. *Appl. Environ. Microbiol.* **1997**, *63*, 3684–3690. [CrossRef]
42. Chivukula, M.; Renganathan, V. Phenolic Azo Dye Oxidation by Laccase from Pyricularia Oryzae. *Appl. Environ. Microbiol.* **1995**, *61*, 4374–4377. [CrossRef]
43. Garcia, F.D.S. Enzimas Oxidorredutases Produzidas por Fungos Filamentosos. Master's Thesis, Universidade de São Paulo, São Paulo, Brasil, 13 June 2018.
44. Chacko, J.T.; Subramaniam, K. Enzymatic degradation of azo dyes-a review. *Int. J. Environ. Sci.* **2011**, *1*, 1250, 2011.
45. Misal, S.A.; Gawai, K.R. Azoreductase: A Key Player of Xenobiotic Metabolism. *Bioresour. Bioprocess.* **2018**, *5*, 17. [CrossRef]
46. Sharma, V.; Upadhyay, L.S.B.; Vasanth, D. Extracellular Thermostable Laccase-Like Enzymes from *Bacillus licheniformis* Strains: Production, Purification and Characterization. *Appl. Biochem. Microbiol.* **2020**, *56*, 420–432. [CrossRef]

47. Kandelbauer, A.; Guebitz, G.M. Bioremediation for the Decolorization of Textile Dyes—A Review. In *Environmental Chemistry*; Lichtfouse, E., Schwarzbauer, J., Robert, D., Eds.; Springer: Berlin/Heidelberg, Germany, 2005; pp. 269–288. ISBN 9783540228608.
48. Paszczynski, A.; Crawford, R.L. Degradation of Azo Compounds by Ligninase from Phanerochaete Chrysosporium: Involvement of Veratryl Alcohol. *Biochem. Biophys. Res. Commun.* **1991**, *178*, 1056–1063. [CrossRef]
49. Ilić Đurđić, K.; Ostafe, R.; Đurđević Đelmaš, A.; Popović, N.; Schillberg, S.; Fischer, R.; Prodanović, R. Saturation Mutagenesis to Improve the Degradation of Azo Dyes by Versatile Peroxidase and Application in Form of VP-Coated Yeast Cell Walls. *Enzym. Microb. Technol.* **2020**, *136*, 109509. [CrossRef] [PubMed]
50. Singh, P.; Iyengar, L.; Pandey, A. Bacterial Decolorization and Degradation of Azo Dyes. In *Microbial Degradation of Xenobiotics*; Singh, S.N., Ed.; Springer: Berlin/Heidelberg, Germany, 2012; pp. 101–133. ISBN 9783642237881.
51. Xu, F.; Mou, Z.; Geng, J.; Zhang, X.; Li, C. Azo Dye Decolorization by a Halotolerant Exoelectrogenic Decolorizer Isolated from Marine Sediment. *Chemosphere* **2016**, *158*, 30–36. [CrossRef] [PubMed]
52. Neifar, M.; Chouchane, H.; Mahjoubi, M.; Jaouani, A.; Cherif, A. *Pseudomonas extremorientalis* BU118: A New Salt-Tolerant Laccase-Secreting Bacterium with Biotechnological Potential in Textile Azo Dye Decolourization. *3 Biotech* **2016**, *6*, 107. [CrossRef] [PubMed]
53. Wang, Y.; Jiang, L.; Shang, H.; Li, Q.; Zhou, W. Treatment of Azo Dye Wastewater by the Self-Flocculating Marine Bacterium Aliiglaciecola Lipolytica. *Environ. Technol. Innov.* **2020**, *19*, 100810. [CrossRef]
54. Zhuang, M.; Sanganyado, E.; Zhang, X.; Xu, L.; Zhu, J.; Liu, W.; Song, H. Azo Dye Degrading Bacteria Tolerant to Extreme Conditions Inhabit Nearshore Ecosystems: Optimization and Degradation Pathways. *J. Environ. Manag.* **2020**, *261*, 110222. [CrossRef] [PubMed]
55. Guadie, A.; Tizazu, S.; Melese, M.; Guo, W.; Ngo, H.H.; Xia, S. Biodecolorization of Textile Azo Dye Using *Bacillus* sp. Strain CH12 Isolated from Alkaline Lake. *Biotechnol. Rep.* **2017**, *15*, 92–100. [CrossRef]
56. Chen, K.-C.; Wu, J.-Y.; Liou, D.-J.; Hwang, S.-C.J. Decolorization of the Textile Dyes by Newly Isolated Bacterial Strains. *J. Biotechnol.* **2003**, *101*, 57–68. [CrossRef]
57. Jadhav, U.U.; Dawkar, V.V.; Ghodake, G.S.; Govindwar, S.P. Biodegradation of Direct Red 5B, a Textile Dye by Newly Isolated Comamonas Sp. UVS. *J. Hazard. Mater.* **2008**, *158*, 507–516. [CrossRef]
58. Asad, S.; Amoozegar, M.A.; Pourbabaee, A.A.; Sarbolouki, M.N.; Dastgheib, S.M.M. Decolorization of Textile Azo Dyes by Newly Isolated Halophilic and Halotolerant Bacteria. *Bioresour. Technol.* **2007**, *98*, 2082–2088. [CrossRef] [PubMed]
59. Shah, M. Evaluation of Aeromonas Spp. In Microbial Degradation and Decolorization of Reactive Black in Microaerophilic—Aerobic Condition. *J. Bioremed. Biodeg.* **2014**, *5*, 246. [CrossRef]
60. Franca, R.D.G.; Vieira, A.; Carvalho, G.; Oehmen, A.; Pinheiro, H.M.; Barreto Crespo, M.T.; Lourenço, N.D. Oerskovia Paurometabola Can Efficiently Decolorize Azo Dye Acid Red 14 and Remove Its Recalcitrant Metabolite. *Ecotoxicol. Environ. Saf.* **2020**, *191*, 110007. [CrossRef] [PubMed]
61. Srinivasan, S.; Sadasivam, S.K. Exploring Docking and Aerobic-Microaerophilic Biodegradation of Textile Azo Dye by Bacterial Systems. *J. Water Process Eng.* **2018**, *22*, 180–191. [CrossRef]
62. Dehghanian, F.; Kay, M.; Kahrizi, D. A Novel Recombinant AzrC Protein Proposed by Molecular Docking and in Silico Analyses to Improve Azo Dye's Binding Affinity. *Gene* **2015**, *569*, 233–238. [CrossRef] [PubMed]
63. Haghshenas, H.; Kay, M.; Dehghanian, F.; Tavakol, H. Molecular Dynamics Study of Biodegradation of Azo Dyes via Their Interactions with AzrC Azoreductase. *J. Biomol. Struct. Dyn.* **2016**, *34*, 453–462. [CrossRef]
64. Srinivasan, S.; Shanmugam, G.; Surwase, S.V.; Jadhav, J.P.; Sadasivam, S.K. In Silico Analysis of Bacterial Systems for Textile Azo Dye Decolorization and Affirmation with Wetlab Studies: General. *Clean Soil Air Water* **2017**, *45*, 1600734. [CrossRef]
65. Sari, I.P.; Simarani, K. Decolorization of Selected Azo Dye by *Lysinibacillus fusiformis* W1B6: Biodegradation Optimization, Isotherm, and Kinetic Study Biosorption Mechanism. *Adsorpt. Sci. Technol.* **2019**, *37*, 492–508. [CrossRef]
66. Joshi, A.U.; Hinsu, A.T.; Kotadiya, R.J.; Rank, J.K.; Andharia, K.N.; Kothari, R.K. Decolorization and Biodegradation of Textile Di-Azo Dye Acid Blue 113 by *Pseudomonas stutzeri* AK6. *3 Biotech* **2020**, *10*, 214. [CrossRef]
67. Du, L.-N.; Li, G.; Zhao, Y.-H.; Xu, H.-K.; Wang, Y.; Zhou, Y.; Wang, L. Efficient Metabolism of the Azo Dye Methyl Orange by Aeromonas Sp. Strain DH-6: Characteristics and Partial Mechanism. *Int. Biodeterior. Biodegrad.* **2015**, *105*, 66–72. [CrossRef]
68. Olukanni, O.D.; Osuntoki, A.A.; Kalyani, D.C.; Gbenle, G.O.; Govindwar, S.P. Decolorization and Biodegradation of Reactive Blue 13 by Proteus Mirabilis LAG. *J. Hazard. Mater.* **2010**, *184*, 290–298. [CrossRef]
69. Mendes, S.; Farinha, A.; Ramos, C.G.; Leitão, J.H.; Viegas, C.A.; Martins, L.O. Synergistic Action of Azoreductase and Laccase Leads to Maximal Decolourization and Detoxification of Model Dye-Containing Wastewaters. *Bioresour. Technol.* **2011**, *102*, 9852–9859. [CrossRef]
70. Thakur, J.K.; Paul, S.; Dureja, P.; Annapurna, K.; Padaria, J.C.; Gopal, M. Degradation of Sulphonated Azo Dye Red HE7B by *Bacillus* sp. and Elucidation of Degradative Pathways. *Curr. Microbiol.* **2014**, *69*, 183–191. [CrossRef]
71. Eskandari, F.; Shahnavaz, B.; Mashreghi, M. Optimization of Complete RB-5 Azo Dye Decolorization Using Novel Cold-Adapted and Mesophilic Bacterial Consortia. *J. Environ. Manag.* **2019**, *241*, 91–98. [CrossRef]
72. Masarbo, R.S.; Niranjana, S.R.; Monisha, T.R.; Nayak, A.S.; Karegoudar, T.B. Efficient Decolorization and Detoxification of Sulphonated Azo Dye Ponceau 4R by Using Single and Mixed Bacterial Consortia. *Biocatal. Biotransform.* **2019**, *37*, 367–376. [CrossRef]

73. Guo, G.; Hao, J.; Tian, F.; Liu, C.; Ding, K.; Xu, J.; Zhou, W.; Guan, Z. Decolorization and Detoxification of Azo Dye by Halo-Alkaliphilic Bacterial Consortium: Systematic Investigations of Performance, Pathway and Metagenome. *Ecotoxicol. Environ. Saf.* **2020**, *204*, 111073. [CrossRef]
74. Guo, G.; Hao, J.; Tian, F.; Liu, C.; Ding, K.; Zhang, C.; Yang, F.; Xu, J. Decolorization of Metanil Yellow G by a Halophilic Alkalithermophilic Bacterial Consortium. *Bioresour. Technol.* **2020**, *316*, 123923. [CrossRef] [PubMed]
75. Guo, G.; Li, X.; Tian, F.; Liu, T.; Yang, F.; Ding, K.; Liu, C.; Chen, J.; Wang, C. Azo Dye Decolorization by a Halotolerant Consortium under Microaerophilic Conditions. *Chemosphere* **2020**, *244*, 125510. [CrossRef] [PubMed]
76. Shah, B.; Jain, K.; Jiyani, H.; Mohan, V.; Madamwar, D. Microaerophilic Symmetric Reductive Cleavage of Reactive Azo Dye—Remazole Brilliant Violet 5R by Consortium VIE6: Community Synergism. *Appl. Biochem. Biotechnol.* **2016**, *180*, 1029–1042. [CrossRef] [PubMed]
77. Khan, Z.; Jain, K.; Soni, A.; Madamwar, D. Microaerophilic Degradation of Sulphonated Azo Dye—Reactive Red 195 by Bacterial Consortium AR1 through Co-Metabolism. *Int. Biodeterior. Biodegrad.* **2014**, *94*, 167–175. [CrossRef]
78. Kannan, D.; Rajan, S.; Murugesan, A.G. Decolourization of azo dye by native microbial consortium. *J. Adv. Sci. Res.* **2019**, *10*, 95–99.
79. Eslami, H.; Shariatifar, A.; Rafiee, E.; Shiranian, M.; Salehi, F.; Hosseini, S.S.; Eslami, G.; Ghanbari, R.; Ebrahimi, A.A. Decolorization and Biodegradation of Reactive Red 198 Azo Dye by a New Enterococcus Faecalis–Klebsiella Variicola Bacterial Consortium Isolated from Textile Wastewater Sludge. *World J. Microbiol. Biotechnol.* **2019**, *35*, 38. [CrossRef] [PubMed]
80. Patel, D.K.; Tipre, D.R.; Dave, S.R. Selection and Development of Efficient Consortia for Decolorization of Metal Complex Dyes. *Toxicol. Environ. Chem.* **2017**, *99*, 252–264. [CrossRef]
81. Chakraborty, J.N. Metal-complex dyes. In *Handbook of Textile and Industrial Dyeing*; Elsevier: Amsterdam, The Netherlands, 2011; pp. 446–465. ISBN 9781845696955.
82. Kelly, P.T.; He, Z. Nutrients Removal and Recovery in Bioelectrochemical Systems: A Review. *Bioresour. Technol.* **2014**, *153*, 351–360. [CrossRef] [PubMed]
83. Cui, M.-H.; Cui, D.; Gao, L.; Cheng, H.-Y.; Wang, A.-J. Analysis of Electrode Microbial Communities in an Up-Flow Bioelectrochemical System Treating Azo Dye Wastewater. *Electrochim. Acta* **2016**, *220*, 252–257. [CrossRef]
84. Mani, V.; Fidal, V.T.; Bowman, K.; Breheny, M.; Chandra, T.S.; Keshavarz, T.; Kyazze, G. Degradation of Azo Dye (Acid Orange 7) in a Microbial Fuel Cell: Comparison Between Anodic Microbial-Mediated Reduction and Cathodic Laccase-Mediated Oxidation. *Front. Energy Res.* **2019**, *7*, 101. [CrossRef]
85. Dai, Q.; Zhang, S.; Liu, H.; Huang, J.; Li, L. Sulfide-Mediated Azo Dye Degradation and Microbial Community Analysis in a Single-Chamber Air Cathode Microbial Fuel Cell. *Bioelectrochemistry* **2020**, *131*, 107349. [CrossRef]
86. Kong, F.; Ren, H.-Y.; Pavlostathis, S.G.; Wang, A.; Nan, J.; Ren, N.-Q. Enhanced Azo Dye Decolorization and Microbial Community Analysis in a Stacked Bioelectrochemical System. *Chem. Eng. J.* **2018**, *354*, 351–362. [CrossRef]
87. Cao, X.; Wang, H.; Zhang, S.; Nishimura, O.; Li, X. Azo Dye Degradation Pathway and Bacterial Community Structure in Biofilm Electrode Reactors. *Chemosphere* **2018**, *208*, 219–225. [CrossRef]
88. Cao, X.; Wang, H.; Li, X.; Fang, Z.; Li, X. Enhanced Degradation of Azo Dye by a Stacked Microbial Fuel Cell-Biofilm Electrode Reactor Coupled System. *Bioresour. Technol.* **2017**, *227*, 273–278. [CrossRef]
89. Fang, Z.; Cheng, S.; Wang, H.; Cao, X.; Li, X. Feasibility Study of Simultaneous Azo Dye Decolorization and Bioelectricity Generation by Microbial Fuel Cell-Coupled Constructed Wetland: Substrate Effects. *RSC Adv.* **2017**, *7*, 16542–16552. [CrossRef]
90. Cui, M.-H.; Cui, D.; Liang, B.; Sangeetha, T.; Wang, A.-J.; Cheng, H.-Y. Decolorization Enhancement by Optimizing Azo Dye Loading Rate in an Anaerobic Reactor. *RSC Adv.* **2016**, *6*, 49995–50001. [CrossRef]
91. Menezes, O.; Brito, R.; Hallwass, F.; Florêncio, L.; Kato, M.T.; Gavazza, S. Coupling Intermittent Micro-Aeration to Anaerobic Digestion Improves Tetra-Azo Dye Direct Black 22 Treatment in Sequencing Batch Reactors. *Chem. Eng. Res. Des.* **2019**, *146*, 369–378. [CrossRef]
92. Assadi, A.; Naderi, M.; Mehrasbi, M.R. Anaerobic–Aerobic Sequencing Batch Reactor Treating Azo Dye Containing Wastewater: Effect of High Nitrate Ions and Salt. *J. Water Reuse Desalination* **2018**, *8*, 251–261. [CrossRef]
93. Kardi, S.N.; Rashid, N.A.A.; Ibrahim, N.; Ahmad, A. Biodegradation of Remazol Black B in Sequential Microaerophilic–Aerobic Operations by NAR-2 Bacterial Consortium. *Environ. Earth Sci.* **2016**, *75*, 1172. [CrossRef]
94. Korenak, J.; Ploder, J.; Trček, J.; Hélix-Nielsen, C.; Petrinic, I. Decolourisations and Biodegradations of Model Azo Dye Solutions Using a Sequence Batch Reactor, Followed by Ultrafiltration. *Int. J. Environ. Sci. Technol.* **2018**, *15*, 483–492. [CrossRef]
95. Wang, Z.; Yin, Q.; Gu, M.; He, K.; Wu, G. Enhanced Azo Dye Reactive Red 2 Degradation in Anaerobic Reactors by Dosing Conductive Material of Ferroferric Oxide. *J. Hazard. Mater.* **2018**, *357*, 226–234. [CrossRef] [PubMed]
96. Sarvajith, M.; Reddy, G.K.K.; Nancharaiah, Y.V. Textile Dye Biodecolourization and Ammonium Removal over Nitrite in Aerobic Granular Sludge Sequencing Batch Reactors. *J. Hazard. Mater.* **2018**, *342*, 536–543. [CrossRef]
97. Al-Hoqani, U.; Zafar, M.; Al Musharafi, S.K.; Mahanty, B.; Behera, S.K. COD Fractionation and Solubility Assessment of Sonicated Waste-activated Sludge. *Environ. Qual. Manag.* **2021**, *1–8*, tqem.21789. [CrossRef]
98. Haddad, M.; Abid, S.; Hamdi, M.; Bouallagui, H. Reduction of Adsorbed Dyes Content in the Discharged Sludge Coming from an Industrial Textile Wastewater Treatment Plant Using Aerobic Activated Sludge Process. *J. Environ. Manag.* **2018**, *223*, 936–946. [CrossRef]

99. Ito, T.; Shimada, Y.; Suto, T. Potential Use of Bacteria Collected from Human Hands for Textile Dye Decolorization. *Water Resour. Ind.* **2018**, *20*, 46–53. [CrossRef]
100. Sarayu, K.; Sandhya, S. Aerobic Biodegradation Pathway for Remazol Orange by *Pseudomonas aeruginosa*. *Appl. Biochem. Biotechnol.* **2010**, *160*, 1241–1253. [CrossRef]
101. Telke, A.A.; Joshi, S.M.; Jadhav, S.U.; Tamboli, D.P.; Govindwar, S.P. Decolorization and Detoxification of Congo Red and Textile Industry Effluent by an Isolated Bacterium *Pseudomonas* sp. SU-EBT. *Biodegradation* **2010**, *21*, 283–296. [CrossRef]
102. Srinivasan, V.; Bhavan, P.S.; Krishnakumar, J. Bioremediation of textile dye effluent by *Bacillus* and *Pseudomonas* spp. *Int. J. Environ. Sci. Technol.* **2014**, *3*, 2215–2224.
103. Bayoumi, M.N.; Al-Wasify, R.S.; Hamed, S.R. Bioremediation of textile wastewater dyes using local bacterial isolates. *Int. J. Curr. Microbiol. App. Sci.* **2014**, *3*, 962–970.
104. Kalathil, S.; Lee, J.; Cho, M.H. Efficient Decolorization of Real Dye Wastewater and Bioelectricity Generation Using a Novel Single Chamber Biocathode-Microbial Fuel Cell. *Bioresour. Technol.* **2012**, *119*, 22–27. [CrossRef]
105. Tara, N.; Arslan, M.; Hussain, Z.; Iqbal, M.; Khan, Q.M.; Afzal, M. On-Site Performance of Floating Treatment Wetland Macrocosms Augmented with Dye-Degrading Bacteria for the Remediation of Textile Industry Wastewater. *J. Clean. Prod.* **2019**, *217*, 541–548. [CrossRef]
106. Hussain, Z.; Arslan, M.; Malik, M.H.; Mohsin, M.; Iqbal, S.; Afzal, M. Treatment of the Textile Industry Effluent in a Pilot-Scale Vertical Flow Constructed Wetland System Augmented with Bacterial Endophytes. *Sci. Total Environ.* **2018**, *645*, 966–973. [CrossRef] [PubMed]
107. Masi, F.; Rizzo, A.; Bresciani, R.; Martinuzzi, N.; Wallace, S.D.; Van Oirschot, D.; Macor, F.; Rossini, T.; Fornaroli, R.; Mezzanotte, V. Lessons Learnt from a Pilot Study on Residual Dye Removal by an Aerated Treatment Wetland. *Sci. Total Environ.* **2019**, *648*, 144–152. [CrossRef]
108. Daniel, D.; Jegathambal, P.; Bevers, B. In Situ Bioremediation of Textile Dye Effluent-Contaminated Soils Using Mixed Microbial Culture. *Int. J. Civ. Eng.* **2019**, *17*, 1527–1536. [CrossRef]
109. Patel, V.R.; Khan, R.; Bhatt, N. Cost-Effective in-Situ Remediation Technologies for Complete Mineralization of Dyes Contaminated Soils. *Chemosphere* **2020**, *243*, 125253. [CrossRef] [PubMed]
110. Tandon, S.A.; Shaikh, S.; Kumar, R. Development and use of soil bacterial consortia for bioremediation of dye polluted soil and municipal waste water. *GJ Biosci. Biotech.* **2014**, *3*, 284–291.
111. Ito, T.; Adachi, Y.; Yamanashi, Y.; Shimada, Y. Long-Term Natural Remediation Process in Textile Dye–Polluted River Sediment Driven by Bacterial Community Changes. *Water Res.* **2016**, *100*, 458–465. [CrossRef] [PubMed]

Article

Combination of Coagulation, Adsorption, and Ultrafiltration Processes for Organic Matter Removal from Peat Water

Muthia Elma [1,2,*], Amalia Enggar Pratiwi [1,2], Aulia Rahma [2], Erdina Lulu Atika Rampun [2], Mahmud Mahmud [3,*], Chairul Abdi [3], Raissa Rosadi [2,4], Dede Heri Yuli Yanto [5] and Muhammad Roil Bilad [6]

1. Department of Chemical Engineering, Faculty of Engineering, Lambung Mangkurat University, Jl. A. Yani KM 36, Banjarbaru 70714, Indonesia; aepratiwi@ulm.ac.id
2. Materials and Membranes Research Group (M²ReG), Lambung Mangkurat University, Jl. A. Yani KM 36, Banjarbaru 70714, Indonesia; arahma@mhs.ulm.ac.id (A.R.); erdinalulu@gmail.com (E.L.A.R.); raissa.rosadi@gmail.com (R.R.)
3. Department of Environment Engineering, Faculty of Engineering, Lambung Mangkurat University, Jl. A. Yani KM 36, Banjarbaru 70714, Indonesia; cabdi@ulm.ac.id
4. Postgraduate Program, Department of Natural Resource and Environmental Management, Lambung Mangkurat University, Jl. A. Yani KM 36, Banjarbaru 70714, Indonesia
5. Research Center for Biomaterials, National Research and Innovation Agency (BRIN), Cibinong Science Center, Cibinong, Jl. Raya Bogor KM 46, Bogor 16911, Indonesia; dede.heri.yuli.yanto@brin.go.id
6. Faculty of Integrated Technologies, Universiti Brunei Darussalam, Gadong BE1410, Brunei; roil.bilad@ubd.edu.bn
* Correspondence: melma@ulm.ac.id (M.E.); mahmud@ulm.ac.id (M.M.)

Abstract: The high content of natural organic matter (NOM) is one of the challenging characteristics of peat water. It is also highly contaminated and contributes to some water-borne diseases. Before being used for potable purposes, peat water must undergo a series of treatments, particularly for NOM removal. This study investigated the effect of coagulation using aluminum sulfate coagulant and adsorption using powdered activated carbon (PAC) as a pretreatment of ultrafiltration (UF) for removal of NOM from actual peat water. After preparation and characterization of polysulfone (Psf)-based membrane, the system's performance was evaluated using actual peat water, particularly on NOM removal and the UF performances. The coagulation and adsorption tests were done under variable dosings. Results show that pretreatment through coagulation–adsorption successfully removed most of the NOM. As such, the UF fouling propensity of the pretreated peat water was substantially lowered. The optimum aluminum sulfate dosing of 175 mg/L as the first pretreatment stage removed up to 75–78% NOM. Further treatment using the PAC-based adsorption process further increased 92–96% NOM removals at an optimum PAC dosing of 120 mg/L. The final UF-PSf treatment reached NOM removals of 95% with high filtration fluxes of up to 92.4 L/(m².h). The combination of three treatment stages showed enhanced UF performance thanks to partial pre-removal of NOM that otherwise might cause severe membrane fouling.

Keywords: coagulation–adsorption; membrane; organic matter; peat water; ultrafiltration; polysulfone

1. Introduction

The supply of high-quality freshwater is a crucial problem in rural areas. In many cases, water resources are of inferior quality (i.e., peat water), making it inconsumable without implementing advanced treatments. Peat water is one of the water sources that are still untapped. It is characteristically acidic (pH 5.9) and high in natural organic matter (NOM), identified using three standard parameters of the non-specific indicator: dissolved organic carbon (DOC, 36.40 mg/L), UV absorbance 254 nm (0.955 1/cm), and organic substances (113.76 mg KMnO₄/L). NOM in peat water may exert odors, aromatization, biological instability, and corrosion of water distribution networks [1]. Conventional water treatments for removal NOM have been widely applied by standalone processes such as coagulation–flocculation and sedimentation [2], activated carbon adsorption [3], and filtration [4]. However, they do not provide optimal treatment for removing NOM.

Previous studies have reported types of water and wastewater treatments that contain high NOM using standalone coagulation, with about 60–70% removal of hydrophobic fraction of NOM and 30–40% of hydrophilic fractions [5–8]. Another work also reported performance of adsorption for NOM removals of up to 98% that were obtained by powdered activated carbon (PAC) with an optimum dosage 500 mg/L [9], which can remove organic materials with a molecular weight (MW) ranging from 0.5–1 to 1–3 kDa. Nevertheless, it could not remove NOM with an MW of <0.5 kDa [10,11]. In addition, the adsorbent may be saturated due to the complete occupation of the adsorption site, while reactivation of the adsorbent results in a complex operation, which may lead performance to decrease [12].

Membrane technology is an advanced treatment process for treating NOM in water, such as wetland or peat water [13–17]. Several studies were reported successful treatment of wetland saline water by pervaporation using silica-based membranes [18–21], wetland saline water by pure silica membrane and organosilica-based membranes [22–28]. Another study showed ultrafiltration (UF) membrane for removal fraction of NOM from peat water [29]. The UF technology is more applicable and better for reducing NOM in water compared with pervaporation. The pervaporation setup is more complex than UF. However, despite NOM's effective removal by the membrane, it is also able to decline membrane performance through membrane fouling [30,31]. Fouling is a major factor that may decrease membrane flux during the separation process, especially the ultrafiltration.

Membrane fouling in peat water treatment is mainly caused by NOM through both the hydrophilic and the hydrophobic fractions [32]. The most common methods to reduce membrane fouling are by altering the physical and chemical properties of the membrane materials by adding additives in the fabrication stage [33–36] or by applying pretreatment of the feed in the operational stages [32,37]. In this study, both coagulation and adsorption were investigated for the first time as a pretreatment of UF for the treatment of real peat water.

This paper reports a preliminary study on NOM removal from peat water by using both aluminum sulfate-based coagulation and PAC-based adsorption as a pretreatment of UF. The polysulfone (PSf) UF membrane was first prepared and characterized. Before being used for the pretreated peat water filtration. The NOM composition in the peat water samples was then characterized. Finally, actual peat water was treated using a series of treatments, namely aluminum sulfate-based coagulation, PAC-based adsorption, and filtration using the developed Psf-UF membrane.

2. Materials and Methods

2.1. Peat Water Characterization

The peat water sample was taken from Banjar Regency, South Kalimantan, Indonesia. Preliminary characterization of peat water included measuring pH using a pH meter (Hanna Hi2211), $KMnO_4$—oxidizable organic substances using the permanganate titrimetric method, and aromatic organic matter absorbance of UV_{254}, and DOC by a total organic carbon analyzer (Shimadzu TOC-L). The permanganate titration method was conducted according to Standard (SNI 06-6989.22-2004). The UV_{254} parameter was measured by a UV visible (UV-1600 Spectrophotometer). On the other hand, DOC was analyzed by high-temperature catalytic oxidation with non-dispersive infrared (NIDR) detection. As a pretreatment, the samples were filtered using Whatman 0.45 μm before being tested by TOC analyzer. Meanwhile, specific UV absorbance ($SUVA_{254}$, L/mg.m) was used to represent TOC normalized aromatic moieties (UV_{254}). Meanwhile, specific UV absorbance ($SUVA_{254}$, L/mg.m) was used to represent TOC normalized aromatic moieties (UV_{254}) by dividing of the UV_{254} with the DOC value.

2.2. Membrane Preparation and Characterization

The dope solution for Psf UF membrane preparation was made using 18 wt.% of Psf (Merck) as the polymer, 64 wt.% dimethylacetamide (DMAc, Merck) as the solvent, and polyethylene glycol (PEG, Merck) PEG 600 as the additive (18 wt.%). According to earlier reports, the membranes were prepared using the phase inversion method [35]. The polymer, solvent, and additive were mixed and stirred until homogeneous. Then, the solution was left idle overnight to release the entrapped bubbles. Subsequently, the dope solution was cast on a glass plate at a wet casting thickness of 165 μm using a casting applicator. The phase inversion was then continued by immersing the cast film into a coagulation bath containing nonsolvent solution comprising of DMAc 35 wt.% and KCl 0.5 wt.% in water.

The hydraulic resistance of the prepared membrane was characterized by measuring the clean water permeability, and a scanning electron microscopy (SEM) was used to determine the membrane PSf morphology and membrane thickness. The pore size of the membranes was determined using image-J software from the surface of the SEM image [38]. The permeability test was conducted by flowing the distilled water on a dead-end system filtration device. The permeate volume was then measured every 5 min intervals for 60 min operation time under different pressures of 1, 1.5, 2, 2.5, and 3 bar.

2.3. Coagulation, Adsorption, and Ultrafiltration

The coagulation tests were done by varying doses of aluminum sulfate (one of the most common coagulants) in a range of 125–250 mg/L using the Jar-test method at adjusting pH 6 (regulated by drop-wise adding 0.1 M NaOH (Merck)) with a working volume of 1.2 L. During the jar test, the coagulant mixture in peat water was stirred at 100 rpm for 1 min, followed by slow stirring at 40 rpm for 20 min and sedimentation for 20 min, according to a protocol reported earlier [39]. The range of the coagulant dosage was defined based on a previous study [31].

The PAC adsorption tests were done using the Jar-test under varying 20–200 mg/L doses. It was carried out for the pretreated peat water through coagulation/flocculation. The feed and PAC (particle size of 100 mesh; surface area of 800 m^2/g, Merck) were mixed with a rotary shaker at 180 rpm for 3 h.

After coagulation and adsorption, the treated supernatant underwent a UF—200 mL of the supernatant was filtered using the developed UF PSf membrane by using a standard dead-end filtration cell (Figure 1) according to a protocol detailed elsewhere [30]. The filtrations were done at variable pressures of 1, 1.5, 2, 2.5, 3 bar for 60 min at room temperature of feed (25 °C), stirred at 50 rpm. A gas compressor generated the pressure, and the permeate volume was collected every 5 min. All of the coagulatin, adsorption, and filtration experiments were done in triplicate.

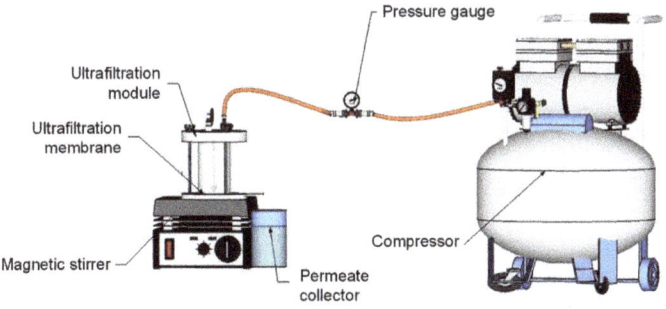

Figure 1. Illustration of the ultrafiltration dead-end experimental setup.

3. Results

3.1. Peat Water Characteristics

The characterizations of peat water were carried out for four periods to monitor the changes of NOM content of the peat water samples, as summarized in Table 1. It shows that the peat water had a neutral pH, similar to previous reports where the pH value on surface water ranged from 5.0–8.1 [31]. The high NOM content was indicated by the DOC values [39,40], the absorbance value of UV_{254}, which are high compared with the results obtained by Kang and Choo [41] and Jeong et al. [42] of $UV_{254} < 0.1$ cm^{-1} for surface water. However, compared with the results obtained elsewhere Herwati, Mahmud, and Abdi [30], Mahmud, Abdi, and Mu'min [31], Saputra [37], Aisyahwalsiah [39] showed the UV_{254} absorbance of the peat water samples was relatively low. Similar to the UV_{254} absorbance value, the $SUVA_{254}$ values of the peat water sample deviated from others (Zularisam et al. [43]). Based on their reports, the $SUVA_{254}$ characteristic of peat water contained a high hydrophobic fraction. The $SUVA_{254}$ values of the peat water sample in this study ranged at 2–3 L/mg.m suggesting the mixture of hydrophobic and hydrophilic substances NOM characteristics, with a large range of MWs. Similar results were found in previous research on surface water that reported $SUVA_{254}$ of < 2 L/mg.m (low hydrophobic character) [41,42].

Table 1. Characteristics of the peat water sample.

No	Parameter	Units	Week I	Week II	Week III	Week IV	Average	STDEV
1	pH		6.3	6.3	6.3	6.3	6.3	0
2	DOC (dissolved organic carbon)	mg/L	36.40	-	-	36.40	36.4	-
3	UV_{254} absorbance	1/cm	0.968	1.005	0.977	0.955	0.976	0.02
4	$KMnO_4$ organic substances	mg $KMnO_4$/L	120.08	126.4	120.08	113.76	120.08	5.16
5	$SUVA_{254}$	L/mg.m	2.659	2.761	2.684	2.624	2.682	0.006

3.2. Characterisation of UF-PSf Membranes

The surface and cross-section SEM images of the prepared membrane are shown in Figure 2. The UF PSf membrane had a tight pore arrangement and sponge-like cross-section morphology without macrovoids (large cavities), which was similar to the membrane structure reported earlier [35]. Based on image-J surface SEM image processing (Figure 3), the surface pore size of the membrane was 0.061 μm, falling under UF range of 0.001–0.1 μm [44].

The polysulfone membrane used in this study had a pore size of less than 0.1 μm (Figure 3). However, from the SEM image it is not possible to determine exactly what the pore size is. However, the pore size distribution can determine by utilized Image-J software following the previous research used digital SEM image data. The result of processing SEM images by Image-J can be seen in Figure 3B,C. The results show the average pore diameter of the membrane is 0.061 μm. Based on literature, the polysulfone membrane used in this work can be categorized as well as ultrafiltration membrane.

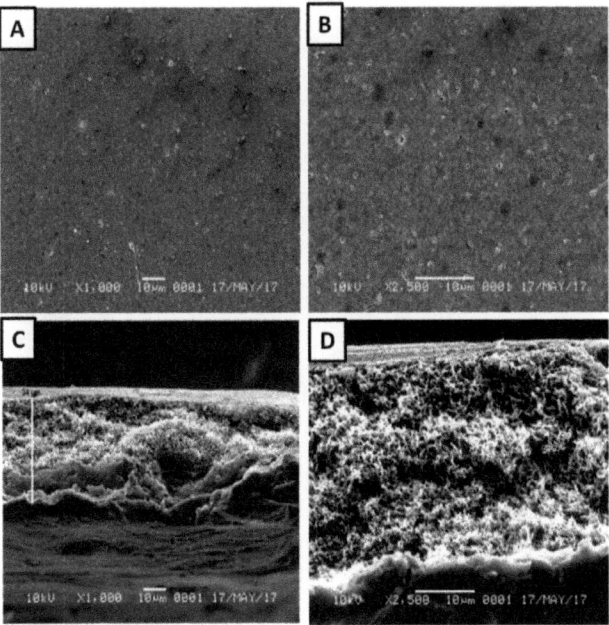

Figure 2. Ultrafiltration polysulfone membrane surface microstructure with (**A**) magnification of 1000× and (**B**) magnification 2500×, and cross-section microstructure with (**C**) magnification of 1000× and (**D**) magnification 2500×.

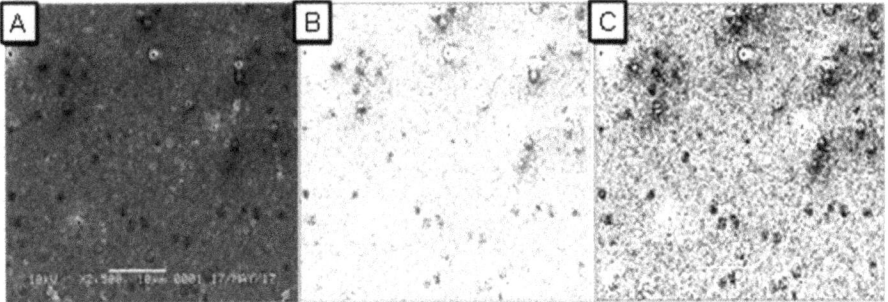

Figure 3. Membrane surface SEM image (2500×) processed with Image-J for pore size determination (**A**) before editing, (**B**) after threshold and (**C**) outline image result.

3.3. Coagulation-Flocculation

Figure 4 shows that for aluminum sulfate dosings of 125–250 mg/L, the removal efficiency of NOM increased from 125 mg/L to 175 mg/L and then a slight decrease until 250 mg/L. Beyond that value, the NOM removal efficiency decreased. The loading restabilization can explain the pattern on the NOM removals as a function of the dosing rate on the addition of $Al_2(SO_4)_3$ coagulant [37].

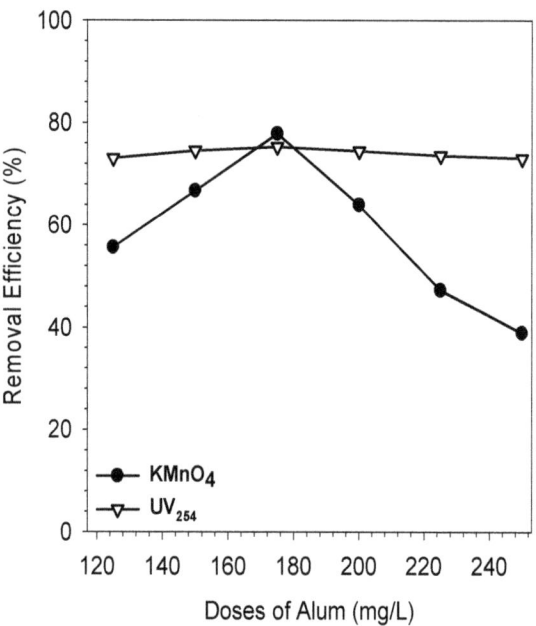

Figure 4. NOM removal rate represented by oxidation with KmnO$_4$ and UV$_{254}$ absorbances as a function of doses of alum in the coagulation–flocculation test.

The oxidation of organic substances by KMnO$_4$ peaked at the optimum dose of 175 mg/L. Meanwhile, a slight decrease in UV$_{254}$ removals was observed. In the process of coagulation–flocculation, the dominant fraction of removed NOM is the one that hydrophobic or with large MWs as detailed elsewhere [31,45]. In addition, according to Suslova et al. [46] that KmnO$_4$ can oxidize various types of organic components irrespective of the MWs.

The NOM removal in the coagulation-flocculation process was achieved optimum at a dose of 175 mg/L corresponding to organic substances KmnO$_4$ and UV$_{254}$ absorbance of 77.78% and 75.24%, respectively. The removal of NOM obtained in this study was higher than the previous studies [47,48]. After adding the coagulant, the coagulation rate decreased, as well as the pH value from 6 to 3.65. It was caused by the reaction of aluminum sulfate with water that produces H$^+$ ions. The acidification of water lowered the coagulation/flocculation efficiencies.

3.4. Coagulation-Adsorption

The PAC adsorption was carried out after the coagulation/flocculation of the raw peat water sample. Figure 5 shows the rate of NOM removal in the PAC adsorption process. The NOM removal rate was higher than the standalone coagulation-flocculation or adsorption processes reported earlier [32,49]. Increased NOM removal at low pH during the PAC adsorption can be attributed due to the low pH of the solution due to the preceding coagulation/flocculation stage [39]. The pH has a significant effect on activated carbon adsorption and the removal efficiency is higher in acidic than in neutral and alkaline conditions. The presence of H$^+$ ions in solution leads to competition between H$^+$ ions and NOM bonding [50].

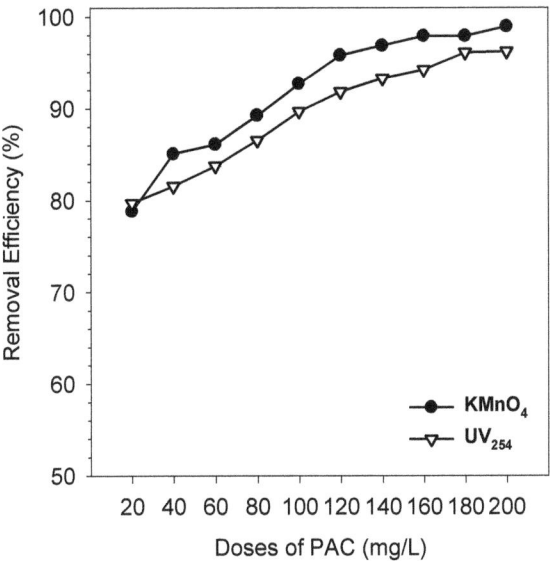

Figure 5. The NOM removal rate represented by organic substances of $KMnO_4$ and UV_{254} absorbances as function of PAC dosages.

The removal rate of $KmnO_4$—oxidizable organic substances in the coagulation–adsorption process is higher than the UV_{254} (representing aromatic moieties). The results were the opposite of the coagulation/flocculation process, which could be attributed to the NOM content with large MW removed in the coagulation–flocculation process. As such, only the low MW NOM remained in the PAC adsorption. In previous research, the adsorption process could remove NOM hydrophilic fractions with small MWs [9–11]. The removal rate of $KMnO_4$ organic substances was higher than the UV_{254} absorbance. These results indicated that $KmnO_4$ organic substances easily oxidized NOM due to their small MWs as stated elsewhere [51].

The best PAC dose was 120 mg/L of PAC, judging from the highest removal of KmnO4 and UV^{254} parameters of >90%. In addition, the efficiency of NOM removal slightly increased up to 120 mg/L. Due to the NOM removal of 120 to 200 mg/L being relatively similar, the 120 mg/L of PAC was chosen because it does not need too much PAC. The removal efficiencies of $KMnO_4$ organic substances and UV_{254} under the optimum dosing were 95.83 and 91.83%, respectively, as shown in Figure 5. The results obtained in this study were in line with others, e.g., Lee et al. [32] and Aisyahwalsiah [39] using PAC as the adsorbant. Nonetheless, higher NOM removals were obtained in this study. After adding PAC, the pH increased as reported by others [49] due to the soluble ash, which is rinsed out of the media during use, and the effluent pH will eventually approach neutral.

3.5. Coagulation-Adsorption-Membrane Experiments

The permeability of pure water (aquadest), pretreatment feed, and non-pretreated feed are shown in Figure 6. Based on the results, pure water permeability was obtained of 38–180 $L/h.m^2$ by the prepared UF polysulfone membrane. This result exhibits the higher water flux of pure water permeability by low transmembrane pressure compared to commercial polysulfone membrane (Merck) of 150–350 $L/h.m^2$ (6–20 bar), which was reported by Adams et al. [52].

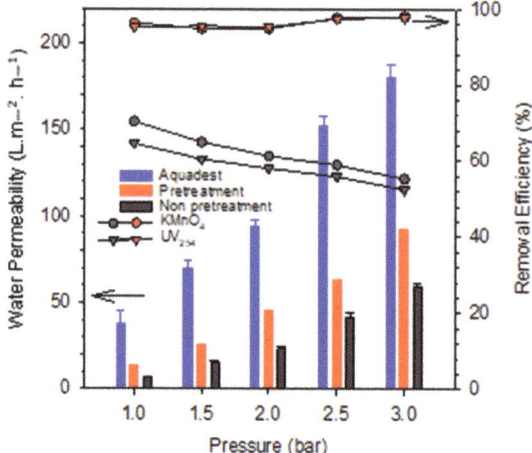

Figure 6. The performance of ultrafiltration of pretreated peat water at different pressures. Aquadest denotes distilled water and represents the permeability of clean water.

Figure 6 shows that the NOM removal efficiency decreased by increasing transmembrane pressure. The results obtained are by previous research, in which the magnitude of NOM rejection is inversely proportional to the applied pressure [53,54]. The deformation of the membrane most probably causes it due to high pressure, which causes membrane compaction that constricts the pore size and the thicker foulant layer that became the secondary filter on top of the PSf membrane.

The NOM removal efficiency reflected from the $KMnO_4$ organic substances was higher than the UV_{254} absorbance obtained in the adsorption process with PAC. The rejection of NOM by membrane was determined by the adequate pore size [53] and an additional dynamic layer formed on the membrane surface.

In addition to the NOM removal rate, water flux value was also an indicator of the optimum pressure. Figure 6 shows that the water flux value was directly proportional to the pressure. The smallest water flux value was obtained at 1 bar of 13.3 L/h.m², and the highest water flux was at 3 bar of 92.5 L/h.m². The permeability of each pressure to percent removal of NOM for $KMnO_4$ organic substances and UV_{254} parameters were determined to determine the optimum pressure. The highest water flux was obtained with the removal rate of $KMnO_4$ and UV_{254} of 94.79 and 94.66%, respectively. The UV_{254} rejection of the polysulfone membrane in this work is extremely high over commercial PSf membrane that was only able to remove about 41% of NOM at 6 bar [52].

The water permeability of treated peat water was smaller than the clean water permeability (Figure 6) due to membrane fouling. However, it was higher by almost two-fold than peat water permeability without pretreatment, which was also similar reported in earlier studies [31,41,54]. It was shown that the pretreatment contributed substantially to reducing the membrane fouling [53].

The permeability decrease in the pretreated peat water filtration can be attributed to the fouling by the residual NOM that escaped from the pretreatment. However, previous works by Kang and Choo [41] and Zhang, et al. [55] ascribed the small water permeability to the use of PAC. The bonding of NOM with PAC particles caused the PAC-NOM particles to become an additional foulant that blocks the membrane pores or forms a cake layer on the membrane surface. In this study, the PAC was separated. Hence the foulant was originated from residual NOM in the feed.

Overall findings suggested that the application of coagulation–adsorption pretreatment of UF is promising to reduce the fouling potential on the feed as indicated by increasing the water permeability value and the removal rate of NOM represented by DOC, $KMnO_4$ organic substance, and absorbance UV_{254}.

In addition, the results obtained were also reinforced with SEM UF-PSF membrane image after treatment. The SEM images in Figure 7 also show the thickness of the UF-PSf membrane (determined from the cross-section SEM image) after being compressed at 3 bars was 85.4 μm. The pore structure of the membrane after passing the feed water was approximately the same as the pristine membrane. It could be seen on the surface of the membrane there is only a thin layer which is thought to be a cake layer. The additional cake layer helped enhance the rejection of NOM and the final quality of the permeate. It is worth noting that a significant difference in thickness was seen from data in Figures 2C and 7B. The high variability was originated from the cutting process.

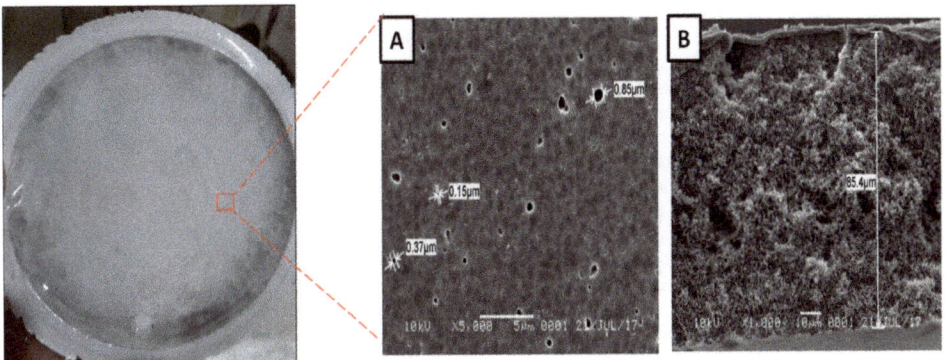

Figure 7. SEM image of used ultrafiltration polysulfone membrane after ultrafiltration process: (**A**) surface section and (**B**) cross-sectional.

4. Conclusions

This study demonstrated the advantages of combining the coagulation–adsorption process and membrane filtration to treat fouling-prone actual peat water. The coagulation-adsorption showed a positive effect as a pretreatment for the ultrafiltration. The pretreated feed showed a lower membrane fouling propensity. The optimum coagulation/flocculation and adsorption condition was at $Al_2(SO_4)_3$ dosing of 175 mg/L and PAC dosing of 120 mg/L, respectively. Higher filtration pressure enhanced the peat water permeability. The optimum pressure on the hybrid process was 3 bar with a permeability value of 92.5 L/m^2.h and an organic removal rate of 95%. The findings highlight the importance of the hybrid system for treating challenging feeds that otherwise proven difficult when applying a standalone system. Moreover, long-term studies are still required to accurately gauge the performance of the hybrid system for treatment of peat water.

Author Contributions: Conceptualization, M.E. and M.M.; methodology, A.E.P.; software, A.E.P.; validation, A.E.P., M.E. and M.M.; formal analysis, A.R.; investigation, A.E.P.; resources, E.L.A.R.; data curation, A.E.P.; writing—original draft preparation, A.E.P.; writing—review and editing, M.E.; visualization, A.E.P.; supervision, M.M., C.A., M.E. and M.R.B.; project administration, A.E.P., R.R., D.H.Y.Y.; funding acquisition, M.E. All authors have read and agreed to the published version of the manuscript.

Funding: This research received no external funding.

Acknowledgments: The authors thank the Engineering Faculty and Materials and Membranes Research Group (M^2ReG), Lambung Mangkurat University for the facilities. Muthia thanks the Applied Research of Universities Grant 2021–2023, Basic Research Grant 2021–2022, and World Class Research Grant 2021–2023 Directorate General of Higher Education, Ministry of Education, Culture, Research, and Technology, Republic of Indonesia.

Conflicts of Interest: The authors declare no conflict of interest.

References

1. Tang, C.; He, Z.; Zhao, F.; Liang, X.; Li, Z. Effects of Cations on The Formation of Ultrafiltration Membrane Fouling Layers When filtering Fulvic Acid. *Desalination* **2014**, *352*, 174–180. [CrossRef]
2. Matilainen, A.; Vepsäläinen, M.; Sillanpää, M. Natural organic matter removal by coagulation during drinking water treatment: A review. *Adv. Colloid Interface Sci.* **2010**, *159*, 189–197. [CrossRef] [PubMed]
3. Menya, E.; Olupot, P.; Storz, H.; Lubwama, M.; Kiros, Y. Production and performance of activated carbon from rice husks for removal of natural organic matter from water: A review. *Chem. Eng. Res. Des.* **2018**, *129*, 271–296. [CrossRef]
4. McMeen, C.R.; Benjamin, M.M. NOM removal by slow sand filtration through iron oxide–coated olivine. *J. Am. Water Work. Assoc.* **1997**, *89*, 57–71. [CrossRef]
5. Zhao, Y.X.; Phuntsho, S.; Gao, B.Y.; Yang, Y.Z.; Kim, J.-H.; Shon, H.K. Comparison of A Novel Polytitanium Chloride Coagulant With Polyaluminium Chloride: Coagulation Performance And Floc Characteristics. *J. Environ. Manag.* **2015**, *147*, 194–202. [CrossRef] [PubMed]
6. Dong, B.-Z.; Chen, Y.; Gao, N.-Y.; Fan, J.-C. Effect of Coagulation Pretreatment on The Fouling of Ultrafiltration Membrane. *J. Environ. Sci.* **2007**, *19*, 278–283. [CrossRef]
7. Rahma, A.; Elma, M.; Mahmud, M.; Irawan, C.; Pratiwi, A.E.; Rampun, E.L.A. Removal of natural organic matter for wetland saline water desalination by coagulation-pervaporation. *J. Kim. Sains Dan Apl.* **2019**, *22*, 85–92. [CrossRef]
8. Elma, M.; Rahma, A.; Pratiwi, A.E.; Rampun, E.L.A. Coagulation as pretreatment for membrane-based wetland saline water desalination. *Asia Pac. J. Chem. Eng.* **2020**, *15*, e2461. [CrossRef]
9. Cui, X.; Choo, K.-H. Reviewer Paper: Natural Organic Matter Removal and Fouling Control in Low-Pressure Membrane Filtration for Water Treatment. *Environ. Eng.* **2014**, *19*, 1–8.
10. Song, Y.; Dong, B.; Gao, N.; Ma, X. Powder Activated Carbon Pretreatment of A Microfiltration Membrane for The Treatment of Surface Water. *Environ. Res. Public Health* **2015**, *12*, 11269–11277. [CrossRef]
11. Liu, S.; Lim, M.; Amal, R. TiO_2-Coated Natural Zeolite: Rapid Humic Acid Adsorption And Effective Photocatalytic Regeneration. *Chem. Eng. Technol.* **2014**, *105*, 46–52. [CrossRef]
12. Gamage, S.M.K.; Sathasivan, A. A Review: Potential And Challenges of Biologically Activated Carbon to Remove Natural Organic Matter in Drinking Water Purification Process. *Chemosphere* **2017**, *167*, 120–138. [CrossRef] [PubMed]
13. Rahma, A.; Elma, M.; Rampun, E.L.A.; Pratiwi, A.E.; Rakhman, A.; Fitriani. Rapid Thermal Processing and Long Term Stability of Interlayer-free Silica-P123 Membranes for Wetland Saline Water Desalination. *Adv. Res. Fluid Mech. Therm. Sci.* **2020**, *71*, 1–9. [CrossRef]
14. Rahma, A.; Elma, M.; Pratiwi, A.E.; Rampun, E.L. Performance of interlayer-free pectin template silica membranes for brackish water desalination. *Membr. Technol.* **2020**, *2020*, 7–11. [CrossRef]
15. Lestari, R.A.; Elma, M.; Rahma, A.; Suparsih, D.; Anadhliyah, S.; Sari, N.L.; Pratomo, D.A.; Sumardi, A.; Lestari, A.E.; Assyaifi, Z.L.; et al. Organo Silica Membranes for Wetland Saline Water Desalination: Effect of membranes calcination temperatures. *E3S Web Conf.* **2020**, *148*, 07006. [CrossRef]
16. Elma, M.; Rampun, E.L.A.; Rahma, A.; Assyaifi, Z.L.; Sumardi, A.; Lestari, A.E.; Saputro, G.S.; Bilad, M.R.; Darmawan, A. Carbon templated strategies of mesoporous silica applied for water desalination: A review. *J. Water Process Eng.* **2020**, *38*, 101520. [CrossRef]
17. Elma, M.; Mujiyanti, D.R.; Ismail, N.M.; Bilad, M.R.; Rahma, A.; Rahman, S.K.; Rakhman, A.; Rampun, E.L.A. Development of Hybrid and Templated Silica-P123 Membranes for Brackish Water Desalination. *Polymers* **2020**, *12*, 2644. [CrossRef]
18. Elma, M.; Riskawati, N.; Marhamah. Silica Membranes for Wetland Saline Water Desalination: Performance and Long Term Stability. *IOP Conf. Ser. Earth Environ. Sci.* **2018**, *175*, 012006. [CrossRef]
19. Elma, M.; Hairullah; Assyaifi, Z.L. Desalination Process via Pervaporation of Wetland Saline Water. *IOP Conf. Ser. Earth Environ. Sci.* **2018**, *175*, 012009. [CrossRef]
20. Elma, M.; Setyawan, H.; Rahma, A.; Pratiwi, A.; Rampun, E.L.A. Fabrication of Interlayer-free P123 Caronised Template Silica Membranes for Water Desalination: Conventional Versus Rapid Thermal Processing (CTP vs RTP) Techniques. *IOP Conf. Ser. Mater. Sci. Eng.* **2019**, *543*, 012076. [CrossRef]
21. Elma, M.; Setyawan, H. Synthesis of Silica Xerogels Obtained in Organic Catalyst via Sol Gel Route. *IOP Conf. Ser. Earth Environ. Sci.* **2018**, *175*, 012008. [CrossRef]
22. Rahman, S.K.; Maimunawaro; Rahma, A.; Syauqiah, I.; Elma, M. Functionalization of hybrid organosilica based membranes for water desalination—Preparation using Ethyl Silicate 40 and P123. *Mater. Today: Proc.* **2020**, *31*, 60–64. [CrossRef]
23. Mawaddah, Y.; Wati, L.S.; Rampun, E.L.A.; Sumardi, A.; Lestari, A.E.; Elma, M. FTIR Studies of the TEOS/TEVS xerogel structure using rapid thermal processing method. *Konversi* **2020**, *9*, 73–78. [CrossRef]
24. Maimunawaro; Rahman, S.K.; Rampun, E.L.A.; Rahma, A.; Elma, M. Deconvolution of carbon silica templated thin film using ES40 and P123 via rapid thermal processing method. *Mater. Today Proc.* **2020**. [CrossRef]

25. Lestari, A.E.; Elma, M.; Rabiah, S.; Rampun, E.L.A.; Rahma, A.; Pratiwi, A.E. Performance of Mesoporous Organo Silica Membrane for Desalination. In *Proceedings of Materials Science Forum*; Trans Tech Publications Ltd.: Bäch, Switzerland, 2020; pp. 285–292.
26. Elma, M.; Rezki, M.R.; Mahmud, M.; Sunardi, S.; Pratiwi, E.N.; Oktaviana, E.N.; Fatimah, S.; Rahma, A. Membran karbon templated silika dari karbon nipah (*Nypa fruticans*) untuk aplikasi desalinasi air rawa asin [Carbon templated silica membranes from nypa carbon (*Nypa fruticans*) applied for wetland saline water desalination]. *J. Ris. Ind. Has. Hutan* **2020**, *12*, 83–92. [CrossRef]
27. Elma, M.; Pratiwi, A.E.; Rahma, A.; Rampun, E.L.A.; Handayani, N. The Performance of Membranes Interlayer-Free Silica-Pectin Templated for Seawater Desalination via Pervaporation Operated at High Temperature of Feed Solution. *Mater. Sci. Forum* **2020**, *981*, 349–355. [CrossRef]
28. Elma, M.; Fitriani; Rakhman, A.; Hidayati, R. Silica P123 Membranes for Desalination of Wetland Saline Water in South Kalimantan. *IOP Conf. Ser. Earth Environ. Sci.* **2018**, *175*, 012007. [CrossRef]
29. Mahmud; Elma, M.; Rampun, E.L.A.; Rahma, A.; Pratiwi, A.E.; Abdi, C.; Rosadi, R. Effect of Two Stages Adsorption as Pre-Treatment of Natural Organic Matter Removal in Ultrafiltration Process for Peat Water Treatment. *Mater. Sci. Forum* **2020**, *988*, 114–121. [CrossRef]
30. Herwati, N.; Mahmud; Abdi, C. Pengaruh pH Air Gambut Terhadap Fouling Membran Ultrafiltrasi. *Jukung J. Tek. Lingkung* **2015**, *1*, 59–73.
31. Mahmud; Abdi, C.; Mu'min, B. Removal Natural Organic Matter (NOM) in Peat Water from Wetland Area by Coagulation-Ultrafiltration Hybrid Process with Pretreatment Two-Stage Coagulation. *J. Wetl. Environ. Manag.* **2013**, *1*, 42–49. [CrossRef]
32. Lee, J.; Vigneswaran, S.; Zhang, Y.; Raj Reddy, R.S.; Liu, Z. Effective Natural Organic Matter Removal In Pond Water by Carbon Nanotube Membrane with Flocculation/Adsorption. *Water Sci. Technol. Water Supply* **2017**, *17*, 1080–1087. [CrossRef]
33. Aryanti, P.T.P.; Joscarita, S.R.; Wardani, A.K.; Subagjo; Ariono, D.; Wenten, I.G. The Influence of PEG 400 and Acetone on Polysulfone Membrane Morphology and Fouling Behaviour. *J. Eng. Technol. Sci.* **2016**, *48*, 135–149. [CrossRef]
34. Wardani, A.K. *Effect of Additives on Polysulfone Based Ultrafiltration Membrane For Peat Water Purification*; Institut Teknologi Bandung: Bandung, Indonesia, 2013; pp. 1–37.
35. Tutriyanti. Synthesis and Characterization of Polysulfone Membranes by Phase Inversion Technique: Effect of Concentration of Impregnant KCl on Coagulant Against Pore Structure of Membranes. Bachelor's Thesis, Universitas Lambung Mangkurat, Banjarbaru, Indonesia, 2017.
36. Ariono, D.; Aryanti, P.T.P.; Subagjo, S.; Wenten, I.G. The Effect of Polymer Concentration on Flux Stability of Polysulfone Membrane. In Proceedings of the International Conference on Engineering Science and Nanotechnology (ICESNANO 2016), Solo, Indonesia, 3–5 August 2016; American Institute of Physics: College Park, MD, USA, 2017; pp. 1–10.
37. Saputra, A.A. Hybrid Coagulation-Ultrafiltration Process in Removing Natural Organic Matter (NOM) in Peat Water: The Effect of Variations in Coagulant Doses on Membrane Fouling. Bachelor's Thesis, Universitas Lambung Mangkurat, Banjarbaru, Indonesia, 2014.
38. Kurniawan, C.; Waluyo, T.B.; Sebayang, P. Particle Size Analysis Using Free Software Image-J. In *Seminar Nasional Fisika*; Pusat Penelitian Fisika-LIPI: Jakarta, Indonesia, 2011; pp. 1–8.
39. Aisyahwalsiah, A. Optimization of Peat Water Treatment Using a Combined Process of Coagulation with Peaty Clay and Activated Carbon Adsorption. Bachelor's Thesis, Universitas Lambung Mangkurat, Banjarbaru, Indonesia, 2013.
40. Mahmud; Notodarmojo, S.; Padmi, T.; Soewondo, P. Adsorpsi Bahan Organik Alami (BOA) Air Gambut Pada Tanah Lempung Gambut Alami dan Teraktivasi: Studi Kesetimbangan Isoterm dan Kinetika Adsorpsi. *INFO-TEKNIK* **2012**, *13*, 28–37.
41. Kang, S.K.; Choo, K.H. Why Does A Mineral Oxide Adsorbent Control Fouling Better than Powdered Activated Carbon in Hybrid Ultrafiltration Powder Treatment? *J. Membr. Sci.* **2010**, *355*, 69–77. [CrossRef]
42. Jeong, K.; Kim, D.G.; Ko, S.O. Adsorption Characteristics of Effluent Organic Matter and Natural Organic Matter by Carbon Based Nanomaterials. *KSCE J. Civ. Eng.* **2016**, *21*, 119–126. [CrossRef]
43. Zularisam, A.W.; Ismail, A.F.; Salim, M.R.; Sakinah, M.; Matsuura, T. Application of Coagulation–Ultrafiltration Hybrid Process for Drinking Water Treatment: Optimization of Operating Conditions Using Experimental Design. *Sep. Purification Technol.* **2009**, *65*, 193–210. [CrossRef]
44. Mulder, M. *Basic Principles of Membrane Technology*; Kluwer Academic Publishers: Dordrecht, The Netherland, 1996.
45. Chih, Y.C.; Chung, Y.W.; Ying, C.C. The Coagulation Characteristics of Humic Acid By Using Acid-Soluble Chitosan, Water-Soluble Chitosan, And Chitosan Coagulant Mixtures. *Environ. Technol.* **2014**, *36*, 1141–1146.
46. Suslova, .; Govorukha, V.; Brovarskaya, .; Matveeva, N.; Tashyreva, H.; Tashyrev, O. Method for Determining Organic Compound Concentration in Biological Systems by Permanganate Redox Titration. *Int. J. Bioautomation* **2014**, *18*, 45–52.
47. Rahman, R.A. Hybrid Coagulation-Ultrafiltration Process in Removal Natural Organic Matter (NOM) in Peat Water: Effect of Coagulants Type and Stirring Speed on Membrane Fouling. Bachelor's Thesis, Universitas Lambung Mangkurat, Banjarbaru, Indonesia, 2014.
48. Fachrozi, M. Ultrafiltration Coagulation Hybrid Process in Removal Natural Organic Matter (NOM) in Peat Water: Effect of Coagulants Type on Membrane Fouling. Bachelor's Thesis, Universitas Lambung Mangkurat, Banjarbaru, Indonesia, 2013.
49. Tomaszewska, M.; Mozia, S.Y.; Morawski, A.W. Removal of Organic Matter by Coagulation Enhanced with Adsorption on PAC. *Desalination* **2004**, *161*, 79–87. [CrossRef]

50. Abdillah, A.I.; Darjito; Khunur, M.M. Pengaruh pH dan Waktu Kontak pada Adsorpsi Ion Logam Cd^{2+} Menggunakan Adsorben Kitin Terikat Silang Glutaraldehid. *Kim. Stud. J.* **2015**, *1*, 826–832.
51. *Oxidation of Organic Molecules by $KMnO_4$*; National Science Foundation University of California, LibreTexts Library: Sacramento, CA, USA, 2015.
52. Adams, F.V.; Nxumalo, E.N.; Krause, R.W.M.; Hoek, E.M.V.; Mamba, B.B. Application of polysulfone/cyclodextrin mixed-matrix membranes in the removal of natural organic matter from water. *Phys. Chem. Earth Parts A/B/C* **2014**, *67–69*, 71–78. [CrossRef]
53. Notodarmojo, S.; Deniva, A. Penurunan Zat Organik dan Kekeruhan Menggunakan Teknologi Membran Ultrafiltrasi dengan Sistem Aliran Dead-End (Studi Kasus: Waduk Saguling, Padalarang). *Proc. ITB Sains Technol. A* **2004**, *36*, 63–82. [CrossRef]
54. Aryanti, P.T.P.; Khoiruddin; Wenten, I.G. Influence of Additives on Polysulfone-Based Ultrafiltration Membrane Performance during Peat Water Filtration. *J. Water Sustain.* **2013**, *3*, 85–96. [CrossRef]
55. Zhang, M.; Li, C.; Benjamin, M.M.; Chang, Y. Fouling and Natural Organic Matter Removal in Adsorbent/Membrane Systems for Drinking Water Treatment. *Environ. Sci. Technol.* **2003**, *37*, 1663–1669. [CrossRef] [PubMed]

Article

Post-Treatment of Palm Oil Mill Effluent Using Immobilised Green Microalgae *Chlorococcum oleofaciens*

Kah Aik Tan [1], Japareng Lalung [1,*], Norhashimah Morad [1], Norli Ismail [1], Wan Maznah Wan Omar [2], Moonis Ali Khan [3], Mika Sillanpää [4] and Mohd Rafatullah [1,*]

Citation: Tan, K.A.; Lalung, J.; Morad, N.; Ismail, N.; Wan Omar, W.M.; Khan, M.A.; Sillanpää, M.; Rafatullah, M. Post-Treatment of Palm Oil Mill Effluent Using Immobilised Green Microalgae *Chlorococcum oleofaciens*. *Sustainability* 2021, 13, 11562. https://doi.org/10.3390/su132111562

Academic Editor: Antonio Zuorro

Received: 5 August 2021
Accepted: 18 October 2021
Published: 20 October 2021

Publisher's Note: MDPI stays neutral with regard to jurisdictional claims in published maps and institutional affiliations.

Copyright: © 2021 by the authors. Licensee MDPI, Basel, Switzerland. This article is an open access article distributed under the terms and conditions of the Creative Commons Attribution (CC BY) license (https://creativecommons.org/licenses/by/4.0/).

[1] School of Industrial Technology, Universiti Sains Malaysia, Gelugor 11800, Malaysia; kangxihuangti@hotmail.com (K.A.T.); nhashim@usm.my (N.M.); norlii@usm.my (N.I.)
[2] School of Biological Sciences, Universiti Sains Malaysia, Gelugor 11800, Malaysia; wmaznah@usm.my
[3] Chemistry Department, College of Science, King Saud University, Riyadh 11451, Saudi Arabia; mokhan@ksu.edu.sa
[4] Department of Biological and Chemical Engineering, Aarhus University, Nørrebrogade 44, 8000 Aarhus, Denmark; mikaesillanpaa@gmail.com
* Correspondence: japareng@usm.my (J.L.); mrafatullah@usm.my (M.R.)

Abstract: Microalgae immobilisation can be a long-term solution for effective wastewater post-treatment. This study was conducted to evaluate the ability of immobilised *Chlorococcum oleofaciens* to remove contaminants from palm oil mill effluent (POME) until it complies with the POME discharge standard. First, the native dominating green microalga was isolated from a polishing POME treatment pond. Then, the microalgae cells were immobilised on sodium alginate beads and cultivated in a lab-scale-treated POME to treat it further. The immobilised microalgae cells demonstrated a high removal of total phosphorus, total nitrogen, ammonia nitrogen, and soluble chemical oxygen demand with 90.43%, 93.51%, 91.26%, and 50.72% of reduction, respectively. Furthermore, the growth rate of the microalgae fitted nicely with the Verhulst logistical model with r^2 of more than 0.99, indicating the model's suitability in modelling the growth. Thus, we concluded that the species can be used for post-treatment of effluents to remove TP, TN, and ammonia nitrogen from palm oil mills until it complies with the POME effluent discharge standard. However, during the process, degradation of the beads occurred and the COD value increased. Therefore, it is not suitable to be used for COD removal.

Keywords: alginate beads; *C. oleofaciens*; immobilisation; green microalgae; POME treatment; post-treatment

1. Introduction

The palm oil industry's contribution to Malaysia's economic growth and rapid development has widely been acknowledged. This industry is rapidly expanding, and Malaysia is the world's second-largest palm oil producer. Unfortunately, the large amounts of byproducts generated during the oil extraction process have contributed to environmental pollution [1]. The wastewater generated by the palm oil extraction process is known as palm oil mill effluent (POME). Total solids (40,500 mg/L), oil and grease (4000 mg/L), chemical oxygen demand (50,000 mg/L), biological oxygen demand (BOD_3) (25,000 mg/L), total nitrogen (TN) (1400 mg/L), and total phosphorus (TP) (150 mg/L) are all present in large concentrations [2]. According to the Malaysian Department of Environment, POME must follow regulatory requirements for BOD (20 mg/L), COD (1000 mg/L), total solid (1500 mg/L), suspended solid (400 mg/L), oil and grease (50 mg/L), and total nitrogen (50 mg/L) before being released into the environment [3]. However, some of these values are often exceeded in the POME discharged into the environment. For example, the COD of POME from Sembilan, Malaysia, was 3250 mg/L [4]. The non-compliance may well be why the discharge limit for COD and TSS was removed in 1982 [3]. Similarly, the COD of POME from Gampong Ujong Lamie, Acheh, Indonesia, was 4177 mg/L exceeding the country's 350 mg/L regulatory discharge limit by more than tenfold [5]. Therefore, there is a need to treat the POME further before releasing it to the environment.

The ponding treatment system, which consists of anaerobic, aerobic, and facultative ponds, is widely used in POME treatment. Because of its ability to handle large quantities of POME with low operating costs, this treatment method is used by around 85% of palm oil mills in Malaysia [6]. However, this treatment system will lead to greenhouse gas emissions, especially the methane gas that causes global warming. In addition, the characteristics of the handled POME using this treatment method do not always follow the requirements of the Malaysian Department of Environment's industrial discharge standard [7]. Therefore, there is a need to improvise the current treatment methods.

Many researchers have recently shifted their focus to wastewater reuse to protect the environment and produce renewable resources such as food, fuel, and feed. Microalgae are thought to be a promising post-treatment option for agro-industrial wastewater because of their ability to use nutrients for growth [8]. While doing so, they could reduce environmental pollution significantly [9]. Value-added compounds can also be extracted from the microalgae, but the extraction method is crucial to determine the quality of the compound [10].

POME has been reported in several studies to have the potential to promote the growth of microalgae because it contains sufficient nutrients, especially TN, TP, and other organic matters [11]. POME has been successfully used to cultivate *Chlamydomonas* sp. [11], *Chlorella sorokiniana* [12], *Botryococcus brauni* [13], and a mixed culture of microalgae [14]. However, Kayombo et al., in 2003, reported that a higher concentration of POME would inhibit microalgae growth due to its biotic and abiotic factors. In addition, the micro-size of microalgae cells adds to the treatment process's difficulty. Therefore, before processed POME is discharged into waterways, microalgae must be harvested. Otherwise, eutrophication, another environmental problem, would occur [15]. One of the difficulties that many researchers face is the isolation or harvesting of microalgae from treated wastewater. Harvesting methods such as flocculation, centrifugation, and filtration have been successful at a laboratory scale, but they require expensive equipment [16]. Immobilisation creates a protective microenvironment for rapid cell development and enables easier harvesting compared to the free-living cells. Various microalgae cells have been successfully immobilised in natural polysaccharide gels or synthetic polymers, enhancing biomass stability and productivity. For example, calcium alginate is one of the most frequently used methods to trap microalgal cells because it does not need heating and is non-toxic [17]. In this research, a dominant native green microalga strain was isolated from a polishing pond of a POME treatment. The strain was used to further treat POME on a laboratory scale. The microalgae cells were immobilised with calcium alginate to study the effects of nutrient load and light penetration on microalgal growth and nutrient removal.

2. Materials and Methods

2.1. Sampling of POME

The raw POME samples were obtained from a palm oil mill holding tank at Nibong Tebal, Pulau Pinang, Malaysia (GPS coordinate: 5°12'33.7" N, 100°29'01.1" E), just before it was discharged into a cooling pond. Sample collection was conducted using high-density polyethylene (HDPE) containers with a capacity of 25 L. All POME samples were kept in a refrigerator at 4 °C to reduce microbial activities. The raw POME collected was pre-treated using two stages of lab-scale treatment: anaerobic and aerobic processes with 50 and 16 days hydraulic retention times (HRTs), respectively. In this analysis, the lab-scale-treated POME (LABT-POME) was used to ensure that the physiochemical characteristics of the POME sample did not change significantly during the experiment.

2.2. Algae Culture Media

Throughout the study, microalgae were grown and cultured using Bold's Basal Medium (BBM). The BBM compositions are shown in Table 1.

Table 1. The Bold's Basal Medium (BBM) composition.

Reagent A	per 400 mL
$NaNO_3$	10.00 g
$MgSO_4.7H_2O$	3.00 g
K_2HPO_4	4.00 g
NaCl	1.00 g
KH_2PO_4	6.00 g
$CaCl_2$	1.00 g
Microelement Stock Solution *	**per 1 L**
$ZnSO_4$	8.82 g
MoO_3	0.71 g
$Co(NO_3)_2.6H_2O$	0.49 g
$MnCl_2$	1.44 g
$CuSO_4.5H_2O$	1.57 g
Solution 1	**per 100 mL**
H_3BO_4	1.14 g
Solution 2	**per 100 mL**
$EDTA.Na_2$	5.00 g
KOH	3.10 g
Solution 3	**per 100 mL**
$FeSO_4.7H_2O$	4.98 g
HCl (Concentrated)	1.00 mL

* autoclave to dissolve before storage.

2.3. Isolation and Cultivation of Green Microalgae

The native green microalgae were isolated from the POME treated in a palm oil mill ponding system (POMST-POME). The sampling site was a polishing pond, as illustrated in Figure 1. The POMST-POME sample was first examined using a Nikon Eclipse E200 light microscope to ensure the presence of microalgae. A total of 50 mL of POMST-POME was added into 250 mL Erlenmeyer flasks before 50 mL of sterilised liquid BBM was added to each flask. All samples were incubated for 14 days at 35 ± 3 °C with continuous agitation at 100 rpm and 32.4 ± 2.7 $\mu mol.m^{-2}s^{-1}$ illuminations. The visible green microalgae cells were withdrawn from each flask, diluted into different series dilutions, and then spread onto the BBM agar plates. These agar plate samples were cultivated under the same condition for another 14 days. Every agar plate with a single colony was examined under a light microscope before the colony was streaked onto a new agar plate for another 14 days of cultivation. This process was repeated several times until a single microalgae species was isolated. Then, the single species of green microalgae was transferred and grown into a liquid BBM at the same cultivation condition. The green microalgae were sub-cultured into a new fresh liquid BBM monthly to maintain the culture.

2.4. Morphological Identification of Microalgae

A light microscope with a magnification of 40× was used to examine the morphology of the isolated green microalgae. Cellular shape, scale, flagella, and other visible characteristics of green microalgae cells were reported and compared with the guidebook [18]. In addition, scanning electron microscopy (SEM) was performed to validate the microalgae cells' three-dimensional shape and scale. Before being examined with SEM, the sample cells were dried and fixed using hexamethyldisilazane (HMDS).

Figure 1. A polishing pond in a palm oil mill located in Nibong Tebal, Penang, from where the green microalgae were collected.

Microalgae Sample Processing Using HMDS Method

Each 1 mL green microalgae cell was withdrawn from a stock culture and centrifuged at 4032 rcf for 15 min. The supernatant of each sample was discarded. The pellet cells were fixed with the McDowell–Trump reagent, prepared in 0.1 M of phosphate buffer, for 24 h. After 24 h, each sample was washed with 0.1 M of phosphate buffer twice by centrifuging and resuspension. The sample was then post-fixed in 1% of Osmium tetraoxide for 1 h. After that, each sample was washed with distilled water, centrifuged, and resuspended twice [19].

After the fixation process, all samples underwent a dehydration process by immersing each sample in ethanol at the concentrations of 50%, 75%, 95%, and 100%, respectively. Each dehydration process was carried out by centrifugation and resuspension of the samples. The reaction time for each procedure was 10 min. In the last step, each sample was immersed in hexamethyldisilazane (HMDS) for another 10 min. Finally, HMDS was decanted from the samples, and the samples were air-dried in a desiccator at ambient temperature for one day. Each of the sample cells was coated with gold and examined under SEM [19].

2.5. Molecular Identification of Algal Strain

One millilitre of green microalgae was taken from each stock culture and centrifuged for two minutes at 11,200 rcf. The supernatant of the green microalgae sample was discarded entirely, and the suspended biomass of the green microalgae was used in the molecular identification process. The genomic DNA from the green microalgae cultures was extracted using Plant DNA Extraction Kits (Vivantis Technologies, Malaysia). The procedures of the extraction works were conducted according to the manufacturer's handbook. The target sequences from the extracted DNA samples were amplified by polymerase chain reaction (PCR). The target genes were amplified with different primers and PCR protocols, as listed in Table 2. The PCR was performed using an Eppendorf Mastercycler® ep, Germany. The sequencing of purified PCR product was carried out by the Centre for Chemical Biology, University Sains Malaysia. The Basic Local Alignment Search Tool (BLAST) was used to analyse the 18S rRNA sequences. The sequence similarities were compared to the available database from the National Center for Biological Information (NCBI).

Table 2. Primer combination, target gene, and PCR protocols used in the present study.

Target Gene	Primer (5′ → 3)	Protocol	Reference
Chloroccocum	**Forward primer** CLO-GEN-S3 (GCATGGAATMRCACGATAGGACTC) **Reverse primer** CLO-GEN-A4 (CGGCATCGTTTATGGTTGGTTGAGACTAC)	Initial: 15 min (95 °C) Denaturation: 30 s (94 °C) Annealing: 90 s (63 °C) Extension: 90 s (72 °C) Elongation: 10 min (72 °C) Total runs: 35 cycles	[20]

2.6. Green Microalgae–Alginate Beads Preparation

The procedure was adopted from Ruiz-Marin and Sánchez-Saavedra (2016) [21]. In brief, 10 mL of the microalgae cells was harvested by centrifugation at 11,200 rcf for 2 min. The pellets were resuspended in 10 mL of autoclaved sterilised water after the microalgae supernatant was discarded. A two-per-cent microalgae–alginate suspension mixture was prepared by mixing each microalgae strain with 10 mL of 4% sterile alginate solution in a 1:1 volume ratio. The mix was vigorously stirred to ensure the uniformity of the solution. As shown in Figure 2, the mixture was then moved into a 10 mL syringe and placed 8–10 cm above a beaker containing a 2% calcium chloride solution. The microalgae–alginate mix was dropped into a calcium chloride solution to form the microalgae–alginate beads immediately. The process created light green spherical beads (as shown in Figure 3a) with around 3 ± 0.5 mm diameter. The beads were left for four hours in the calcium chloride solution at room temperature to harden. Next, the microalgae–alginate beads were washed with a 0.85% sterile sodium chloride solution followed by autoclaved distilled water to remove any residuals. The same process was used to create blank alginate beads as control, but distilled water was used instead of condensed microalgae cells. Figure 3b shows the morphology of the blank beads.

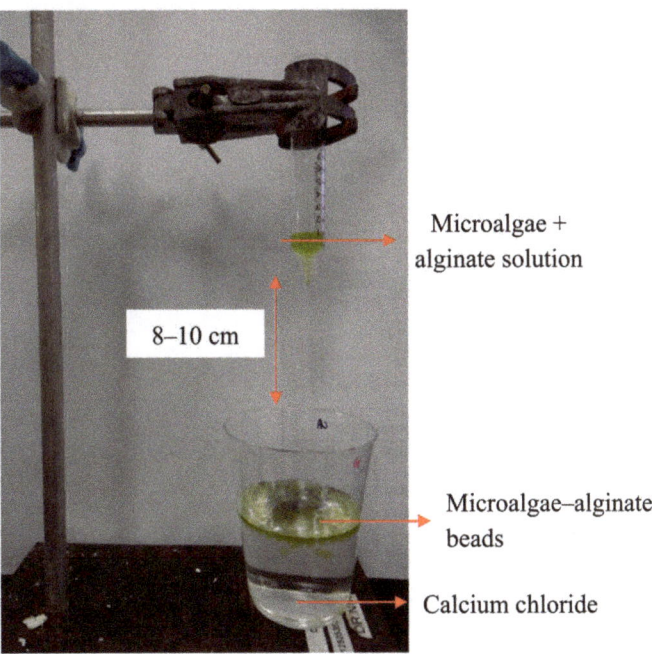

Figure 2. Experimental work for microalgae beads preparation.

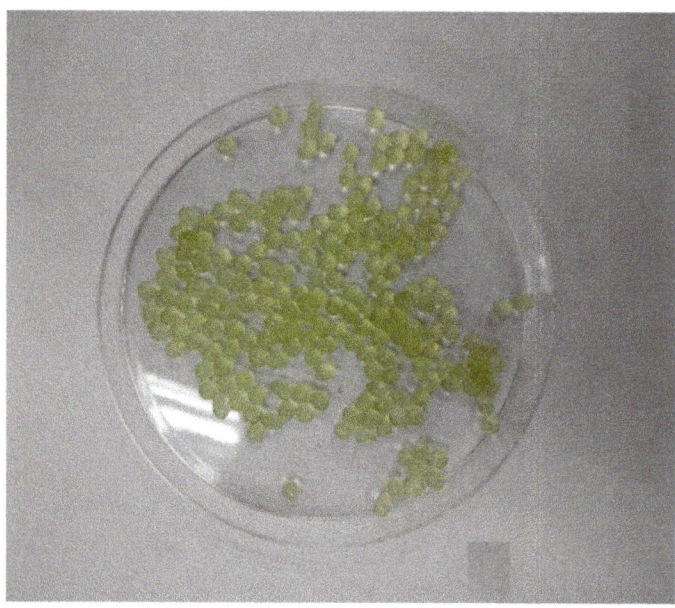

(a)

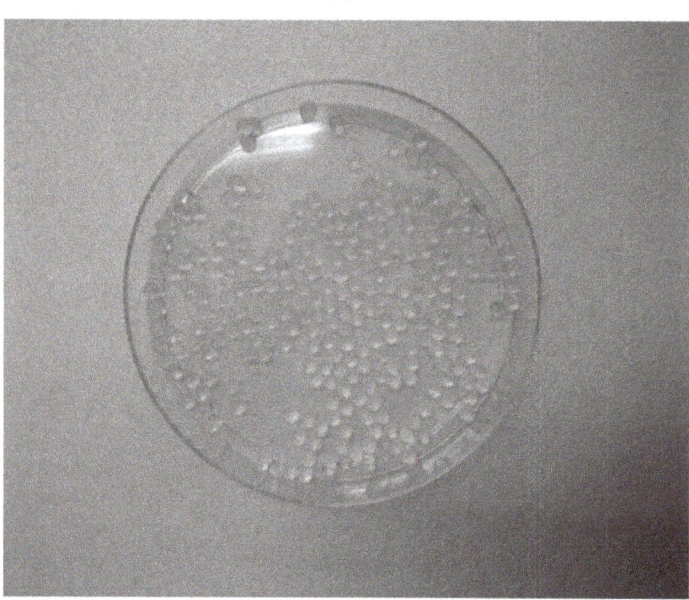

(b)

Figure 3. The product of (**a**) microalgae–alginate beads and (**b**) blank alginate beads.

2.7. Experimental Setup

The microalgae–alginate beads were prepared with the following microalgae–alginate to calcium chloride concentration ratios: 2:2, 2:4, 2:6, 4:2, 4:4, 4:6, 6:4, and 6:6. All of the beads were placed into Erlenmeyer flasks containing 150 mL of LABT-POME. The flasks were placed at (35 ± 3 °C) with 12 h of light and 12 h of darkness intervals. All samples were shaken on an orbital shaker model KJ-201B, Xiangtian China, with a speed of 100 rpm at 32.4 ± 2.7 µmol.m^{-2}s^{-1} light intensity. After ten days, ten beads from each sample were removed to determine their hardness and diameter. A texture analyser (TA.XT*Plus*; U.S.A.) was used to calculate the beads' hardness, while a digital calliper (TPI 3C350-NB; China) was used to determine their diameter. For the post-treatment LABT-POME study, the microalgae–alginate beads were prepared using the selected best pair of ratios (sodium alginate: calcium chloride). All of the microalgae–alginate beads were placed in a 250 mL Erlenmeyer flask containing 150 mL of LABT-POME. Three sets of control runs were conducted in this study: a blank control run (BLK-CTRL), autoclaved blank control (AUTOBLK-CTRL), and actual control (ACT-CTRL). BLK-CTRL was prepared with addition of blank beads into the LABT-POME sample. AUTOBLK-CTRL was prepared by mixing blank beads with a sterilised LABT-POME sample. ACT-CTRL was prepared by the addition of microalgae–alginate beads into a sterile LABT-POME sample. All samples were placed at 35 ± 3 °C with continuous agitation at 100 rpm and a light intensity of 32.4 ± 2.7 µmol.m^{-2}s^{-1}. For the growth study, ten beads from each sample flask were withdrawn every two days and re-dissolved for chlorophyll-a extraction (details are further discussed in Section 2.8.2). At the same time, 10 mL of LABT-POME was collected from each flask to test for reducing or removing TN, TP, ammonia nitrogen, and SCOD. The analysis of TN, TP, ammonia nitrogen, and SCOD was according to the HACH standard methods. All the experimental works were ended when the beads started to degrade. All the experimental runs were carried out in triplicates.

2.8. Analytical Methods

2.8.1. Measurement of Substrate Reduction

The percentage of substrate reduction was calculated using Equation (1) below,

$$p = \frac{S_o - S_e}{S_o} \times 100\% \tag{1}$$

where p is the percentage removal; S_o is the initial substrate concentration (mg/L); and S_e is the concentration of substrate at equilibrium state (mg/L).

2.8.2. Chlorophyll-a Extraction and Quantification

Ten microalgae–alginate beads were collected from each sample every two days. To obtain cells trapped in the beads, they were dissolved in 10 mL of 0.1 M tri-sodium citrate dihydrate. The Chlorophyll-a was extracted from the microalgae cells. In brief, each flask had 1 mL of green microalgae pipetted into a 1.5 mL centrifuge tube. All sample cells were centrifuged at 11,200 rcf for 2 min. Next, each sample cell's supernatant was discarded and resuspended in 1 mL of methanol. Then, all sample cells were vortexed for 5 s and held at 4 °C in the dark for an hour. The sample cells were then centrifuged for 2 min at 11,200 rcf. Finally, UV-spectrophotometer analysis was performed on the supernatant from each sample cell at 649 nm and 665 nm wavelengths. Equation (2) was used to calculate the chlorophyll-a concentration of green microalgae [22].

$$Chl. = -8.0962 A_{652} + 16.169 A_{665} \tag{2}$$

2.8.3. Growth Rate Study

The doubling time of the green microalgae was calculated using Equation (3) [23,24]:

$$t_d = \frac{\ln 2}{\mu} \quad (3)$$

where t_d is the doubling time of microalgae (d), and μ is the specific growth rate of microalgae (d^{-1}).

The specific growth rate of the microalgae was determined using Equation (4) [22,23]:

$$\mu = \frac{\ln x_2 - \ln x_1}{t_2 - t_1} \quad (4)$$

where μ is the specific growth rate (d^{-1}); x_1 and x_2 are chlorophyll-a (µg/L); and t_1 and t_2 are the time within the exponential phase (d).

2.8.4. Growth Kinetic Study

Verhulst logistical model was used to determine the dynamic growth of microalgae in the present study [25]. The equation as below:

$$X = \frac{X_o e^{\mu t}}{1 - (\frac{X_o}{X_m})(1 - e^{\mu t})} \quad (5)$$

where X is the chlorophyll-a concentration of green microalgae in the time-course (g/L); X_o is the initial biomass concentration of green microalgae (g/L); X_m is the biomass concentration at equilibrium state (g/L); t is the time taken (d); and μ is the maximum specific growth rate of green microalgae (d^{-1}).

2.8.5. Quantification of Substrate Consumption Rate

The substrate consumption rate of immobilised microalgae was calculated using Equation (6) [21]:

$$r = \frac{S_o - S_i}{t_i - t_o} \quad (6)$$

where r is the consumption rate of substrate (mg/L.d); S_o is the initial concentration of substrate (mg/L); S_i is the concentration of substrate at time t_i (mg/L); and t_o and t_i are the time within the exponential phase (d).

3. Results and Discussion

3.1. Identification of Green Microalgae

Table 3 shows the picture and morphological characteristics of the green microalgae. The microalgae cells were unicellular and green in colour. As a result, it is classified as a Chlorophyte [18]. Table 3 shows that the average size of microalgae cells is between 10 and 12 µm. The identities of the microalgae cells were suspected based on their visible characteristics, which are described in Table 3. The morphology and features of these microalgae led to *Chlorococcum* sp. as the potential genus. Table 4 lists the most related species for this green microalga using the DNA sequence in the NCBI method. As shown in Table 4, the 18rRNA result for this green microalgae showed 99–100% similarity. Therefore, due to morphology and molecular identification, the microalgal species is most likely *C. oleofaciens*.

3.2. Selection of the Suitable Beads

The size of the microalgae–alginate beads shrank over time. As a result, some experimental work was conducted to increase the strength of the microalgae–alginate beads to achieve a longer-lasting duration of beads in the LABT-POME before conducting the immobilisation studies of *C. oleofaciens* for POME post-treatment. The beads were reinforced using three typical concentration ratios of calcium chloride (2%, 4%, and 6%) to microalgae–alginate (2%, 4%, and 6%). Figure 4 shows the hardness of the beads made at various percentages of calcium chloride to microalgae–alginate concentration ratios. All of the beads were inoculated for ten days in LABT-POME at the same time. Throughout the cultivation time, the growth of the strain and the bead degradation were measured. Table 5 summarises all of the data obtained.

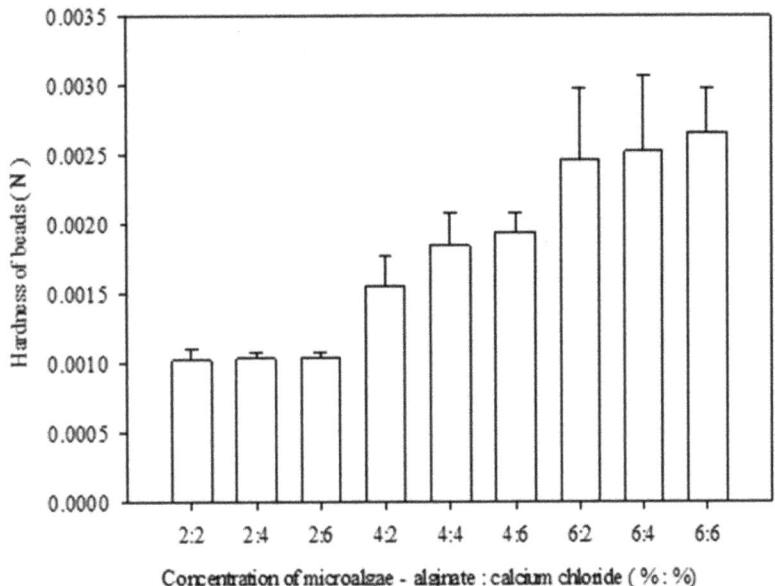

Figure 4. Hardness of microalgae–alginate beads prepared from different volume ratio pairs of microalgae–alginate to calcium chloride.

Figure 4 shows the increased hardness of beads with the increase in the ratio of microalgae–alginate to calcium chloride. The 6:6 ratio produced the hardest beads, and the size diameter deteriorated the least throughout the cultivation period, as shown in Table 5. In addition, *C. oleofaciens* had a low chlorophyll-a concentration average (2.309 µg/mL). These data indicate that the strain did not develop well in the 6:6 ratios because the higher alginate concentration hindered nutrient transport from the LABT-POME into the beads. As a result, the species' proliferation in the microalgae–alginate beads was slowed. The research work from Banerjee et al. (2019) supports this [26]. They reported that increasing alginate concentration would reduce the pore size of the alginate bead wall, leading to restrictive uptake of nutrients or other macro-molecules.

Table 3. The morphological characteristics of the green microalgae species viewed under light and scanning electron microscopy.

Microscopic Image (40× Magnification)	SEM Image	Cell Size (μm)	Visible Characteristics	Possible Genus
		10–12	green in colour, sphere-shaped cells, contain numerous sac-like organelles	*Chlorococcum* sp.

Table 4. The DNA sequence identification for the green microalgae species.

NCBI BLAST System Result							The most Possible Species
Description	Max score	Total score	Query cover	E value	Ident	Accession	
Chlorococcum oleofaciens strain Ru-1-1 small subunit ribosomal RNA gene, partial sequence; internal transcribed spacer 1	1821	1821	100%	0.0	99%	MH702751.1	*C. oleofaciens*
Chlorococcum tatrense gene for 18S ribosomal RNA, partial sequence	1821	1821	100%	0.0	99%	AB936290.1	
Chlorococcum oleofaciens strain CAMU MZ-Ch4 small subunit ribosomal RNA gene, partial sequence	1821	1821	100%	0.0	99%	MG491216.1	
Chlorococcum sphacocccum strain ACSSI 188 small subunit ribosomal RNA gene, partial sequence	1821	1821	100%	0.0	99%	MG982207.1	
Chlorococcum oleofaciens strain ACSSI 208 small subunit ribosomal RNA gene, partial sequence	1821	1821	100%	0.0	99%	MG491516.1	

Table 5. Comparison of the strength of microalgae–alginate beads and chlorophyll-a of *C. oleofaciens* before and after cultivation in LABT-POME.

Microalgae–Alginate:CaCl$_2$ (% conc.:% conc.)	Initial Diameter of Beads (mm)	Diameter of Beads after 10 Days (mm)	Initial Chlorophyll-a (µg/mL)	Chlorophyll-a after 10 Days (µg/mL)
2:2	3.5 ± 0.05	1.2 ± 0.03	0.501	5.881
2:4	3.3 ± 0.05	1.8 ± 0.05	0.493	5.679
2:6	3.5 ± 0.03	2.0 ± 0.07	0.512	5.654
4:2	3.3 ± 0.05	2.5 ± 0.08	0.502	5.021
4:4	3.4 ± 0.02	3.0 ± 0.06	0.495	4.978
4:6	3.5 ± 0.02	3.1 ± 0.02	0.522	4.889
6:2	3.3 ± 0.03	3.1 ± 0.09	0.506	2.899
6:4	3.4 ± 0.04	3.3 ± 0.01	0.483	2.333
6:6	3.5 ± 0.01	3.4 ± 0.01	0.517	2.309

The immobilised cells had a high average of chlorophyll-a concentration. The bead size decreased by more than 65% for the ratio pairs of 2:2, 2:4, and 2:6 after ten days of cultivation in LABT-POME. This finding shows that the microalgae–alginate beads made using these ratios were inappropriate for LABT-POME treatment because they deteriorated quickly. As a result, the 4:4 ratio was chosen as the best ratio for preparing microalgae–alginate beads. Table 5 shows that after ten days of cultivation, the size of the beads decreased by approximately 10%, and the chlorophyll-a concentration average from immobilised *C. oleofaciens* increased to 4.978 µg/mL from 0.495 µg/mL. This result proved that *C. oleofaciens* could be grown in certain proportions in the beads. Hence, a ratio of 4:4 was selected for further study.

3.3. Growth Study of Immobilised C. oleofaciens in LABT-POME

Figure 5 shows the growth of immobilised *C. oleofaciens* over 16 days. When cultivating the immobilised stain in the LABT-POME sample, the lag phase took two days to develop. In the ACT-CTRL sample, the lag phase was developed on the first day. After the lag time, both the ACT-CTRL and LABT-POME growth curves of immobilised cell cultivation showed a sharp increase with no stationary phase. The size of microalgae–alginate beads decreases as the cultivation time increases. Due to this, this study only measured the growth of immobilised cells for up to 16 days. Consequently, stationary phases were absent from both growth curves (Figure 5). Over the 16 days, the size of microalgae–alginate beads in both the ACT-CTRL and LABT-POME samples was decreased by approximately 12% and 30%, respectively. The microalgae–alginate beads degraded due to the higher pH and TP content in LABT-POME. When cultivated in the LABT-POME study, the size of microalgae–alginate beads shrank the most. Since alginate is an organic compound, other microorganisms in the LABT-POME may be causing the beads to shrink [27]. On day 14, the deterioration of the beads caused the microalgae cell to leak from the beads, resulting in lower chlorophyll-a concentrations from immobilised cells cultivated in the LABT-POME sample (35.776 µg/mL) relative to the ACT-CTRL sample (51.435 µg/mL). This result demonstrates that immobilising microalgae can concentrate a significant amount of biomass that can be used as a byproduct. The research on the immobilisation of *Synechococcus* sp. produced similar results. Immobilised cells, which developed more biomass than free cell cultures, were immobilised in chitosan to protect the cell walls from NaOH toxicity [28]. Therefore, similar to the previous study, the cells grew faster when immobilised with alginate beads than free cell culture.

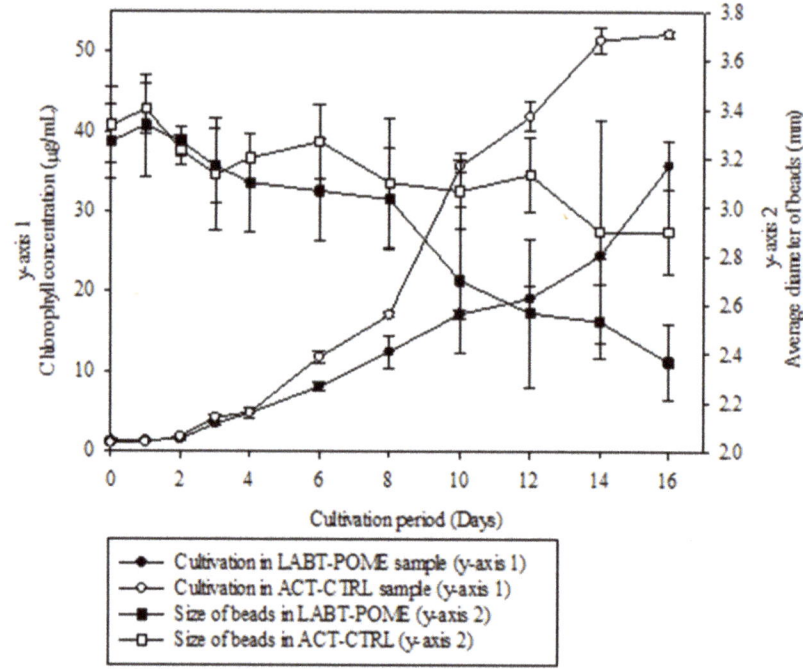

Figure 5. Average chlorophyll-a concentration of immobilised *C. oleofaciens* in LABT-POME.

3.4. Removal of TP from LABT-POME

BLK-CTRL has existing POME bacteria, and the bacteria in AUTOBLK-CTRL were eliminated. The result of this study shows that nutrient removal was not affected by the presence of the bacteria. Figure 6 shows the absorption of TP by immobilised *C. oleofaciens* for 16 days. The TP removal curves for both the ACT-CTRL and LABT-POME samples decreased over the 16 days. The species took approximately eight days to eliminate more than 90% of TP from the LABT-POME in an immobilised state with an ingestion rate of 8.6950 mg/L.d. The higher TP concentration, on the other hand, can hinder the deterioration of the beads. The Na^+ ions exchange ions with Ca^{2+} ions, loosening the structure of the beads. Calcium phosphate was then formed by the reaction of phosphate ions with Ca^{2+}, which increased the turbidity of the LABT-POME [29]. As a result, the diameter of the beads shrank throughout the 16-day cultivation cycle. The removal of TP by free-living cells was more than 90% within 16 days of cultivation [30]. However, immobilised cells could remove more than 90% of TP from POME within just 6–8 days of cultivation. This result indicates that the efficiency of immobilised cells is higher than free-living cells. Therefore, the TP removal was probably done not only by algal cells but also by the beads. However, the removal of TP by beads was not evaluated.

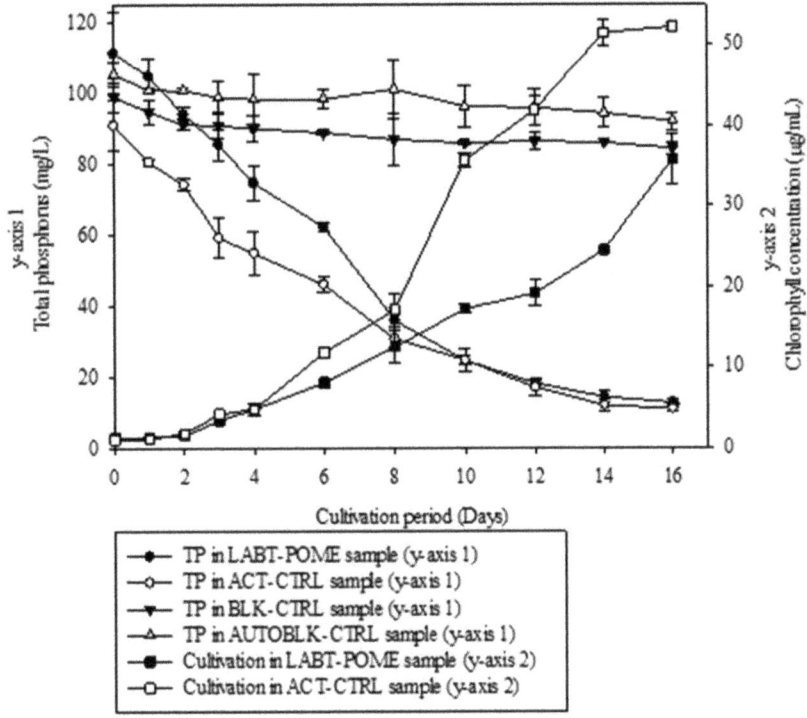

Figure 6. Removal of total phosphorus from LABT-POME using immobilised *C. oleofaciens*.

3.5. Removal of Ammonia Nitrogen and TN from LABT-POME

Figure 7a,b depict TN and ammonia nitrogen elimination from LABT-POME over 16 days using immobilised cells. In both the ACT-CTRL and LABT-POME samples, TN and ammonia nitrogen concentrations decreased over the 16-day cultivation period, as shown in Figure 7a,b. By referring to Figure 7a,b, the immobilised cells performed a high rate of removal TN (25.5536 mg/L.d) and ammonia nitrogen (7.3857 mg/L.d). A similar result was reported by Liu et al. (2012), which concluded that immobilised *Chlorella sorokiniana* GXNN 01 removed more nitrogen from synthetic wastewater than free-living cells under three different conditions: autotrophic, heterotrophic, and 23 microaerobic cultivation over six days [30]. Within 6–8 days of the cultivation date, the immobilised cells extracted over 95% of TN and ammonia nitrogen from ACT-CTRL and LABT-POME samples. However, the free-living cells took a longer time (about 16 days) to achieve more than 90% of TN and ammonia nitrogen removal [30]. This result shows that immobilised cells are more efficient than free-living cells. When cells are contained inside the alginate, it is shielded from being ingested by other microorganisms. Furthermore, since nutrients can diffuse into the beads for *C. oleofaciens* to absorb, the immobilisation produces a micro-environment for the species to develop [27]. Therefore, consistent with the previous study, the TN and ammonia nitrogen can be efficiently removed by microalgae immobilised with alginate.

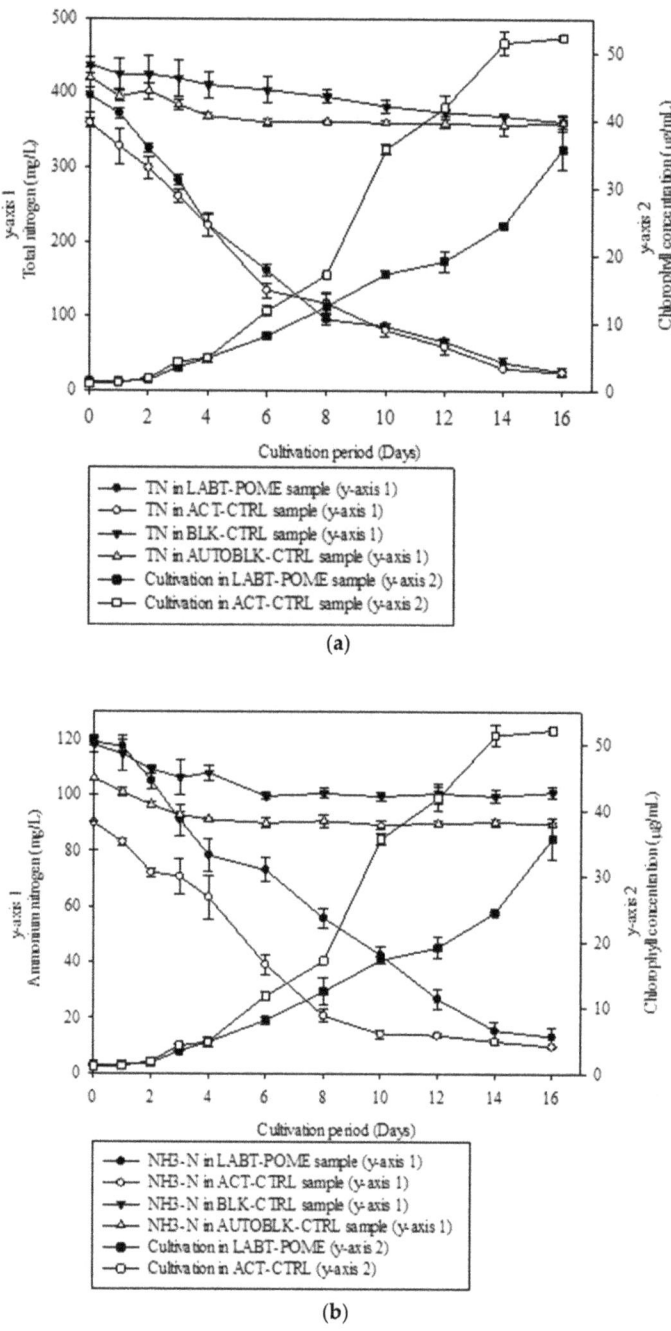

Figure 7. Removal of (**a**) total nitrogen and (**b**) ammonia nitrogen from LABT-POME using immobilised *C. oleofaciens*.

3.6. Reduction of Soluble COD (SCOD) from LABT-POME

Figure 8 shows the reduction of SCOD from LABT-POME over 16 days using immobilised cells. In the first ten days, the SCOD removal pattern is shown to be dramatically reduced. Within ten days of cultivation, immobilised cells were able to reduce SCOD by about 60%. Both the ACT-CTRL and LABT-POME samples had significantly higher SCOD concentrations after day 10. The breakdown of microalgae–alginate beads contributes to the rise in SCOD. After the degradation of the beads, some functional groups such as the carboxyl in alginate can contribute SCOD into the LABT-POME [29]. Hence, the SCOD curve showed a slight increase over the cultivation period. The 60% removal is lower than the 70% removal by *Nannochloropsis* sp. microalgae [4]. However, the POME used by Emparan et al., 2020, [4] was filtered or diluted. In contrast, the raw POME used in the present study was not diluted but pre-treated using two-stage, anaerobic and aerobic, lab-scale treatment. Therefore, considering the dilution factor, the total COD removal in this study was higher than that of Emparan et al., 2020, [4]. Therefore if this lab-scale treatment can be upscaled, it can be used to treat POME directly, without dilution.

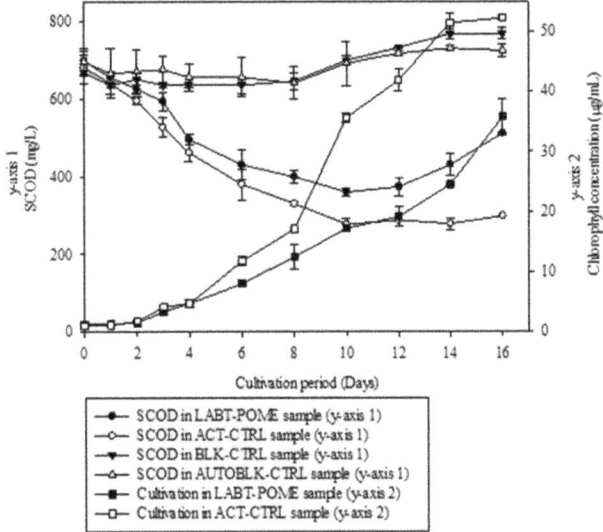

Figure 8. Reduction of SCOD from LABT-POME using immobilised *C. oleofaciens*.

The SigmaPlot®12.5 programme was used to adapt the growth data for immobilised cells to the Verhulst logistic model. The kinetic model coefficients are shown in Table 6. The data fitting curves to the Verhulst logistic model are shown in Figure 9. For growth in immobilised form, the p-value was less than 0.05. Both systems had an r^2 of 0.99. As a result, the Verhulst logistic model was accurate in describing the growth of immobilised cells. Similar to Azlin Suhaida et al.'s (2018) research work, the microalgae growth was also well fitted with the Verhulst logistic model [31]. The growth rate of immobilised cells was 0.4867 d^{-1}. The high growth rate of immobilised cells could be because cells are protected from being consumed by other microorganisms in POME. Therefore, the maximum chlorophyll-a concentration, X_m obtained from immobilised cells at the stationary state, was 36.9460 µg/mL.

Table 6. Values of kinetic coefficients.

Microalgae	X_o (µg/mL)	X_m (µg/mL)	µ (d^{-1})	r^2	p-Value
Immobilised C. oleofaciens	0.7034	36.9460	0.4867	0.9887	<0.0001

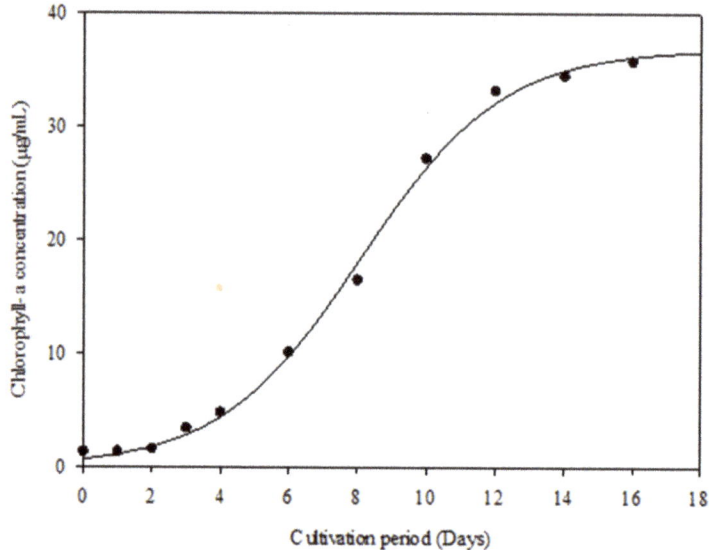

Figure 9. Fits of the normalised logistic model to the relative reproductive rates of immobilised *C. oleofaciens*.

4. Conclusions

The findings of this report suggest that immobilised *C. oleofaciens* showed good performance in reducing or extracting substrates (TN, ammonia nitrogen, TP, and SCOD) from POME samples. Furthermore, the growth of the immobilised cells was well suited by the Verhulst logistical model, with r^2 values of >0.95, indicating the model's suitability in modelling growth. One of the traditional methods for separating microalgae from treated wastewater used in this study was the immobilisation of microalgae. However, one of the key challenges in the post-treatment of POME in the industry is the deterioration of the beads. As a result, future research should look into the strength of the beads to find longer-lasting beads for POME phycoremediation.

Author Contributions: Conceptualisation, K.A.T., J.L., N.M., N.I. and W.M.W.O.; writing—original draft preparation, K.A.T.; writing—review and editing, J.L., M.A.K., M.S. and M.R.; supervision, J.L., N.M., N.I., W.M.W.O. and M.R.; funding acquisition, M.A.K., M.S. and M.R. All authors have read and agreed to the published version of the manuscript.

Funding: Moonis Ali Khan acknowledge the financial support through the Researchers Supporting Project number (RSP-2021/345), King Saud University, Riyadh, Saudi Arabia, and Japareng Lalung acknowledge the financial support from Universiti Sains Malaysia in the form of Research University Grant No. 1001/PTEKIND/811253. The APC was funded by Universiti Sains Malaysia, Malaysia.

Institutional Review Board Statement: Not applicable.

Informed Consent Statement: Not applicable.

Data Availability Statement: The data presented in this study are available on request from the corresponding author.

Acknowledgments: Moonis Ali Khan acknowledge the financial support through the Researchers Supporting Project number (RSP-2021/345), King Saud University, Riyadh, Saudi Arabia, and Japareng Lalung acknowledge the financial support from Universiti Sains Malaysia in the form of Research University Grant No. 1001/PTEKIND/811253.

Conflicts of Interest: The authors declare no conflict of interest.

References

1. Parthasarathy, S.; Mohammed, R.R.; Fong, C.M.; Gomes, R.L.; Manickam, S. A novel hybrid approach of activated carbon and ultrasound cavitation for the intensification of palm oil mill effluent (POME) polishing. *J. Clean. Prod.* **2016**, *112*, 1218–1226. [CrossRef]
2. Ahmad, A.; Buang, A.; Bhat, A.H. Renewable and sustainable bioenergy production from microalgal co-cultivation with palm oil mill effluent (POME): A review. *Renew. Sustain. Energy Rev.* **2016**, *65*, 214–234. [CrossRef]
3. EQA. *Environmental Quality Act (Prescribed Premises) (Crude Palm Oil) Regulations*; Department of Environment: Putrajaya, Malaysia, 1974; Volume 93.
4. Emparan, Q.; Jye, Y.S.; Danquah, M.K.; Harun, R. Cultivation of Nannochloropsis sp. microalgae in palm oil mill effluent (POME) media for phycoremediation and biomass production: Effect of microalgae cells with and without beads. *J. Water Process. Eng.* **2020**, *33*, 101043.
5. Zulfahmi, I.; Ravika, N.; Feizia, H.; Lina, R.; Muliari, M.; Kizar, A.S.; Mohammad, M.R. Phytoremediation of Palm Oil Mill Effluent (POME) Using Water Spinach (Ipomoea Aquatica Forsk). *Environ. Technol. Innov.* **2021**, *21*, 101260. [CrossRef]
6. Ahmad, A.; Krimly, M.Z. Palm oil mill effluent treatment process evaluation and fate of priority components in an open and closed digestion system. *Curr. World Environ.* **2014**, *9*, 321–330. [CrossRef]
7. Wu, T.Y.; Mohammad, A.W.; Jahim, J.M.; Anuar, N. Pollution control technologies for the treatment of palm oil mill effluent (POME) through end-of-pipe processes. *J. Environ. Manag.* **2010**, *91*, 1467–1490. [CrossRef]
8. Delrue, F.; Álvarez-Díaz, P.D.; Fon-Sing, S.; Fleury, G.; Sassi, J.-F. The Environmental Biorefinery: Using Microalgae to Remediate Wastewater, a Win-Win Paradigm. *Energies* **2016**, *9*, 132. [CrossRef]
9. Goswami, R.K.; Mehariya, S.; Verma, P.; Lavecchia, R.; Zuorro, A. Microalgae-based biorefineries for sustainable resource recovery from wastewater. *J. Water Process. Eng.* **2021**, *40*, 101747. [CrossRef]
10. Mehariya, S.; Fratini, F.; Lavecchia, R.; Zuorro, A. Green extraction of value-added compounds form microalgae: A short review on natural deep eutectic solvents (NaDES) and related pre-treatments. *J. Environ. Chem. Eng.* **2021**, *9*, 105989. [CrossRef]
11. Ding, G.T.; Yaakob, Z.; Takkriff, M.S.; Salihon, J.; Rahaman, M.S.A. Biomass production and nutrients removal by a newly-isolated microalgal strain Chlamydomonas sp in palm oil mill effluent (POME). *Int. J. Hydrogen Energy* **2016**, *41*, 4888–4895. [CrossRef]
12. Khalid, A.A.H.; Yaakob, Z.; Abdullah, S.R.S.; Takriff, M.S. Growth improvement and metabolic profiling of native and commercial Chlorella sorokiniana strains acclimatised in recycled agricultural wastewater. *Bioresour. Technol.* **2016**, *247*, 930–939. [CrossRef]
13. Nur, M.M.A.; Setyoningrum, T.M.; Budiaman, I.G.S. Potency of Botryococcus braunii cultivated on palm oil mill effluent (POME) wastewater as the source of biofuel. *Environ. Eng. Res.* **2018**, *22*, 1–8.
14. Babu, A.; Katam, K.; Gundupalli, M.P.; Bhattacharyya, D. Nutrient removal from wastewater using microalgae: A kinetic evaluation and lipid analysis. *Water Environ. Res.* **2018**, *90*, 520–529. [CrossRef]
15. Kayombo, S.; Mbwette, T.S.A.; Katima, J.H.Y.; Jorgensen, S.E. Effects of substrate concentrations on the growth of heterotrophic bacteria and algae in secondary facultative ponds. *Water Res.* **2003**, *37*, 2937–2943. [CrossRef]
16. Sanchez-Galvis, E.M.; Cardenas-Gutierrez, I.Y.; Contreras-Ropero, J.E.; Garcia-Martinez, J.B.; Barajas-Solano, A.F.; Zuorro, A. An innovative low-cost equipment for electro-concentration of microalgal biomass. *Appl.Sci.* **2020**, *10*, 4841. [CrossRef]
17. Richmond, A. *Handbook of Microalgal Culture: Biotechnology and Applied Phycology*; John Wiley & Sons: Hoboken, NJ, USA, 2008.
18. Bellinger, E.G.; Sigee, D.C. *Freshwater Algae: Identification, Enumeration and Use as Bioindicators*; John Wiley & Sons. Ltd.: Chichester, UK, 2015.
19. Glauert, A.M. Fixation, Dehydration and Embedding of Biological Specimens. In *Practical Methods in Electron Microscopy*; Elsevier Science: Amsterdam, The Netherlands, 1984.
20. Fiera, C. Detection of food in the gut content of Heteromurusnitidus (Hexapoda:Collembola) by DNA/PCR-based molecular analysis. *North-West. J. Zool.* **2014**, *10*, 67–73.
21. Ruiz-Güereca, D.A.; Sánchez-Saavedra, M.d.P. Growth and phosphorus removal by Synechococcus elongatus co-immobilized in alginate beads with Azospirillum brasilense. *J. Appl. Phycol.* **2016**, *28*, 1501–1507. [CrossRef]
22. Ritchie, R.J. Consistent Sets of Spectrophotometric Chlorophyll Equations for Acetone, Methanol and Ethanol Solvents. *Photosynth. Res.* **2006**, *89*, 27–41. [CrossRef] [PubMed]
23. Delgadillo-Mirquez, L.; Lopes, F.; Taidi, B.; Pareau, D. Nitrogen and phosphate removal from wastewater with a mixed microalgae and bacteria culture. *Biotechnol. Rep.* **2016**, *11*, 18–26. [CrossRef]
24. Ermis, H.; Altınbaş, M. Determination of biokinetic coefficients for nutrient removal from anaerobic liquid digestate by mixed microalgae. *Environ. Boil. Fishes* **2018**, *31*, 1773–1781. [CrossRef]
25. Peleg, M.; Corradini, M.; Normand, M.D. The logistic (Verhulst) model for sigmoid microbial growth curves revisited. *Food Res. Int.* **2007**, *40*, 808–818. [CrossRef]

26. Banerjee, S.; Balakdas Tiwade, P.; Sambhav, K.; Banerjee, C.; Kumar Bhaumik, S. Effect of alginate concentration in wastewater nutrient removal using alginate-immobilised microalgae beads: Uptake kinetics and adsorption studies. *Biochem. Eng. J.* **2019**, *149*, 1–12. [CrossRef]
27. de-Bashan, L.E.; Bashan, Y. Immobilised microalgae for removing pollutants: Review of practical aspects. *Bioresour. Technol.* **2010**, *101*, 1611–1627. [CrossRef]
28. Aguilar-May, B.; del Pilar Sánchez-Saavedra, M.; Lizardi, J.; Voltolina, D. Growth of Synechococcus sp. immobilised in chitosan with different times of contact with NaOH. *J. Appl. Phycol.* **2007**, *19*, 181–183.
29. Bajpai, S.; Sharma, S. Investigation of swelling/degradation behaviour of alginate beads crosslinked with Ca^{2+} and Ba^{2+} ions. *React. Funct. Polym.* **2004**, *59*, 129–140. [CrossRef]
30. Liu, K.; Li, J.; Qiao, H.; Lin, A.; Wang, G. Immobilisation of Chlorella sorokiniana GXNN 01 in alginate for removal of N and P from synthetic wastewater. *Bioresour. Technol.* **2012**, *114*, 26–32. [CrossRef] [PubMed]
31. Azlin Suhaida, A.; Nurain Atikah Che, A.; Noor Illi Mohamad, P.; Amanatuzzakiah Abdul, H.; Faridah, Y.; Suzana, Y. Chlorella vulgaris logistic growth kinetics model in high concentrations of aqueous ammonia. *IIUM Eng. J.* **2018**, *19*, 1–9.

Review

Insights into Solar Disinfection Enhancements for Drinking Water Treatment Applications

Abdassalam A. Azamzam [1], Mohd Rafatullah [1,*], Esam Bashir Yahya [1], Mardiana Idayu Ahmad [1], Japareng Lalung [1], Sarah Alharthi [2], Abeer Mohammad Alosaimi [2] and Mahmoud A. Hussein [3,4]

[1] Division of Environmental Technology, School of Industrial Technology, Universiti Sains Malaysia, Penang 11800, Malaysia; azamzamabdassalam@student.usm.my (A.A.A.); esamyahya@student.usm.my (E.B.Y.); mardianaidayu@usm.my (M.I.A.); japareng@usm.my (J.L.)
[2] Department of Chemistry, Faculty of Science, Taif University, P.O. Box 11099, Taif 21944, Saudi Arabia; sarah.alharthi@tu.edu.sa (S.A.); a.alosaimi@tu.edu.sa (A.M.A.)
[3] Chemistry Department, Faculty of Science, King Abdulaziz University, P.O. Box 80923, Jeddah 21589, Saudi Arabia; maabdo@kau.edu.sa
[4] Chemistry Department, Faculty of Science, Assiut University, Assiut 71516, Egypt
* Correspondence: mrafatullah@usm.my; Tel.: +6-046-532-111; Fax: +6-046-56375

Abstract: Poor access to drinking water, sanitation, and hygiene has always been a major concern and a main challenge facing humanity even in the current century. A third of the global population lacks access to microbiologically safe drinking water, especially in rural and poor areas that lack proper treatment facilities. Solar water disinfection (SODIS) is widely proven by the World Health Organization as an accepted method for inactivating waterborne pathogens. A significant number of studies have recently been conducted regarding its effectiveness and how to overcome its limitations, by using water pretreatment steps either by physical, chemical, and biological factors or the integration of photocatalysis in SODIS processes. This review covers the role of solar disinfection in water treatment applications, going through different water treatment approaches including physical, chemical, and biological, and discusses the inactivation mechanisms of water pathogens including bacteria, viruses, and even protozoa and fungi. The review also addresses the latest advances in different pre-treatment modifications to enhance the treatment performance of the SODIS process in addition to the main limitations and challenges.

Keywords: water treatment; solar disinfection; pre-treatment; mechanism; SODIS system

1. Introduction

Access to safe drinking water is essential to health, which has been always a major concern as one of the basic human health fundamental rights worldwide. More than 2 billion people around the world lack proper sanitation and hygiene in terms of drinking water [1]. Treatment techniques for drinking water in many undeveloped and developing countries are either insufficient or inaccessible, and thus, millions lack access to safe water service. However, it has been reported that roughly 144 million mainly rely on surface water as their main source for drinking, cooking, and daily usage, resulting in half a million deaths annually, caused by diarrheal symptoms alone apart from other waterborne diseases [2]. In many urban areas, the governments supply households with clean water, after intensive treatment with a combination of traditional treatment techniques including chlorination, filtration, ultraviolet irradiation (UV irradiation), ozonation, flocculation, Fenton and photo-Fenton approach, etc. However, with the significant increase in world population, these methods may not adequately address the requirement of all households for drinking water, especially in the poor villages.

The past few years witnessed the development of many cost-effective and proper water treatment technologies, using sustainable and eco-friendly principals instead of chemical treatment processes [3]. Conventional SODIS has been used in many isolated regions in Africa and south-east Asia as a household water treatment method [3,4]. The utilization of solar power in water treatment has recently gained tremendous attention due to its accessibility, cost-effectiveness, and availability in most undeveloped countries. Despite the effectiveness and low costs of the most currently applied water treatment approaches such as ozonation, chlorination, and advanced oxidation processes they have major drawbacks including the generation of potential harmful byproducts, decomposition of organic pollutants, and/or inactivation instead of killing of water pathogenic microorganisms [5–8]. Different approaches have been used for solar disinfection, such as the filling of transparent polyethylene bottles with untreated water and then exposing them to sunlight for variable time, depending on the climate [9]. This simple procedure was found to be sufficient for killing the bacteria and other water microorganisms and considered a successful approach in different regions with illuminated periods such as Cameroon, India, Senegal, and South Africa [10]. However, conventional SODIS technology has been associated with several drawbacks, including the small and limited volume of water bottles, as efficiency of the process mainly depends on various factors including the initial level of water contamination, solar irradiance, water turbidity, time of day, and the atmospheric conditions [11]. The required time for some microbial inactivation was reviewed recently by Malato et al. [12] and reported to range from 20 min to 8 h depending on the microbe. The long period of time and unavailable direct solar power on some days are some of the known limitations of the SODIS process. However, great advances have been made in this field to enhance efficiency and overcome such issues, which can be observed from the marked accelerated increase of the number of publications each year regarding the utilization of solar disinfection of water as presented in Figure 1.

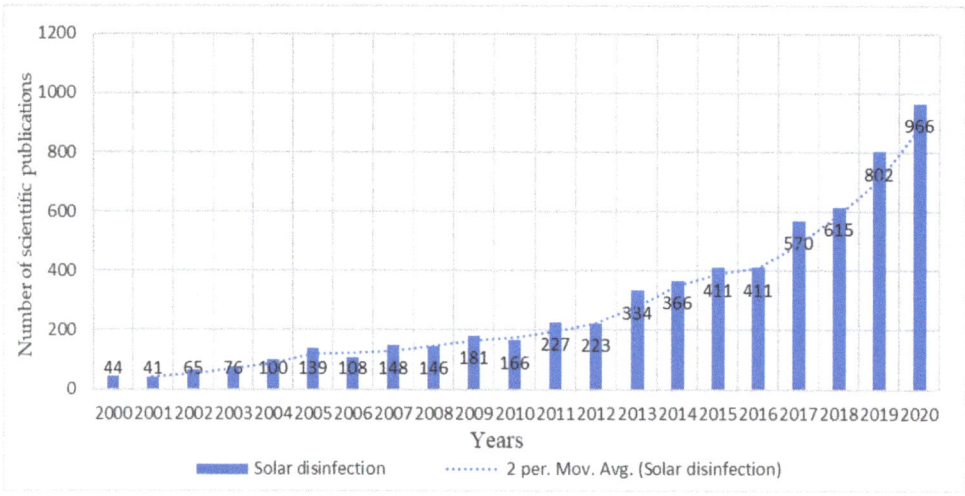

Figure 1. The number of scientific publications related to solar disinfection for the past 20 years (search done through Science Direct database on 4 September 2021 by using the keywords "Solar disinfection").

Many of these publications reviewed the mechanism of solar disinfection of microbes [13], modeling approaches to make SODIS faster [14], and photocatalytic enhancement of solar disinfection [11]. Previous reviews either discuss a particular side of the process or focus on a specific approach such as TiO_2 photocatalysts [15], ozone and photocatalytic processes [16], or the application of SODIS [17]. Chaúque et al. [18] recently identified the main limitations associated with the conventional SODIS process and reviewed its potential and challenges for large-scale application as a drinking water treatment strategy. This review presents a comprehensive discussion of the SODIS process, its inactivation mechanism of different water microorganisms, and the factors affecting the disinfection process. Different pre-treatment techniques to enhance the disinfection efficiency of SODIS process including the use of chemical agents such as photocatalysts as well as using synthetic and natural coagulants are also discussed. Finally, the main challenges and health precautions of consuming water after its exposure to such treatment systems are covered, presenting the latest health assessment in this regard.

2. Water Treatment Approaches

Several methods have been developed for water treatment depending on the initial level of water quality, the type of pollution, and the availability. The applicability of these approaches has always been a great challenge in many countries, taking into consideration the cost of installation, safety, and the effectiveness. However, most of these methods are designed to have two major phases: a solid–liquid separation phase, which runs as a pretreatment step, necessarily followed by a disinfection phase (such as SODIS) [19]. The pre-treatment phase could be physical, chemical, or biological approaches, which are applied to reduce the water turbidity, resulting from dusts, oil, and/or grease. The disinfection phase is used to eliminate the microorganisms using UV or solar treatments to meet the national or international requirements [20]. Despite the cost and applicability, pretreatment approaches can be classified into three approaches including physical, chemical, or biological approaches. Physical approaches are mostly used as pretreatment steps, which include filtration, adsorption, etc. In some cases, physical approaches may be sufficient for water treatment, but in terms of water microorganisms, further steps may be required. Chemical approaches such as coagulation, precipitation, hydrolysis, etc. are not favorable in terms of drinking water due to the potential health effect for the chemical agents.

2.1. Physical Approaches

The physical water treatment approaches typically consist of filtration techniques that involve using any particle size-based separation such as screens and cross flow filtration membranes, in addition to adsorption, sedimentation, and distillation techniques [21]. Physical treatment approaches are typically used in numerous undeveloped and developing countries as a pretreatment method for removing larger suspended material from water. Thines et al. [22] stated two main types of adsorption for different pollutants including physisorption and chemisorption. However, physisorption occurs due to the attraction of the organic pollutant and adsorbent by van der Waals force, which is a weak force (0.4–4.0 kJ/mol) resulting from electrical interactions between two or more atoms or molecules that are very close to each other, compared with the strong chemical bond such as hydrogen bonds, ionic interactions, and hydrophobic interactions (15, 20, and <40 kJ/mol, respectively) that may result between the surface of the adsorbent and the organic pollutant in chemisorption. Refer to Table 1 for a summary of conventional physical approaches for water treatment.

Table 1. Illustration of different physical approaches for water treatment.

Physical Approach	The Principle	Ref.
Adsorption	The adsorbent attracts organic and inorganic pollutants, leading to their accumulation onto its surface and eventually precipitate.	[23]
Filtration	Passing of water through a permeable membrane containing small pores able to trap the desired pollutant.	[24]
Sedimentation	Removing the suspended solids from water by means of gravity or separation.	[25]

2.2. Chemical Approaches

The chemical water treatment approach involves using chemicals in an array of procedures to expedite purification or separation of the water pollutant. Chemical approaches consist of various reactions that eventually lead to neutralization and hydrolyzing the water pollutants into harmless chemicals [26]. Salimi et al. [27] stated that coagulation and advanced oxidation processes are the main chemical methods in water treatment processes, which also include ozonation and Fenton treatment. These approaches are usually followed by photocatalysis and/or membrane techniques. The chlorination technique has been also used prior to powdered activated carbon filtration as a two-step water treatment approach [28]. Advanced oxidation is a technique that consists of using oxidizing agents for the oxidation of the pollutant [29]. Various advantages have been reported to this technique such as rapid reaction rates compared to similar techniques, not requiring a large area for processing the water, and less retention time [30].

3. Solar Disinfection and Water Treatment

Solar water disinfection, or simply SODIS, is a water disinfection process widely known in many African and Asian countries as a simple and low-cost treatment approach to eliminate the pathogenic microorganisms from drinking water [31]. The SODIS process is not a recent technology, numerous studies give descriptions of this approach in ancient India over 2000 years ago, with the water being placed in open trays outside in the sun [32]. However, the exact role and the mechanism of microbial disinfection was discovered during the 20th century when Aftim Acra et al. conducted and published a similar work explaining how sunlight disinfects water microorganisms [33,34]. Since then, a significant number of studies have been conducted revealing the great potential of SODIS in inactivating pathogenic microorganisms in water [35,36].

3.1. Mechanisms of Solar Disinfection

Recent investigation revealed that the microbial disinfection in the SODIS process is caused by the effect of two main factors, namely increasing the temperature resulting from light exposure and UV (A and B) of the solar rays [37]. Most UVC is absorbed while the rays pass through the atmosphere and the germicidal effect of UVB is significantly stronger than that of UVA (100 and 1000 times). UVB was found to pass through the atmosphere and reach natural water bodies, leading to a significant disinfection effect due to the induction of mutations and possibly apoptosis as well as imminent cell death [38]. Different studies have confirmed the effect of high temperature resulting from solar exposure on different water microorganisms, especially its synergistic effect with UV radiation that comes with solar radiation, which accelerates the disinfection process [39]. The short wavelengths of UVB (280–320 nm) are highly absorbable by the nucleic acid of living organisms within the water, causing severe damage in the genetic material and eventually causing their disinfection [40]. Numerous investigations have recently been conducted to confirm the ability and the mechanisms of sunlight to inactivate different standard microorganisms including bacteria, viruses, and fungi, in addition to protozoa and some helminths [3,41–43].

3.1.1. Bacterial Disinfection

Bacterial cells contain regulators of reactive oxygen species (ROS) to naturalize any byproduct resulting from cellular oxidative metabolism or imbalances that may be generated within the bacterial cells in addition to withstanding the production of reactive oxygen species by enzyme auto-oxidation [44]. *Escherichia coli* is known to be the most frequently studied species, as pathogenic waterborne bacteria and the main pathogen indicators in drinking water [45]. Solar light was found to affect the endogenous photosensitizers in bacterial cells in addition to ROS production upon exposure to solar light. Reactive oxygen species have been reported to cause cellular damage by initiating a variety of cellular reactions such as intracellular oxidation [46]. Berney et al. [47] investigated the inactivation of this species using the SODIS system and revealed that bacterial disinfection was mainly caused by the massive disrupting of cellular function. Upon solar exposure, efflux pump activity and the synthesis of ATP energy compounds cease shortly in the initial minutes and accelerated by time, leading to a reduction in glucose uptake and gradual loss of membrane potential. Finally, the bacterial cell membrane becomes highly permeable, and thus the inactivation occurs [48]. Catalase enzyme is the most known defense line for most bacteria in addition to Ahp Alkyl hydroperoxide reductase, glutathione reductase (GR), hydroperoxidases (HPI, HPII), and superoxide dismutases (FeSOD, MnSOD) [49]. Decomposition of hydrogen peroxide was found to be caused by a catalase enzyme, while scavenging the normally production of hydrogen peroxide in *E. coli* is the function of Ahp Alkyl hydroperoxide reductase enzyme [50]. Many studies stated that H_2O_2 itself is not harmful to bacterial DNA; however, some studies reported that oxidation of adenine can be caused by H_2O_2 production. H_2O_2 may also engulf and eliminate the adverse effect of hydroxyl radical and thus, its accumulation could be a sign of cell survival [51]. Gomes et al. [52] stated that when the bacterial cells are exposed to solar light, several reactions occur as consecutive events following complex mechanisms, which may be inhibited in some bacteria by the synergistic effect of ROS and the action of light. However, prolonged exposure to UVB was found to have direct effect on catalase functions in *E. coli* cells, and therefore enhance H_2O_2 accumulation [53]. UVA irradiation is responsible for excess production of singlet oxygen, which is an essential factor in bacterial cytotoxicity and gene expression through excitation of porfyrins and other chromophoric substances [54]. Catalase (CAT) and superoxide dismutase (SOD) are another two key enzymes in bacterial disinfection by solar light action. The reduction in these enzymes was found to be even worse than the over accumulation of ROS inside the bacterial cell. Some previous studies proposed that the excess H_2O_2 accumulation inside the cell followed by the subsequent reactions could lead to genetic mutations and possible cellular apoptosis, which is also accelerated by solar light [55].

3.1.2. Viral Disinfection

Viruses are obligate intracellular parasites, do not have a cell membrane, and are present in many environments including water. The endogenic inactivation process of these microorganisms is less important compared to the bacteria [48]. It has been reported that virus inactivation by the SODIS process occurs by damaging the viral genome by reactive oxygen species resulting from dissolved photosensitizers in the water. However, blocking of the UVB rays was found to prevent direct inactivation of viruses in PET bottles, and the inactivation was said to be dependent on initial water quality [56]. In a recent investigation, Sagripanti et al. [57] estimated the potential of solar radiation in COVID-19 inactivation as double-stranded viruses compared with previous reports of single-stranded RNA viruses. The authors found that double-stranded viruses should be inactivated relatively faster than single-stranded upon exposure to solar light. In contrast to bacteria, viruses are known for their simple structure that consists of protein capsid surrounding their genetic material. Some viruses contain endogenous chromophores, able to absorb the visible range of UVA light. Kohn et al. [58] stated that indirect photo-inactivation of viruses in aquas media can entirely depends on many factors (even in low concentrations), including exogenous presence, organic matter, nitrate, nitrite, and the constituents of iron-containing solution, which can significantly aid in sunlight absorption and thus produce reactive oxygen species such as H_2O_2, O_2^-, O_2, and OH^-. Studies evaluating the effect of SODIS process against different viruses that are commonly present in water are scarce. Many viruses including bacteriophage, bovine rotavirus, and somatic phage were completely inactivated at three logarithm unit reduction in solar exposure in less than 3 h [59,60]. Safapour and Metcalf [61] reported that Enterobacteria phage T2 was viable (able to infect the bacterial cells) at 62 °C, partially inactivated at 65 °C, and completely inactivated at 70 °C in 1.5 h. In a different study, viable FRNA coliphages were detected in SODIS reactors fitted with reflectors (increasing the water temperature by an additional 8–10 °C to 64–75 °C) even though *E. coli* was easily disinfected under identical conditions [62]. These viruses have the ability to resist the SODIS process by developing several protection mechanisms. In a similar investigation, Harding and Schwab [63] were also unable to reduce the viable murine noro-virus populations by using solar based

UVA light, revealing the potential of these microorganisms to develop resistance to such conventional treatment systems.

3.1.3. Disinfection of Other Water Microbes

Other water microorganisms including fungi, protozoa, and some helminths can be found in water sources and may realistically be classed as serious waterborne human pathogens. Most fungal disinfection studies using the SODIS process have concentrated on plant fungi [64]. However, Lonnen et al. [65] demonstrated that *Candida albicans* is readily inactivated after 6 h of solar exposure. Many protozoan pathogens commonly present in water such as *Cryptosporidium* spp., *Giardia lamblia*, and amoebae were reported to be more resistant to the SODIS process than bacteria and viruses [66]. Many waterborne protists possess, during several phases of their life, a thick-walled chitinous-cysts or oocysts; these strong structures protect them from undesirable conditions including the conventional SODIS process as well as other forms of disinfection such as chlorination and even boiling [67]. García-Gil et al. [48] reported that disinfection of protozoa and fungi is done through endogenous photo-inactivation mechanisms, which is similar to viral disinfection (through the action of UVB). However, it has been reported that impure water can enhance exogenous indirect damage, and thus the inactivation of bacteria as well as viruses, but not for other microorganisms such as protozoa. Exogenous indirect damage for protozoa and some fungi is generally negligible as they are characterized by a thick resistant oocyst-wall [68]. Inactivation of protozoa is a multifaceted mechanism depending on many factors such as the microbial species and its physiological state in addition to time of exposure and radiation wavelength. Natural solar light was found to be enough to deactivate many protozoa such as *Cryptosporidium parvum protozoon* by direct genome damage produced by the UVB radiation [69].

3.2. SODIS Effectiveness and Impact

SODIS is a simple process that depends on few factors associated with the initial quality of water to be treated, the container, and the weather in terms of solar irradiation, and weather in terms of temperature. The effectiveness of SODIS has been confirmed for various types of water microorganisms, including bacteria, fungi, viruses, and protozoa [18]. In order to achieve safe inactivation of water microorganisms, a significant number of studies reported the importance of gathering both the optical and thermal effect resulting from solar radiation to accelerate the inactivation process [39,70]. The synergistic effect of thermal and optical effect rays mainly depends on the initial water quality and the type of container in addition to the seasonal and environmental factors.

3.2.1. Initial Water Quality and Type of Container

The SODIS process involves placing the water (untreated or partially treated) into a specific container in most cases made from plastic or glass and then exposing it to direct sunlight for a certain period of time before drinking [1]. Glass containers considered more transparent and can be used to provide better sunlight penetration, but they have been linked with many drawbacks including blocking UV radiation, high expenses, fragility, and weight, making use of polyethylene terephthalate (PET) or polyethylene (PE) [71,72]. Turbidity of water was also found to affect the permeability of solar light; Chauque et al. [73] found that in water with less than 1 NTU, the water temperature reached 55 °C, which was found to be enough for the inactivation of *Acanthamoeba castellanii* cysts, *Salmonella typhimurium*, *E. coli*, *Enterococcus faecalis*, and *P. aeruginosa* in only 0.5 min. However, with turbid water (50 NTU). the same inactivation results were only achieved when the temperature reached 60 °C [73]. Amirsoleimani and Brion [74] stated that the inactivation of *E. coli* markedly decreased (from 5 to 1 log) with the increase of water turbidity from 0 to 200 NTU. The authors followed the same protocol and found that more than 4-log inactivation of two Gram-positive bacteria (*Staphylococcus epidermidis* and *Staphylococcus aureus*) was achieved. Pretreatment of water to obtain minimal turbidity is important to achieve the synergy of the thermal and optical effects of solar irradiation and thus better water disinfection. Microbial disinfection could be achieved by any or both of these effects, depending on exposure time, the type of microorganism, and the intensity. Even enough heat without solar radiation or in high turbidity may kill the bacteria, the presence of solar light and less turbidity will significantly enhance and speed up the disinfection process even with the thermal effect of solar power.

3.2.2. Environmental Factors

Being an environment-dependent process, SODIS is highly dependent on the weather and solar intensity, which is the major set-back in many countries. Prolonged periods are required in cloudy weather or during cold season to achieve satisfactory water disinfection. Luzi et al. [75] stated that on a clear sunny day, 6 h of exposure is enough compared with cloudy conditions. The effectiveness of SODIS was found to be the best in locations with significant amounts of strong sunlight during midday, mostly located around the equator [76]. Various studies have been conducted to overcome the weather issue in term of partially sunny or during cold season and enhance SODIS; Sommer et al. [77] suggested using darker containers or painting the underside of them with black to enhance the thermal disinfection. However, this could increase the water temperature, but it will prevent the penetration of solar irradiation through the water and minimize the action of endogenous photo-inactivation (through the action of UVB) [62]. Other studies reported using concentrators or reflectors to increase solar ray intensity and enhance water exposure [78,79]; using such method could significantly reduce the required exposure time and those utilize the partial sunny weather. In monsoon and winter seasons, it has been reported that log reduction value > 4 can be achieved for an exposure time to sunlight of around 17 h due to the significant low temperature in some countries, which may require the use of reflective reactors [39]. However, in hot seasons, only 6 to 8 h were found to be enough to achieve log reduction value of over 5 even by using the conventional PET reactors [17,39].

3.2.3. Resistance of Water Microorganisms

Despite the high sensitivity of most waterborne microorganisms to solar disinfection (thermal and optical effect) and them being easily inactivated by different intensities and different time, some studies have pointed out the ability of certain pathogens to develop resistance and adaptation to such conditions [80]. It has been reported that these pathogens, which include many forms of protozoa (such as *Cryptosporidium parvum* and *Acanthamoeba* spp.) and bacterial spores (such as *Bacillus subtilis* spores) are also more likely to resist conventional disinfection processes such as chlorination and ozonation [81,82]. Although SODIS has proven its ability to inactivate or considerably reduce the viability of most important and common waterborne disease such as *E. coli*, total and fecal coliforms as reported in many studies, the more recalcitrant the pathogen is, the higher doses of solar radiation and longer the exposure time will be required for its inactivation. Fiorentino et al. [83] stated that chlorination with only 1 mg/L was found to be more effective against the vegetative and non-spore-forming bacteria, particularly *E. coli*, than the use of H_2O_2 (50 mg/L)+ SODIS. However, the same authors found that chlorination was less effective in controlling the regrowth of the same bacteria; the percentage of *E. coli* in SODIS treated samples significantly decreased with the increase in incubation time, which was the opposite in the chlorinated samples [83]. SODIS was effective against chlorine-resistant microorganisms. Table 2 presents a summary of the literature regarding drinking water pathogens and their profile of resistance to SODIS.

Table 2. The effect of SODIS against chlorine-resistant microorganisms.

Microbial Species	Treatment Condition	Optical Effect (W/m^2)	Thermal Effect	Time	Inactivation	Ref
Acanthamoeba castellanii (cysts)	Simulated sunlight conditions	550 (UV B)	≤45	6 h	2.2-log	[66]
Cryptosporidium parvum (oocysts)	Simulated sunlight conditions + TiO$_2$	870 and 200 (UV B and A respectively)	40 °C	6 h	5.9-log	[84]
Acanthamoeba castellanii (cysts)	NaCl and NaOCl	243 (UV C)	55 °C	2 h	6-log	[85]
Cryptosporidium parvum (oocysts)	Simulated sunlight conditions + Cl$_2$	-	25 °C	1 h	>2-log	[86]
Bacillus subtilis spores	Simulated sunlight + TiO$_2$	870 and 200 (UV B and A respectively)	40 °C	8 h	1.1-log	[65]

3.3. Enhancement Approaches for Solar Disinfection of Water

The past few years witnessed a significant number of studies aiming to enhance the efficiency of the SODIS process and overcome its limitations by using a photocatalyst, pretreatment of water, or using continuous flow-based systems.

3.3.1. Enhancement of SODIS Using Photocatalyst

Advanced oxidation processes have recently been used as a promising option for drinking water treatment as well as simultaneous mineralization of organic matter [87]. A photo-assisted Fenton process has been successfully coupled with SODIS for the enhancement of bacterial inactivation. This process combines three main factors including Fe^{2+}, H_2O_2, and light [88]. Non-selective ROSs resulting from the consumption of H_2O_2 are the main effective factors in this deactivation process, which was found to inactivate most of pathogenic bacteria present in water. Shekoohiyan et al. [89] fabricated an iron oxide-based film to enhance the process of solar disinfection. The film was placed inside the plastic bottles (PET bottles) for better bactericidal capacity. The authors were able to enhance SODIS and reduce the exposure time by 60% in addition to significantly eliminating microbial regrowth. Titanium dioxide nanoparticles have been used as a photocatalytic water treatment to degrade organic pollutants and eliminate the pathogenic microorganisms [90]. Owing to its safety and non-toxicity, availability, and low cost, as well as its high photocatalytic activity, titanium dioxide (TiO$_2$) has gained the attention of many scientists although there are some remaining concerns about the separation of the nanoparticles from treated water. Metal oxide nanoparticles have been reported to display photocatalytic activity, resulting in the elimination of microbial growth. Zinc oxide nanoparticles possess many desirable properties such as its unique surface reactivity that is attached to many active sites, making them emergent and efficient nanophotocatalysts as compared to titanium dioxide [91]. However, some studies have concluded that nano-sized zinc oxide exhibited toxicity even at lower doses [92–94]. Pasupuleti et al. [95] reported that the incidences of microscopic lesions in liver, pancreas, heart, and stomach treated rates were higher in lower doses of nano-sized zinc oxide compared to higher doses, while for micro-sized zinc oxide, the incidences of the above lesions were higher in rats treated with a high dose. The inability to control the dose of zinc oxide in poor villages is still a great challenge that has restricted its usage. Reddy et al. [96] investigated silver-loaded TiO$_2$ nanoparticles supported on hydroxyapatite to enhance solar disinfection using *E. coli* bacteria in aqueous media, which showed 100% killing of *E. coli* within only 2 min. The separation of titanium dioxide is still a major challenge, as this photocatalyst is mostly used in slurry systems, thus the separation will increase the cost of the treatment process. However, immobilization of the photocatalyst on specific carriers such as films can overcome the extra costs of doing the separation step, especially for the household and treatment of small volumes of water.

It can also be a possible future water treatment approach in industrialized countries using renewable solar energy [97]. Photocatalytic materials have the advantage of the ability to be deposited on flexible substrates as coating agents, due to their high flexibility nature. Numerous studies reported the fabrication of photocatalytic coatings on a wide range of substrates for water treatment and other applications [98,99]. It is well known that upon using TiO_2, both hydroxyl and superoxide radicals are generated, which act as a catalyst under direct sunlight (UVA). ROSs including HO and O_2^- generated from TiO_2, which are able to oxidize and convert most of the organic molecules to simpler forms or induce mineralization (complete oxidization to CO_2 and H_2O) [100]. H_2O_2 and singlet oxygen are other oxidative species that may be generated, which also react together with hydroxyl radicals and superoxide, resulting in bacterial deactivation due to the excessive damage of their cell membrane and genetic material [101].

Carbon nanoparticles have recently received great attention among scientists, due to their hydrophobic nature and high surface area of nanoparticles [102]. In addition to their adsorption role, carbonaceous nanomaterials have photothermal or photocatalytic properties, making them highly desired in water treatment applications. The antibacterial activity of these materials is based on the photothermal heat produced by the nanoparticles when exposed to sunlight (photon energy) [103]. Many studies revealed that the physical interaction between carbon nanoparticles and the microorganisms can be crucial for solar disinfection; exposing polyethylene terephthalate bottles of contaminated water for a certain period of time can lead to 100% killing of all the microorganisms [104]. Malato et al. [12] reported that the use of photocatalysts was found to significantly shorten the required time for total disinfection. Carbon nanoparticles induce physical damage to bacterial cells as a result of their interaction, which can be desirable and helpful for the SODIS process. Upon the exposure of contaminated water in PET bottles to direct sunlight for a certain period of time, carbon nanomaterials were found to enhance the description of bacterial cell membrane. Owing to the generation of photo-thermal heat, when these nanoparticles are exposed to sunlight, there is a similar mechanism to cancerous cells destruction [105]. In a recent investigation, Maddigpu et al. [106] prepared composites from chitosan and carbon nanoparticles by the solution-casting approach and studied its inactivation efficiency towards *E. coli* bacteria under sunlight. The authors reported that the composite significantly enhanced SODIS as compared to the control with higher bactericidal efficiency. In a different study, Larlee [107] investigated using a low-tech photocatalyst for solar water disinfection and color removal. The authors used different bare clays as photocatalysts for the SODIS process, and revealed the ability to inactivate *E. coli* within only 1 h of sunlight exposure, suggesting bare clays as cheap material for potential photocatalysts for SODIS. Refer to Table 3 for a summary of literature regarding the enhancement of SODIS using different photocatalysts.

Table 3. Summary of the literature regarding the enhancement of SODIS using different photocatalyst.

Photocatalyst	Microorganisms Tested	Enhancing the Inactivation Rate	Time	Condition	Ref.
Composite $TiO_2/SiO_2/Au$ films	*E. coli*, total coliforms, and *Enterococci*	1.5, 1.3, and 1.6 fold survival decrease	24 to 72 h	Under natural solar radiation	[108]
photo-Fenton system ($Fe^{3+}/H_2O_2/hv$)	*E. coli*	Increased by 355%	240 min	Under natural solar radiation and natural pH	[109]

Table 3. Cont.

Photocatalyst	Microorganisms Tested	Enhancing the Inactivation Rate	Time	Condition	Ref.
ZnO Nanorods	E. coli	2.5-fold survival decrease after	180 min	Under simulated sunlight of low intensity	[110]
TiO$_2$	Cryptosporidium parvum	Enhanced SODIS elimination from 81.3% to 98.3%	8 h	Cloudy solar irradiance	[111]
TiO$_2$/SiO$_2$ thin films	Vibrio spp., Enterococci, and E. coli	Enhanced SODIS elimination by 27%	80 min	Under natural solar light	[108]
Fe$_2$O$_3$-TiO$_2$-based nanoparticles	Vibrio fischeri	99.4% inactivation efficiency	240 min	Visible light irradiation at different temperatures	[112]
Nano-structured ZnO	E. coli	15% higher disinfection efficiency	15 min	Under actual sunlight	[113]
TiO$_2$ nanoparticles and ZnO/TiO$_2$	Enterobacter, Klebsiella and pseudomonas	Reduction of bacteria from 3.6×10^8 to 1.63×10^4 CFU	-	UV lamp and solar radiation	[114]
TiO$_2$ nanoparticles	E. coli	Total inactivation (6.5-log) compared with 4-log without TiO$_2$	240 min	Under solar conditions	[72]
Boron-doped ZnO	E. coli and Enterococcus sp.	Total inactivation (3-log) compared with 1-log of ZnO	180 min	Under solar simulated irradiation.	[115]
Fe^{3+} and H$_2$O$_2$	E. coli and Klebsiella pneumoniae	Complete inactivation of E. coli, but K. pneumoniae decreased only 1-log.	350 min	Under simulated solar light	[116]

3.3.2. Coagulant Pre-Treatment to Solar Disinfection

Most of water bodies possess a certain level of turbidity, resulting from dissolved and suspended organic materials as well as microorganisms, which are able to block or limit the penetration of sunlight through the untreated water during solar disinfection [93]. Thus, many scientists suggested pre-treatment of turbid water to less than 30 NTU before using SODIS [117]. Numerous methods have been used to reduce turbidity including a variety of filtration techniques, gravity settling, centrifuge force, and the use of natural and synthetic coagulants/adsorbents [118]. The use of natural coagulants followed by a filtration process is highly preferable compared to synthetic coagulation agents due to cytotoxicity concerns. In a recent investigation, Keogh et al. [119] used *Moringa oleifera* as a flocculating agent to reduce turbidity as a pre-treatment for SODIS. The authors reported that this flocculating agent was able to reduce the turbidity of water samples, leaving a bio-active sludge layer in the bottoms of the bottles. After 24 h of *Moringa oleifera* treatment, the authors reported impressive reduction in water microorganisms compared with untreated turbid controls. In a different study, an aqua lens coupled with the natural coagulant (*Moringa oleifera*) was found to be effective as an eco-friendly and safe treatment approach for households, which showed great potential in enhancing the efficiency of solar disinfection [120]. Natural coagulants that release lower sludge quantity are preferable as the post-separation costs will be reduced and the natural alkalinity of water is retained during the process of the water treatment [121]. The seed powder of *Strychnos potatorum* is another effective natural coagulant, which was tested by Arafat and Mohamed [122]. The authors reported 90–99% removal of water bacteria after using this natural coagulant in addition to its coagulation role. *Carica papaya* seeds are another natural coagulant obtained from tropical trees and characterized by cystine protease proteins (water-soluble and positively charged proteins), which are also utilized as a putative natural coagulant [123]. In a study done by Unnisa et al. [124], *Carica papaya* seeds were used as a pre-treatment for SODIS to reduce water turbidity and inactivate coliform bacteria. The authors reported 100% removal efficiency for turbidity even at the minimum concentrations of natural coagulant in only 30 min. The time for SODIS was reduced to 2 h, which was enough to achieve 100% removal of coliform bacteria from the water. Limited studies have been done on coupling natural coagulant with the SODIS process. Table 4 presents a summary of studies of using natural coagulants to reduce the water turbidity and coupling them with the SODIS process.

Table 4. Summary of the literature on using coagulant pretreatment for SODIS process.

Coagulant	Initial Turbidity	Enhanced Turbidity	Microorganisms Tested	Inactivation Rate	Time	Condition	Ref
Powdered *Moringa oleifera* seeds	200 NTU	28.5 NTU	*E. coli*	6-log	6 h	Direct sunlight	[119]
Opuntia cochenillifera	111 NTU	7.83 NTU	*E. coli*	2.86-log	12 h	Under natural sunlight	[125]
Maerua subcordata or *Moringa stenopetala*	150 NTU	10 NTU	Fecal coliforms	Complete removal	4 h	Under natural sunlight	[126]
Carica papaya seeds	60 NTU	0 NTU	*E. coli* and coliforms	Complete removal	2 h	Under natural sunlight	[124]
Moringa oleifera	150 NTU	3 NTU	Total and fecal coliforms	99% removal	6 h	Under natural sunlight	[127]
Artemisia annua	-	-	*E. coli, B. subtilis,* and *E. faecalis*	6-log	6 h	Under natural sunlight	[128]

3.4. Continuous Flow SODIS System

In the photocatalytic disinfection process using normal sunlight, it was found that using photocatalysts (semiconductors) with suitable optical band-gap for the generation of reactive oxygen species in the presence of solar-light can significantly enhance the inactivation of microorganisms [35,129]. Several solar light/visible light metallic and even non-metallic photocatalysts have been investigated and proven to have a significant photocatalytic potential at laboratory scale [130,131]. Zhang et al. [132] stated that the operational costs can be reduced by immobilizing the visible light photocatalysts to avoid the requirement of photocatalyst separation phase. Mbonimpa et al. [38] developed a continuous-flow reactor based on SODIS by ultraviolet B radiation. For the reactor, the authors used a quartz tube with an internal diameter of 2.25 cm and placed the tube in the focus of a solar collector of compound parabolic with 125 cm long and 42 cm opening. Significant reduction in *E. coli* count was achieved, which was believed to be due to the exposure of the raw water for less than an hour and the marked increase in water temperature. Chauque et al. [73] developed a novel continuous-flow system that combines both the effect of optical and thermal solar radiation with the recirculation of water. The authors designed the system to have both disinfection mechanisms of SODIS using parabolic trough concentrator to increase water temperature with a thermo-absorbent tube in the focus of the concentrator. Furthermore, the UV irradiator of solar power was a Fresnel collector, which joined the parabolic trough with the concentrator, and then the authors placed tubular quartz reactors (where the water circulates) in the focus of the concentrators. The authors reported that water temperature reached 70 °C, which then allowed it to flow through the irradiator for several minutes and thus full inactivation of all cysts as well as water bacteria was achieved [73]. In another recent study, Roshith et al. [133] reported a novel technique for a continuous flow water disinfection process by using red phosphorus-based photocatalyst. The authors immobilized crystalline red phosphorus in the inner walls of quartz capillary tube using a solid-state approach. The process of continuous water treatment was achieved under direct sunlight using the set-up as an optofluidic reactor and continuous flow photocatalytic (Figure 2). The reactor with the immobilized photocatalyst was able to reduce 99.99% of the *E. coli* in only 14 min when it was tested under direct sunlight. Furthermore, the authors reported that no visible colonies were found from the water samples of 28 min exposure time, which confirms the high efficiency of this process and its promising potential for commercial scale-up.

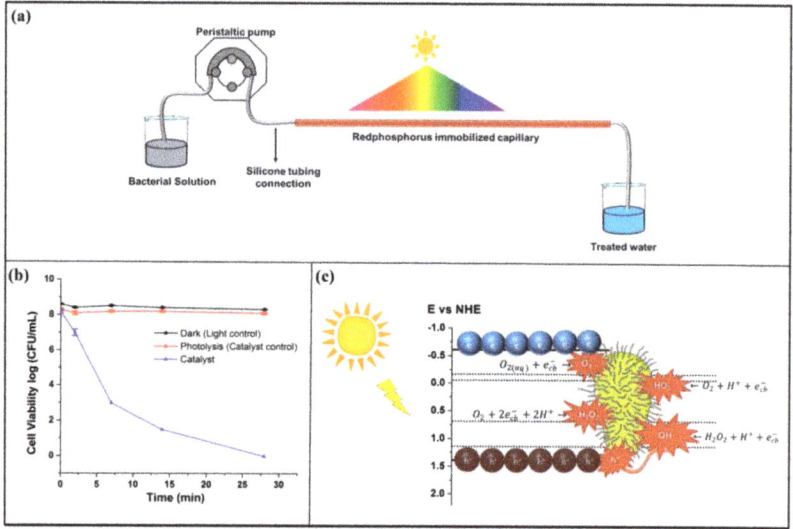

Figure 2. Schematic illustrations of continuous-flow solar photocatalytic disinfection: (**a**) the overall disinfection process, (**b**) cell viability, and (**c**) the mechanism of photocatalytic disinfection. Adapted from Roshith et al. [133].

Numerous designs of SODIS-based continuous flow water purifications systems have recently been investigated, which were based on different solar radiation collectors able to increase water temperature in a continuous flow manner. Domingos et al. [134] developed a continuous-flow SODIS based water disinfection system. This system was composed of a double reflection disk-shaped collector (3.8 and 1.3 m in diameter). The two solar disks were able to collect and concentrate the radiation and focus it in the aluminum block (the reactor) surrounding the tubular channels for water-flow. The authors were able to achieve 65 °C water temperatures on average days of solar radiation with a flow rate of 63 L per hour. However, this temperature was able to inactivate *E. coli* and other waterborne pathogens. Yildiz [135] stated that the raw water should be filtered to reduce its turbidity and to remove larger particulate matter prior to entering the raw water tank in the continuous flow water purifications systems. Amara et al. [136] fabricated a solar-based continuous flow water purification system by using a 1.3 m diameter parabolic solar disk. The authors attached the focus of the concentrator to a fluid filled chamber, in addition to using a copper tube (14 mm diameter) to permit fluid-to-fluid heat exchange. By using this system, the authors were able to significantly enhance SODIS performance by increasing the temperature of raw water by more than 8.5 °C (from 52 °C to more than 60 °C) at a flow rate of 1 L per minute. In a different study, Dobrowsky et al. [137] investigated the water disinfection efficiency of a closed coupled purification system able to significantly reduce the number of water microorganisms. Owing to the metal-based system, the authors were able to significantly enhance the water heating, and reported reduction of *E. coli*, heterotrophic bacteria, and total coliforms.

4. Limitations and Challenges of SODIS

Natural sunlight only contains 5% of UV, which may not enough to disinfect all the water pathogens even using the photocatalysts. Therefore, most of the solar spectrum is not utilized by these wide band-gap photocatalysts [138]. Another limitation that has been reported with photocatalytic processes is that the particles of the photocatalysts are suspended in the reaction as these processes are normally conducted in the batch mode, forming a slurry of suspended particles of used catalyst, which has to be removed eventually using a post-treatment step such as centrifugation or filtration [139]. Using an external UV source and/or a post-treatment step may significantly impede the cost effectiveness of the SODIS process [140]. Numerous studies aimed to overcome these issues in the past few years, aiming to develop an effective visible light photocatalyst able to ensure that the photocatalysis process worked under normal conditions of SODIS without the need for any external light sources [141,142]. The efficacy and mechanisms of SODIS against *E. coli* have been extensively studied since the 1980s; other types of pathogens are yet to be studied. In this regard, there are no available predictive approaches for the expected effectiveness of the SODIS process worldwide. Most published studies conducted on batch experiments suggest that the ability of microorganisms to adapt to solar may change the statistics of disinfection potential. Although SODIS is proven to be an effective water treatment process in many low-income communities with promotional strategies, many concerns have been raised recently regarding the conventional SODIS process including the fear of leaching in plastic bottles, cloudy weather and latitude, water turbidity, and community acceptance.

4.1. Leaching in Plastic Bottles and Potential Adverse Health Effects

The use of polyethylene terephthalate bottles in the SODIS system in low-income communities raises a public concern of potential chemical release at high temperatures. The use of plastic bottles to expose the water to sunlight, which causes elevation in water temperature, raises health concerns about the potential generation of photoproducts or migration of organic compound from the plastic bottles to treated water, which have been linked with many chronic and genetic disorders [143,144]. A limited number of research studies assessed the adverse health effects and the potential toxicity of consuming the water from SODIS process [145,146]. The Ames test was carried out to assess the potential cytotoxicity of the water resulting from the SODIS process [147,148]. However, the authors did not report any significant cytotoxicity in the water. The evaluation of advanced oxidative processes (AOP), as most of the studies claim the toxic effect of plastic, was carried out either on higher temperatures than sunlight or the direct evaluation of PET plastic [149]. Westerhoff et al. [150] reported potential health risks from using water after SODIS treatment, due to the plasticizers and carcinogenic compounds that could leach out into the water inside the bottles at a higher temperature. The same authors found exceeded carcinogenic risk levels in SODIS-treated water, which came from the plasticizer di(2-ethylhexyl)adipate (DEHA). In a different study, Schmid et al. [151] evaluated the potential transfer of any organic compounds from the PET bottles to treated water during the SODIS process. The authors exposed the bottles to solar power for 17 h with a total residence time of 48 h and reported similar values of plasticizer in commercially produced bottled water, revealing a minimum toxicological risk, and a safety factor of 8.5, for consuming such water. However, other researchers recommended replacing the plastic bottles at least every six months to avoid or at least minimize the effects of their wear and tear. Biopolymeric plastic based bottles could be used as a safer option than the petroleum-based ones. A significant number of studies have proven the biocompatibility and safety of using biopolymer-based materials in different applications [152–154]. We believe that biopolymer-based water bottles could be a potential solution to remove the health risk concern of conventional plastic and to minimize plastic pollution and maintain sustainability. Despite the higher leaching potential of biopolymeric bottles compared with the conventional one, their natural base makes them a highly preferable and safer option.

4.2. Inclement Weather and Latitude

In the SODIS process, the exposure time of the bottles is mainly dependent on the weather conditions and the location, which may not be available all the time in many countries. Luzi et al. [75] stated that it is recommended to expose the contaminated water for at least 6 h on sunny days in tropical countries, suggesting leaving the bottles to have a full day in possible cloudy interrupting weather, which is uncommonly found in many countries. However, according to the Swiss Federal Institute, the top population countries around the world, including China, India, Bangladesh, Indonesia, Kenya, and Nigeria, are located in the region considered to be the most suitable regions for SODIS [155]. The weather may not be suitable all the time, but as discussed earlier, using SODIS enhancers could minimize the exposure time and thus increase the rate of water treatment. The use of an external source may increase the costs, but in a large-scale water treatment project, this could be a good option especially in the regions that usually have cloudy interrupting weather.

4.3. Water Turbidity and Community Acceptance

Water turbidity is another factor that significantly affects the SODIS process; turbid water by different suspended particles was found to block solar light from reaching the microorganisms within the water, and thus, SODIS is unsuitable in areas with water turbidity of >30 NTU, unless the water was pre-treated or filtered [119]. After treatment, drinking water should have <1 NTU, but 5 NTU is satisfactory in many rural water facilities. Successful implementation of a SODIS system requires behavior change as well as alteration to daily life, which is important to reduce health risks among the residents of particular communities. Sommers M. [156] stated that SODIS promotion depends on many factors, which must be taken into serious consideration including risks, attitudes, norms, and ability, which can all significantly impact the efficiency of promotion. Adopting an at-home water treatment systems such as SODIS based is still challenging, as it's associated with poverty as well as lack of resources. However, a community tends to be less likely to adopt it than if it was associated with positive emotions, which is the attitude of many rural areas especially in Africa and southeast Asia [157]. A pre-filter for turbid water and unsuitable weather requires an additional step/s for the water such as pre-treatment filtration and external UV treatment, which some individuals may find to be less user friendly. Furthermore, community acceptance and preconceived notions are further hurdles that the SODIS process faces in many countries. More studies in this regard should be conducted to spread knowledge about the effectiveness and safety of such treatment.

5. Future Prospective

Great advances made in recent decades in using SODIS process for water disinfection have proven that this process is an effective and potentially inexpensive water treatment method providing safe drinking water. The modification of the SODIS process using different pre-treatment steps such as coagulant or the use of photocatalysts has significantly enhanced the performance of the SODIS process which has become able to eliminate most of water pathogens and thus reduce the prevalence and even mortality of waterborne diseases. Many communities around the world lacking safe drinking water and still depending on the conventional SODIS may benefit from such modification techniques for better water quality in shorter time. Future generations could witness more utilization of solar energy in water treatment applications even in developed countries. The direct conversion of solar energy into electricity using photo-voltaic effects has become priority for many countries as a more sustainable source of electricity than electricity powered by fossil fuels [158]. However, these electric power stations could be developed to treat water and produce energy at the same time. Continuous flow systems for SODIS have been already achieved; a thin network of tubes can be exposed to solar power in these stations to continuously provide the water in addition to the energy in a large-scale base for developed and developing countries. SODIS-based continuous flow systems could significantly increase the productivity of treated water compared with a conventional SODIS process (patch-based approach) using bottles. Traditional SODIS has been used in most cases when sufficient resources of water are not available, especially in rural areas as we mentioned earlier, for affording more expensive treatments approaches such as filtration, chlorination, etc. The future of SODIS will not be only based on bottles, it could be developed to become a faster and more productive continuous flow system. However, in many poor areas all over the world, household-level SODIS could be their available option for many years. Based on our previous discussion, a collective of several studies could be combined together to obtain the desired goal in terms of effectiveness under affordable costs. The modification techniques of SODIS may not be affordable for a household-based system, given that if they cannot afford using the conventional techniques such as filtration and chlorination, it is extremely unlikely that they would be able to afford SODIS additives such as titanium dioxide. Realistically speaking, the SODIS process could someday be applied in the large-scale treatment of drinking water, taking into consideration the benefits of different modification and enhancement techniques that have been applied by researchers. Further economic evaluation studies are highly needed that can address the costs of applying such projects and their potential compared with conventional techniques.

Author Contributions: Conceptualization, A.A.A., E.B.Y. and M.R.; writing—original draft preparation, A.A.A. and E.B.Y.; writing—review and editing, M.I.A., J.L., S.A., A.M.A., M.A.H. and M.R.; supervision, M.I.A., J.L. and M.R.; funding acquisition, M.R. All authors have read and agreed to the published version of the manuscript.

Funding: The authors would like to express their appreciation to Ministry of Higher Education Malaysia for Fundamental Research Grant Scheme with Project Code: FRGS/1/2019/STG07/USM/02/12. The APC was funded by Universiti Sains Malaysia, Malaysia.

Institutional Review Board Statement: Not applicable.

Informed Consent Statement: Not applicable.

Data Availability Statement: Not applicable.

Acknowledgments: The authors would like to express their appreciation to Universiti Sains Malaysia, Malaysia for providing the support and research facilities for this study.

Conflicts of Interest: The authors declare that they have no known competing financial interests or personal relationships that could have appeared to influence the work reported in this paper.

References

1. Cowie, B.E.; Porley, V.; Robertson, N. Solar disinfection (SODIS) provides a much underexploited opportunity for researchers in photocatalytic water treatment (PWT). *ACS Catal.* **2020**, *10*, 11779–11782. [CrossRef]
2. Pichel, N.; Vivar, M.; Fuentes, M. The problem of drinking water access: A review of disinfection technologies with an emphasis on solar treatment methods. *Chemosphere* **2019**, *218*, 1014–1030. [CrossRef] [PubMed]
3. McGuigan, K.G.; Conroy, R.M.; Mosler, H.-J.; du Preez, M.; Ubomba-Jaswa, E.; Fernandez-Ibanez, P. Solar water disinfection (SODIS): A review from bench-top to roof-top. *J. Hazard. Mater.* **2012**, *235*, 29–46. [CrossRef] [PubMed]
4. Polo-López, M.; Fernández-Ibáñez, P.; Ubomba-Jaswa, E.; Navntoft, C.; García-Fernández, I.; Dunlop, P.; Schmid, M.; Byrne, J.; McGuigan, K.G. Elimination of water pathogens with solar radiation using an automated sequential batch CPC reactor. *J. Hazard. Mater.* **2011**, *196*, 16–21. [CrossRef] [PubMed]

5. Singer, P.C.; Obolensky, A.; Greiner, A. DBPs in chlorinated North Carolina drinking waters. *J. Am. Water Work. Assoc.* **1995**, *87*, 83–92. [CrossRef]
6. Nasuhoglu, D.; Isazadeh, S.; Westlund, P.; Neamatallah, S.; Yargeau, V. Chemical, microbial and toxicological assessment of wastewater treatment plant effluents during disinfection by ozonation. *Chem. Eng. J.* **2018**, *346*, 466–476. [CrossRef]
7. Aguilar, S.; Rosado, D.; Moreno-Andrés, J.; Cartuche, L.; Cruz, D.; Acevedo-Merino, A.; Nebot, E. Inactivation of a wild isolated Klebsiella pneumoniae by photo-chemical processes: UV-C, UV-C/H2O2 and UV-C/H2O2/Fe3+. *Catal. Today* **2018**, *313*, 94–99. [CrossRef]
8. Levchuk, I.; Rueda-Márquez, J.; Suihkonen, S.; Manzano, M.; Sillanpää, M. Application of UVA-LED based photocatalysis for plywood mill wastewater treatment. *Sep. Purif. Technol.* **2015**, *143*, 1–5. [CrossRef]
9. Nalwanga, R.; Quilty, B.; Muyanja, C.; Fernandez-Ibañez, P.; McGuigan, K.G. Evaluation of solar disinfection of *E. coli* under Sub-Saharan field conditions using a 25L borosilicate glass batch reactor fitted with a compound parabolic collector. *Sol. Energy* **2014**, *100*, 195–202. [CrossRef]
10. Helali, S.; Polo-López, M.I.; Fernández-Ibáñez, P.; Ohtani, B.; Amano, F.; Malato, S.; Guillard, C. Solar photocatalysis: A green technology for *E. coli* contaminated water disinfection. Effect of concentration and different types of suspended catalyst. *J. Photochem. Photobiol. A Chem.* **2014**, *276*, 31–40. [CrossRef]
11. Byrne, J.A.; Fernandez-Ibanez, P.A.; Dunlop, P.S.; Alrousan, D.; Hamilton, J.W. Photocatalytic enhancement for solar disinfection of water: A review. *Int. J. Photoenergy* **2011**, *2011*. [CrossRef]
12. Malato, S.; Fernández-Ibáñez, P.; Maldonado, M.I.; Blanco, J.; Gernjak, W. Decontamination and disinfection of water by solar photocatalysis: Recent overview and trends. *Catal. Today* **2009**, *147*, 1–59. [CrossRef]
13. Castro-Alférez, M.; Polo-López, M.I.; Fernández-Ibáñez, P. Intracellular mechanisms of solar water disinfection. *Sci. Rep.* **2016**, *6*, 1–10.
14. García-Gil, Á.; García-Muñoz, R.A.; McGuigan, K.G.; Marugán, J. Solar Water Disinfection to Produce Safe Drinking Water: A Review of Parameters, Enhancements, and Modelling Approaches to Make SODIS Faster and Safer. *Molecules* **2021**, *26*, 3431. [CrossRef]
15. Demirel, C.S.U.; Birben, N.C.; Bekbolet, M. A comprehensive review on the use of second generation TiO2 photocatalysts: Microorganism inactivation. *Chemosphere* **2018**, *211*, 420–448. [CrossRef]
16. Gomes, J.; Matos, A.; Gmurek, M.; Quinta-Ferreira, R.M.; Martins, R.C. Ozone and photocatalytic processes for pathogens removal from water: A review. *Catalysts* **2019**, *9*, 46. [CrossRef]
17. Zhang, Y.; Sivakumar, M.; Yang, S.; Enever, K.; Ramezanianpour, M. Application of solar energy in water treatment processes: A review. *Desalination* **2018**, *428*, 116–145. [CrossRef]
18. Chaúque, B.J.M.; Rott, M.B. Solar disinfection (SODIS) technologies as alternative for large-scale public drinking water supply: Advances and challenges. *Chemosphere* **2021**, *281*, 130754. [CrossRef] [PubMed]
19. Li, F.; Wichmann, K.; Otterpohl, R. Review of the technological approaches for grey water treatment and reuses. *Sci. Total Environ.* **2009**, *407*, 3439–3449. [CrossRef] [PubMed]
20. Chibowski, E.; Szcześ, A. Magnetic water treatment—A review of the latest approaches. *Chemosphere* **2018**, *203*, 54–67. [CrossRef] [PubMed]
21. Cho, Y.I.; Lane, J.; Kim, W. Pulsed-power treatment for physical water treatment. *Int. Commun. Heat Mass Transf.* **2005**, *32*, 861–871. [CrossRef]
22. Thines, R.; Mubarak, N.; Nizamuddin, S.; Sahu, J.; Abdullah, E.; Ganesan, P. Application potential of carbon nanomaterials in water and wastewater treatment: A review. *J. Taiwan Inst. Chem. Eng.* **2017**, *72*, 116–133. [CrossRef]
23. Panahi, Y.; Mellatyar, H.; Farshbaf, M.; Sabet, Z.; Fattahi, T.; Akbarzadehe, A. Biotechnological applications of nanomaterials for air pollution and water/wastewater treatment. *Mater. Today Proc.* **2018**, *5*, 15550–15558. [CrossRef]
24. Anjum, M.; Miandad, R.; Waqas, M.; Gehany, F.; Barakat, M. Remediation of wastewater using various nano-materials. *Arab. J. Chem.* **2019**, *12*, 4897–4919. [CrossRef]
25. Qian, Y.; Chen, Y.; Hu, Y.; Hanigan, D.; Westerhoff, P.; An, D. Formation and control of C-and N-DBPs during disinfection of filter backwash and sedimentation sludge water in drinking water treatment. *Water Res.* **2021**, *194*, 116964. [CrossRef] [PubMed]
26. Meghwal, K.; Agrawal, R.; Kumawat, S.; Jangid, N.K.; Ameta, C. Chemical and Biological Treatment of Dyes. In *Impact of Textile Dyes on Public Health and the Environment*; IGI Global: Hershey, PA, USA, 2020; pp. 170–204.
27. Salimi, M.; Esrafili, A.; Gholami, M.; Jafari, A.J.; Kalantary, R.R.; Farzadkia, M.; Kermani, M.; Sobhi, H.R. Contaminants of emerging concern: A review of new approach in AOP technologies. *Environ. Monit. Assess.* **2017**, *189*, 1–22. [CrossRef] [PubMed]
28. Li, W.; Wu, R.; Duan, J.; Saint, C.P.; van Leeuwen, J. Impact of prechlorination on organophosphorus pesticides during drinking water treatment: Removal and transformation to toxic oxon byproducts. *Water Res.* **2016**, *105*, 1–10. [CrossRef]
29. Wang, J.; Zhuan, R. Degradation of antibiotics by advanced oxidation processes: An overview. *Sci. Total Environ.* **2020**, *701*, 135023. [CrossRef]
30. Marican, A.; Durán-Lara, E.F. A review on pesticide removal through different processes. *Environ. Sci. Pollut. Res.* **2018**, *25*, 2051–2064. [CrossRef]
31. Sciacca, F.; Rengifo-Herrera, J.A.; Wéthé, J.; Pulgarin, C. Dramatic enhancement of solar disinfection (SODIS) of wild *Salmonella* sp. in PET bottles by H2O2 addition on natural water of Burkina Faso containing dissolved iron. *Chemosphere* **2010**, *78*, 1186–1191. [CrossRef]

32. Downes, A. Researches on the effect of light upon bacteria and other organisms. *Proc. R. Soc. Lond.* **1877**, *26*, 488–500.
33. Acra, A.; Karahagopian, Y.; Raffoul, Z.; Dajani, R. Disinfection of oral rehydration solutions by sunlight. *Lancet* **1980**, *2*, 1257–1258. [CrossRef]
34. Acra, A.; Jurdi, M.; Mu'Allem, H.; Karahagopian, Y.; Raffoul, Z. Sunlight as disinfectant. *Lancet* **1989**, *333*, 280. [CrossRef]
35. Porley, V.; Chatzisymeon, E.; Meikap, B.C.; Ghosal, S.; Robertson, N. Field testing of low-cost titania-based photocatalysts for enhanced solar disinfection (SODIS) in rural India. *Environ. Sci. Water Res. Technol.* **2020**, *6*, 809–816. [CrossRef]
36. Rommozzi, E.; Giannakis, S.; Giovannetti, R.; Vione, D.; Pulgarin, C. Detrimental vs. beneficial influence of ions during solar (SODIS) and photo-Fenton disinfection of *E. coli* in water:(Bi) carbonate, chloride, nitrate and nitrite effects. *Appl. Catal. B Environ.* **2020**, *270*, 118877. [CrossRef]
37. Giannakis, S. Analogies and differences among bacterial and viral disinfection by the photo-Fenton process at neutral pH: A mini review. *Environ. Sci. Pollut. Res.* **2018**, *25*, 27676–27692. [CrossRef] [PubMed]
38. Mbonimpa, E.G.; Vadheim, B.; Blatchley III, E.R. Continuous-flow solar UVB disinfection reactor for drinking water. *Water Res.* **2012**, *46*, 2344–2354. [CrossRef] [PubMed]
39. Vivar, M.; Pichel, N.; Fuentes, M.; López-Vargas, A. Separating the UV and thermal components during real-time solar disinfection experiments: The effect of temperature. *Sol. Energy* **2017**, *146*, 334–341. [CrossRef]
40. Valero, P.; Giannakis, S.; Mosteo, R.; Ormad, M.P.; Pulgarin, C. Comparative effect of growth media on the monitoring of *E. coli* inactivation and regrowth after solar and photo-Fenton treatment. *Chem. Eng. J.* **2017**, *313*, 109–120. [CrossRef]
41. Gárcia-Fernández, I.; Polo-López, M.I.; Oller, I.; Fernandez-Ibanez, P. Bacteria and fungi inactivation using Fe3+/sunlight, H2O2/sunlight and near neutral photo-Fenton: A comparative study. *Appl. Catal. B Environ.* **2012**, *121*, 20–29. [CrossRef]
42. Giannakis, S.; Ruales-Lonfat, C.; Rtimi, S.; Thabet, S.; Cotton, P.; Pulgarin, C. Castles fall from inside: Evidence for dominant internal photo-catalytic mechanisms during treatment of Saccharomyces cerevisiae by photo-Fenton at near-neutral pH. *Appl. Catal. B Environ.* **2016**, *185*, 150–162. [CrossRef]
43. Polo, D.; García-Fernández, I.; Fernández-Ibañez, P.; Romalde, J.L. Hepatitis A Virus Disinfection in Water by Solar Photo–Fenton Systems. *Food Environ. Virol.* **2018**, *10*, 159–166. [CrossRef]
44. Imlay, J.A. Cellular defenses against superoxide and hydrogen peroxide. *Annu. Rev. Biochem.* **2008**, *77*, 755–776. [CrossRef]
45. Gill, L.; Price, C. Preliminary observations of a continuous flow solar disinfection system for a rural community in Kenya. *Energy* **2010**, *35*, 4607–4611. [CrossRef]
46. Giannakis, S.; Voumard, M.; Rtimi, S.; Pulgarin, C. Bacterial disinfection by the photo-Fenton process: Extracellular oxidation or intracellular photo-catalysis? *Appl. Catal. B Environ.* **2018**, *227*, 285–295. [CrossRef]
47. Berney, M.; Weilenmann, H.-U.; Egli, T. Flow-cytometric study of vital cellular functions in *Escherichia coli* during solar disinfection (SODIS). *Microbiology* **2006**, *152*, 1719–1729. [CrossRef] [PubMed]
48. García-Gil, Á.; Pablos, C.; García-Muñoz, R.A.; McGuigan, K.G.; Marugán, J. Material selection and prediction of solar irradiance in plastic devices for application of solar water disinfection (SODIS) to inactivate viruses, bacteria and protozoa. *Sci. Total Environ.* **2020**, *730*, 139126. [CrossRef]
49. Hoerter, J.D.; Arnold, A.A.; Kuczynska, D.A.; Shibuya, A.; Ward, C.S.; Sauer, M.G.; Gizachew, A.; Hotchkiss, T.M.; Fleming, T.J.; Johnson, S. Effects of sublethal UVA irradiation on activity levels of oxidative defense enzymes and protein oxidation in Escherichia coli. *J. Photochem. Photobiol. B Biol.* **2005**, *81*, 171–180. [CrossRef] [PubMed]
50. Mahaseth, T.; Kuzminov, A. Potentiation of hydrogen peroxide toxicity: From catalase inhibition to stable DNA-iron complexes. *Mutat. Res. Rev. Mutat. Res.* **2017**, *773*, 274–281. [CrossRef]
51. Brudzynski, K.; Abubaker, K.; Miotto, D. Unraveling a mechanism of honey antibacterial action: Polyphenol/H2O2-induced oxidative effect on bacterial cell growth and on DNA degradation. *Food Chem.* **2012**, *133*, 329–336. [CrossRef]
52. Gomes, A.; Asad, L.; Felzenszwalb, I.; Leitão, A.; Silva, A.; Guillobel, H.; Asad, N. Does UVB radiation induce SoxS gene expression in *Escherichia coli* cells? *Radiat. Environ. Biophys.* **2004**, *43*, 219–222. [CrossRef]
53. Yahya, E.; Almashgab, A.M.; Abdulsamad, M.A.; Allaq, A.A.; Alqadhi, A.M.; Garatem, F.M.; Aljundi, S.S. Evaluation the Effect of Microwave Radiation on Gram Positive and Negative Bacteria. *J. Chem. Nutr. Biochem.* **2021**, *2*, 39–45.
54. Tyrrell, R.M.; Pourzand, C.A.; Brown, J.; Hejmadi, V.; Kvam, V.; Ryter, S.; Watkin, R. Cellular studies with UVA radiation: A role for iron. *Radiat. Prot. Dosim.* **2000**, *91*, 37–39. [CrossRef]
55. Feng, L.; Peillex-Delphe, C.; Lü, C.; Wang, D.; Giannakis, S.; Pulgarin, C. Employing bacterial mutations for the elucidation of photo-Fenton disinfection: Focus on the intracellular and extracellular inactivation mechanisms induced by UVA and H2O2. *Water Res.* **2020**, *182*, 116049. [CrossRef] [PubMed]
56. Carratalà, A.; Dionisio Calado, A.; Mattle, M.J.; Meierhofer, R.; Luzi, S.; Kohn, T. Solar disinfection of viruses in polyethylene terephthalate bottles. *Appl. Environ. Microbiol.* **2016**, *82*, 279–288. [CrossRef] [PubMed]
57. Sagripanti, J.L.; Lytle, C.D. Estimated inactivation of coronaviruses by solar radiation with special reference to COVID-19. *Photochem. Photobiol.* **2020**, *96*, 731–737. [CrossRef] [PubMed]
58. Kohn, T.; Nelson, K.L. Sunlight-mediated inactivation of MS2 coliphage via exogenous singlet oxygen produced by sensitizers in natural waters. *Environ. Sci. Technol.* **2007**, *41*, 192–197. [CrossRef]
59. Wegelin, M.; Canonica, S.; Mechsner, K.; Fleischmann, T.; Pesaro, F.; Metzler, A. Solar water disinfection: Scope of the process and analysis of radiation experiments. *Aqua* **1994**, *43*, 154–169.

60. Davies-Colley, R.; Craggs, R.; Park, J.; Sukias, J.; Nagels, J.; Stott, R. Virus removal in a pilot-scale "advanced" pond system as indicated by somatic and F-RNA bacteriophages. *Water Sci. Technol.* **2005**, *51*, 107–110. [CrossRef]
61. Safapour, N.; Metcalf, R.H. Enhancement of solar water pasteurization with reflectors. *Appl. Environ. Microbiol.* **1999**, *65*, 859–861. [CrossRef] [PubMed]
62. Rijal, G.; Fujioka, R. Use of reflectors to enhance the synergistic effects of solar heating and solar wavelengths to disinfect drinking water sources. *Water Sci. Technol.* **2004**, *48*, 481–488. [CrossRef]
63. Harding, A.S.; Schwab, K.J. Using limes and synthetic psoralens to enhance solar disinfection of water (SODIS): A laboratory evaluation with norovirus, *Escherichia coli*, and MS2. *Am. J. Trop. Med. Hyg.* **2012**, *86*, 566–572. [CrossRef]
64. Sichel, C.; De Cara, M.; Tello, J.; Blanco, J.; Fernández-Ibáñez, P. Solar photocatalytic disinfection of agricultural pathogenic fungi: Fusarium species. *Appl. Catal. B Environ.* **2007**, *74*, 152–160. [CrossRef]
65. Lonnen, J.; Kilvington, S.; Kehoe, S.; Al-Touati, F.; McGuigan, K. Solar and photocatalytic disinfection of protozoan, fungal and bacterial microbes in drinking water. *Water Res.* **2005**, *39*, 877–883. [CrossRef]
66. Heaselgrave, W.; Kilvington, S. The efficacy of simulated solar disinfection (SODIS) against Ascaris, Giardia, Acanthamoeba, Naegleria, Entamoeba and Cryptosporidium. *Acta Trop.* **2011**, *119*, 138–143. [CrossRef] [PubMed]
67. McGuigan, K.; Méndez-Hermida, F.; Castro-Hermida, J.; Ares-Mazás, E.; Kehoe, S.; Boyle, M.; Sichel, C.; Fernández-Ibáñez, P.; Meyer, B.; Ramalingham, S. Batch solar disinfection inactivates oocysts of Cryptosporidium parvum and cysts of Giardia muris in drinking water. *J. Appl. Microbiol.* **2006**, *101*, 453–463. [CrossRef] [PubMed]
68. Nelson, K.L.; Boehm, A.B.; Davies-Colley, R.J.; Dodd, M.C.; Kohn, T.; Linden, K.G.; Liu, Y.; Maraccini, P.A.; McNeill, K.; Mitch, W.A. Sunlight-mediated inactivation of health-relevant microorganisms in water: A review of mechanisms and modeling approaches. *Environ. Sci. Process. Impacts* **2018**, *20*, 1089–1122. [CrossRef] [PubMed]
69. Liu, Y.; Dong, S.; Kuhlenschmidt, M.S.; Kuhlenschmidt, T.B.; Drnevich, J.; Nguyen, T.H. Inactivation mechanisms of Cryptosporidium parvum oocysts by solar ultraviolet irradiation. *Environ. Sci. Water Res. Technol.* **2015**, *1*, 188–198. [CrossRef]
70. Castro-Alférez, M.; Polo-López, M.I.; Marugán, J.; Fernández-Ibáñez, P. Mechanistic modeling of UV and mild-heat synergistic effect on solar water disinfection. *Chem. Eng. J.* **2017**, *316*, 111–120. [CrossRef]
71. Walker, D.C.; Len, S.-V.; Sheehan, B. Development and evaluation of a reflective solar disinfection pouch for treatment of drinking water. *Appl. Environ. Microbiol.* **2004**, *70*, 2545–2550. [CrossRef]
72. Dunlop, P.; Ciavola, M.; Rizzo, L.; Byrne, J. Inactivation and injury assessment of *Escherichia coli* during solar and photocatalytic disinfection in LDPE bags. *Chemosphere* **2011**, *85*, 1160–1166. [CrossRef]
73. Chauque, B.J.M.; Benetti, A.D.; Corcao, G.; Silva, C.E.; Goncalves, R.F.; Rott, M.B. A new continuous-flow solar water disinfection system inactivating cysts of Acanthamoeba castellanii, and bacteria. *Photochem. Photobiol. Sci.* **2021**, *20*, 123–137. [CrossRef] [PubMed]
74. Amirsoleimani, A.; Brion, G.M. Solar disinfection of turbid hygiene waters in Lexington, KY, USA. *J. Water Health* **2021**, *19*, 642–656. [CrossRef]
75. Luzi, S.; Tobler, M.; Suter, F.; Meierhofer, R. *SODIS Manual: Guidance on Solar Water Disinfection*; Eawag: Duebendorf, Switzerland, 2016.
76. Arzu, T.; Sevil, Ç. Drinking water disinfection by solar radiation. *Environ. Ecol. Res.* **2017**, 400–408. [CrossRef]
77. Sommer, B.; Marino, A.; Solarte, Y.; Salas, M.; Dierolf, C.; Valiente, C.; Mora, D.; Rechsteiner, R.; Setter, P.; Wirojanagud, W. SODIS-an emerging water treatment process. *Aqua* **1997**, *46*, 127–137.
78. Kehoe, S.; Joyce, T.; Ibrahim, P.; Gillespie, J.; Shahar, R.; McGuigan, K. Effect of agitation, turbidity, aluminium foil reflectors and container volume on the inactivation efficiency of batch-process solar disinfectors. *Water Res.* **2001**, *35*, 1061–1065. [CrossRef]
79. Mani, S.K.; Kanjur, R.; Singh, I.S.B.; Reed, R.H. Comparative effectiveness of solar disinfection using small-scale batch reactors with reflective, absorptive and transmissive rear surfaces. *Water Res.* **2006**, *40*, 721–727. [CrossRef]
80. Boyle, M.; Sichel, C.; Fernández-Ibáñez, P.; Arias-Quiroz, G.; Iriarte-Puná, M.; Mercado, A.; Ubomba-Jaswa, E.; McGuigan, K. Bactericidal effect of solar water disinfection under real sunlight conditions. *Appl. Environ. Microbiol.* **2008**, *74*, 2997–3001. [CrossRef] [PubMed]
81. Soliman, A.; El-Adawy, A.; Abd El-Aal, A.A.; Elmallawany, M.A.; Nahnoush, R.K.; Abd Eiaghni, A.R.; Negm, M.S.; Mohsen, A. Usefulness of sunlight and artificial UV radiation versus chlorine for the inactivation of Cryptosporidium oocysts: An in vivo animal study. *Open Access Maced. J. Med Sci.* **2018**, *6*, 975. [CrossRef]
82. Hijnen, W.; Beerendonk, E.; Medema, G.J. Inactivation credit of UV radiation for viruses, bacteria and protozoan (oo) cysts in water: A review. *Water Res.* **2006**, *40*, 3–22. [CrossRef] [PubMed]
83. Fiorentino, A.; Ferro, G.; Alferez, M.C.; Polo-López, M.I.; Fernández-Ibañez, P.; Rizzo, L. Inactivation and regrowth of multidrug resistant bacteria in urban wastewater after disinfection by solar-driven and chlorination processes. *J. Photochem. Photobiol. B Biol.* **2015**, *148*, 43–50. [CrossRef] [PubMed]
84. Méndez-Hermida, F.; Castro-Hermida, J.; Ares-Mazas, E.; Kehoe, S.; McGuigan, K.G. Effect of batch-process solar disinfection on survival of Cryptosporidium parvum oocysts in drinking water. *Appl. Environ. Microbiol.* **2005**, *71*, 1653–1654. [CrossRef]
85. Chaúque, B.J.; Rott, M.B. Photolysis of sodium chloride and sodium hypochlorite by ultraviolet light inactivates the trophozoites and cysts of Acanthamoeba castellanii in the water matrix. *J. Water Health* **2021**, *19*, 190–202. [CrossRef]
86. Zhou, P.; Di Giovanni, G.D.; Meschke, J.S.; Dodd, M.C. Enhanced inactivation of Cryptosporidium parvum oocysts during solar photolysis of free available chlorine. *Environ. Sci. Technol. Lett.* **2014**, *1*, 453–458. [CrossRef]

87. Villegas-Guzman, P.; Giannakis, S.; Rtimi, S.; Grandjean, D.; Bensimon, M.; De Alencastro, L.F.; Torres-Palma, R.; Pulgarin, C. A green solar photo-Fenton process for the elimination of bacteria and micropollutants in municipal wastewater treatment using mineral iron and natural organic acids. *Appl. Catal. B Environ.* **2017**, *219*, 538–549. [CrossRef]
88. Giannakis, S.; López, M.I.P.; Spuhler, D.; Pérez, J.A.S.; Ibáñez, P.F.; Pulgarin, C. Solar disinfection is an augmentable, in situ-generated photo-Fenton reaction—Part 1: A review of the mechanisms and the fundamental aspects of the process. *Appl. Catal. B Environ.* **2016**, *199*, 199–223. [CrossRef]
89. Shekoohiyan, S.; Rtimi, S.; Moussavi, G.; Giannakis, S.; Pulgarin, C. Enhancing solar disinfection of water in PET bottles by optimized in-situ formation of iron oxide films. From heterogeneous to homogeneous action modes with H2O2 vs. O2–Part 1: Iron salts as oxide precursors. *Chem. Eng. J.* **2019**, *358*, 211–224. [CrossRef]
90. Fagan, R.; McCormack, D.E.; Dionysiou, D.D.; Pillai, S.C. A review of solar and visible light active TiO2 photocatalysis for treating bacteria, cyanotoxins and contaminants of emerging concern. *Mater. Sci. Semicond. Process.* **2016**, *42*, 2–14. [CrossRef]
91. Baruah, S.; Pal, S.K.; Dutta, J. Nanostructured zinc oxide for water treatment. *Nanosci. Nanotechnol. Asia* **2012**, *2*, 90–102. [CrossRef]
92. Abbasalipourkabir, R.; Moradi, H.; Zarei, S.; Asadi, S.; Salehzadeh, A.; Ghafourikhosroshahi, A.; Mortazavi, M.; Ziamajidi, N. Toxicity of zinc oxide nanoparticles on adult male Wistar rats. *Food Chem. Toxicol.* **2015**, *84*, 154–160. [CrossRef]
93. Wong, S.W.; Leung, P.T.; Djurišić, A.; Leung, K.M. Toxicities of nano zinc oxide to five marine organisms: Influences of aggregate size and ion solubility. *Anal. Bioanal. Chem.* **2010**, *396*, 609–618. [CrossRef]
94. Sahu, D.; Kannan, G.; Vijayaraghavan, R.; Anand, T.; Khanum, F. Nanosized zinc oxide induces toxicity in human lung cells. *Int. Sch. Res. Not.* **2013**, *2013*. [CrossRef]
95. Pasupuleti, S.; Alapati, S.; Ganapathy, S.; Anumolu, G.; Pully, N.R.; Prakhya, B.M. Toxicity of zinc oxide nanoparticles through oral route. *Toxicol. Ind. Health* **2012**, *28*, 675–686. [CrossRef]
96. Reddy, M.P.; Venugopal, A.; Subrahmanyam, M. Hydroxyapatite-supported Ag–TiO2 as *Escherichia coli* disinfection photocatalyst. *Water Res.* **2007**, *41*, 379–386. [CrossRef]
97. Loeb, S.K.; Alvarez, P.J.; Brame, J.A.; Cates, E.L.; Choi, W.; Crittenden, J.; Dionysiou, D.D.; Li, Q.; Li-Puma, G.; Quan, X. The technology horizon for photocatalytic water treatment: Sunrise or sunset? *Environ. Sci. Technol.* **2019**, *53*, 2937–2947. [CrossRef] [PubMed]
98. Hatamie, A.; Khan, A.; Golabi, M.; Turner, A.P.; Beni, V.; Mak, W.C.; Sadollahkhani, A.; Alnoor, H.; Zargar, B.; Bano, S. Zinc oxide nanostructure-modified textile and its application to biosensing, photocatalysis, and as antibacterial material. *Langmuir* **2015**, *31*, 10913–10921. [CrossRef] [PubMed]
99. Yan, B.; Wang, Y.; Jiang, X.; Liu, K.; Guo, L. Flexible photocatalytic composite film of ZnO-microrods/polypyrrole. *ACS Appl. Mater. Interfaces* **2017**, *9*, 29113–29119. [CrossRef] [PubMed]
100. Monteagudo, J.M.; Durán, A.; San Martín, I.; Acevedo, A.M. A novel combined solar pasteurizer/TiO2 continuous-flow reactor for decontamination and disinfection of drinking water. *Chemosphere* **2017**, *168*, 1447–1456. [CrossRef]
101. Venieri, D.; Fraggedaki, A.; Kostadima, M.; Chatzisymeon, E.; Binas, V.; Zachopoulos, A.; Kiriakidis, G.; Mantzavinos, D. Solar light and metal-doped TiO2 to eliminate water-transmitted bacterial pathogens: Photocatalyst characterization and disinfection performance. *Appl. Catal. B Environ.* **2014**, *154*, 93–101. [CrossRef]
102. Chalew, T.E.A.; Ajmani, G.S.; Huang, H.; Schwab, K.J. Evaluating nanoparticle breakthrough during drinking water treatment. *Environ. Health Perspect.* **2013**, *121*, 1161–1166. [CrossRef]
103. Liu, X.; Wang, M.; Zhang, S.; Pan, B. Application potential of carbon nanotubes in water treatment: A review. *J. Environ. Sci.* **2013**, *25*, 1263–1280. [CrossRef]
104. Keogh, M.B.; Castro-Alférez, M.; Polo-López, M.; Calderero, I.F.; Al-Eryani, Y.; Joseph-Titus, C.; Sawant, B.; Dhodapkar, R.; Mathur, C.; McGuigan, K.G. Capability of 19-L polycarbonate plastic water cooler containers for efficient solar water disinfection (SODIS): Field case studies in India, Bahrain and Spain. *Sol. Energy* **2015**, *116*, 1–11. [CrossRef]
105. Neumann, O.; Feronti, C.; Neumann, A.D.; Dong, A.; Schell, K.; Lu, B.; Kim, E.; Quinn, M.; Thompson, S.; Grady, N. Compact solar autoclave based on steam generation using broadband light-harvesting nanoparticles. *Proc. Natl. Acad. Sci. USA* **2013**, *110*, 11677–11681. [CrossRef] [PubMed]
106. Maddigpu, P.R.; Sawant, B.; Wanjari, S.; Goel, M.; Vione, D.; Dhodapkar, R.S.; Rayalu, S. Carbon nanoparticles for solar disinfection of water. *J. Hazard. Mater.* **2018**, *343*, 157–165. [CrossRef] [PubMed]
107. Larlee, S.M. Low-Tech Photocatalysts for Solar Water Disinfection (SODIS). Ph.D. Thesis, University of Toronto, Toronto, ON, Canada, 2017.
108. Levchuk, I.; Kralova, M.; Rueda-Márquez, J.J.; Moreno-Andrés, J.; Gutiérrez-Alfaro, S.; Dzik, P.; Parola, S.; Sillanpää, M.; Vahala, R.; Manzano, M.A. Antimicrobial activity of printed composite TiO2/SiO2 and TiO2/SiO2/Au thin films under UVA-LED and natural solar radiation. *Appl. Catal. B Environ.* **2018**, *239*, 609–618. [CrossRef]
109. Spuhler, D.; Rengifo-Herrera, J.A.; Pulgarin, C. The effect of Fe2+, Fe3+, H2O2 and the photo-Fenton reagent at near neutral pH on the solar disinfection (SODIS) at low temperatures of water containing *Escherichia coli* K12. *Appl. Catal. B Environ.* **2010**, *96*, 126–141. [CrossRef]
110. Achouri, F.; Merlin, C.; Corbel, S.; Alem, H.; Mathieu, L.; Balan, L.; Medjahdi, G.; Ben Said, M.; Ghrabi, A.; Schneider, R. ZnO nanorods with high photocatalytic and antibacterial activity under solar light irradiation. *Materials* **2018**, *11*, 2158. [CrossRef]

111. Méndez-Hermida, F.; Ares-Mazás, E.; McGuigan, K.G.; Boyle, M.; Sichel, C.; Fernández-Ibáñez, P. Disinfection of drinking water contaminated with Cryptosporidium parvum oocysts under natural sunlight and using the photocatalyst TiO2. *J. Photochem. Photobiol. B Biol.* **2007**, *88*, 105–111. [CrossRef]
112. Baniamerian, H.; Safavi, M.; Alvarado-Morales, M.; Tsapekos, P.; Angelidaki, I.; Shokrollahzadeh, S. Photocatalytic inactivation of Vibrio fischeri using Fe2O3-TiO2-based nanoparticles. *Environ. Res.* **2018**, *166*, 497–506. [CrossRef]
113. Danwittayakul, S.; Songngam, S.; Sukkasi, S. Enhanced solar water disinfection using ZnO supported photocatalysts. *Environ. Technol.* **2020**, *41*, 349–356. [CrossRef]
114. Ángel-Hernández, B.; Hernández-Aldana, F.; Osorio, G.P.; Gutiérrez-Arias, J.M. Municipal wastewater treatment by photocatalysis. *Rev. Mex. Ing. Química* **2021**, *20*, Cat2438.
115. Núñez-Salas, R.E.; Rodríguez-Chueca, J.; Hernández-Ramírez, A.; Rodríguez, E.; de Lourdes Maya-Treviño, M. Evaluation of B-ZnO on photocatalytic inactivation of *Escherichia coli* and *Enterococcus* sp. *J. Environ. Chem. Eng.* **2021**, *9*, 104940. [CrossRef]
116. Alvear-Daza, J.J.; García-Barco, A.; Osorio-Vargas, P.; Gutiérrez-Zapata, H.M.; Sanabria, J.; Rengifo-Herrera, J.A. Resistance and induction of viable but non culturable states (VBNC) during inactivation of *E. coli* and Klebsiella pneumoniae by addition of H2O2 to natural well water under simulated solar irradiation. *Water Res.* **2021**, *188*, 116499. [CrossRef] [PubMed]
117. Meierhofer, R.; Wegelin, M. *Solar Water Disinfection: A Guide for the Application of SODIS*; Eawag: Dübendorf, Switzerland, 2002.
118. Xia, X.; Lan, S.; Li, X.; Xie, Y.; Liang, Y.; Yan, P.; Chen, Z.; Xing, Y. Characterization and coagulation-flocculation performance of a composite flocculant in high-turbidity drinking water treatment. *Chemosphere* **2018**, *206*, 701–708. [CrossRef]
119. Keogh, M.B.; Elmusharaf, K.; Borde, P.; McGuigan, K.G. Evaluation of the natural coagulant Moringa oleifera as a pretreatment for SODIS in contaminated turbid water. *Sol. Energy* **2017**, *158*, 448–454. [CrossRef]
120. Lamore, Y.; Beyene, A.; Fekadu, S.; Megersa, M. Solar disinfection potentials of aqua lens, photovoltaic and glass bottle subsequent to plant-based coagulant: For low-cost household water treatment systems. *Appl. Water Sci.* **2018**, *8*, 1–9. [CrossRef]
121. Anwar, F.; Rashid, U. Physico-chemical characteristics of Moringa oleifera seeds and seed oil from a wild provenance of Pakistan. *Pak. J. Bot* **2007**, *39*, 1443–1453.
122. Arafat, M.; Mohamed, S. Preliminary study on efficacy of leaves, seeds and bark extracts of Moringa oleifera in reducing bacterial load in water. *Int. J. Adv. Res.* **2013**, *1*, 124–130.
123. Kristianto, H.; Kurniawan, M.A.; Soetedjo, J.N. Utilization of papaya seeds as natural coagulant for synthetic textile coloring agent wastewater treatment. *Int. J. Adv. Sci. Eng. Inf. Technol.* **2018**, *8*, 2071–2077. [CrossRef]
124. Unnisa, S.A.; Bi, S.Z. Carica papaya seeds effectiveness as coagulant and solar disinfection in removal of turbidity and coliforms. *Appl. Water Sci.* **2018**, *8*, 1–8. [CrossRef]
125. Freitas, B.L.S.; Sabogal-Paz, L.P. Pretreatment using Opuntia cochenillifera followed by household slow sand filters: Technological alternatives for supplying isolated communities. *Environ. Technol.* **2019**, 2783–2794. [CrossRef]
126. Megersa, M.; Beyene, A.; Ambelu, A.; Triest, L. Coupling extracts of plant coagulants with solar disinfection showed a complete inactivation of faecal coliforms. *Clean Soil Air Water* **2019**, *47*, 1700450. [CrossRef]
127. Marobhe, N.J.; Sabai, S.M. Treatment of drinking water for rural households using Moringa seed and solar disinfection. *J. Water Sanit. Hyg. Dev.* **2021**. [CrossRef]
128. Muñoz-Restrepo, M.; Orrego, L.V.; Muñoz-Arango, D.C.; Lozano-Andrade, C.N.; Escobar-Restrepo, M.C.; Arcos-Arango, Y.; Lutgen, P.; Mejia-Ruiz, R. Microbicidal effect of solar radiation (SODIS) combined with Artemisia annua. *Dyna* **2014**, *81*, 71–76. [CrossRef]
129. Yu, R.; Zhang, S.; Chen, Z.; Li, C. Isolation and application of predatory Bdellovibrio-and-like organisms for municipal waste sludge biolysis and dewaterability enhancement. *Front. Environ. Sci. Eng.* **2017**, *11*, 10. [CrossRef]
130. Xia, D.; Shen, Z.; Huang, G.; Wang, W.; Yu, J.C.; Wong, P.K. Red phosphorus: An earth-abundant elemental photocatalyst for "green" bacterial inactivation under visible light. *Environ. Sci. Technol.* **2015**, *49*, 6264–6273. [CrossRef] [PubMed]
131. Ansari, S.A.; Ansari, M.S.; Cho, M.H. Metal free earth abundant elemental red phosphorus: A new class of visible light photocatalyst and photoelectrode materials. *Phys. Chem. Chem. Phys.* **2016**, *18*, 3921–3928. [CrossRef]
132. Zhang, Y.; Li, Y.; Ni, D.; Chen, Z.; Wang, X.; Bu, Y.; Ao, J.P. Improvement of BiVO4 photoanode performance during water photo-oxidation using Rh-doped SrTiO3 perovskite as a co-catalyst. *Adv. Funct. Mater.* **2019**, *29*, 1902101.
133. Roshith, M.; Pathak, A.; Kumar, A.N.; Anantharaj, G.; Saranyan, V.; Ramasubramanian, S.; Babu, T.S.; Kumar, D.V.R. Continuous flow solar photocatalytic disinfection of *E. coli* using red phosphorus immobilized capillaries as optofluidic reactors. *Appl. Surf. Sci.* **2021**, *540*, 148398. [CrossRef]
134. Domingos, M.; Sanchez, B.; Vieira-da-Motta, O.; Samarão, S.S.; Canela, M.C. A new automated solar disc for water disinfection by pasteurization. *Photochem. Photobiol. Sci.* **2019**, *18*, 905–911. [CrossRef]
135. Yildiz, B.S. Performance assessment of modified biosand filter with an extra disinfection layer. *J. Water Supply Res. Technol. Aqua* **2016**, *65*, 266–276. [CrossRef]
136. Amara, S.; Baghdadli, T.; Knapp, S.; Nordell, B. Legionella disinfection by solar concentrator system. *Renew. Sustain. Energy Rev.* **2017**, *70*, 786–792. [CrossRef]
137. Dobrowsky, P.; Carstens, M.; De Villiers, J.; Cloete, T.; Khan, W. Efficiency of a closed-coupled solar pasteurization system in treating roof harvested rainwater. *Sci. Total Environ.* **2015**, *536*, 206–214. [CrossRef] [PubMed]

138. Kang, X.; Berberidou, C.; Galeckas, A.; Bazioti, C.; Sagstuen, E.; Norby, T.; Poulios, I.; Chatzitakis, A. Visible Light Driven Photocatalytic Decolorization and Disinfection of Water Employing Reduced TiO2 Nanopowders. *Catalysts* **2021**, *11*, 228. [CrossRef]
139. Odling, G.; Robertson, N. Bridging the gap between laboratory and application in photocatalytic water purification. *Catalysis Sci. Technol.* **2019**, *9*, 533–545. [CrossRef]
140. Li, X.; Lin, H.; Chen, X.; Niu, H.; Liu, J.; Zhang, T.; Qu, F. Dendritic α-Fe 2 O 3/TiO 2 nanocomposites with improved visible light photocatalytic activity. *Phys. Chem. Chem. Phys.* **2016**, *18*, 9176–9185. [CrossRef] [PubMed]
141. Habibi-Yangjeh, A.; Mousavi, M.; Nakata, K. Boosting visible-light photocatalytic performance of g-C3N4/Fe3O4 anchored with CoMoO4 nanoparticles: Novel magnetically recoverable photocatalysts. *J. Photochem. Photobiol. A Chem.* **2019**, *368*, 120–136. [CrossRef]
142. Asadzadeh-Khaneghah, S.; Habibi-Yangjeh, A.; Seifzadeh, D. Graphitic carbon nitride nanosheets coupled with carbon dots and BiOI nanoparticles: Boosting visible-light-driven photocatalytic activity. *J. Taiwan Inst. Chem. Eng.* **2018**, *87*, 98–111. [CrossRef]
143. Baiguini, A.; Colletta, S.; Rebella, V. Materials and articles intended to come into contact with food: Evaluation of the rapid alert system for food and feed (RASFF) 2008–2010. *Ig. Sanita Pubblica* **2011**, *67*, 293–305.
144. Yahya, E.B.; Alqadhi, A.M. Recent trends in cancer therapy: A review on the current state of gene delivery. *Life Sci.* **2021**, 119087. [CrossRef] [PubMed]
145. Groh, K.J.; Muncke, J. In vitro toxicity testing of food contact materials: State-of-the-art and future challenges. *Compr. Rev. Food Sci. Food Saf.* **2017**, *16*, 1123–1150. [CrossRef]
146. Yahya, E.B.; Alfallous, K.A.; Wali, A.; Hameid, S.; Zwaid, H. Growth rate and antibiotic sensitivity effect of some natural and petroleum based materials on Staphylococcus aureus. *Int. J. Res. Appl. Sci. Biotechnol.* **2020**, *7*, 7–11. [CrossRef]
147. Monarca, S.; De Fusco, R.; Biscardi, D.; De Feo, V.; Pasquini, R.; Fatigoni, C.; Moretti, M.; Zanardini, A. Studies of migration of potentially genotoxic compounds into water stored in PET bottles. *Food Chem. Toxicol.* **1994**, *32*, 783–788. [CrossRef]
148. Ubomba-Jaswa, E.; Fernández-Ibáñez, P.; McGuigan, K.G. A preliminary Ames fluctuation assay assessment of the genotoxicity of drinking water that has been solar disinfected in polyethylene terephthalate (PET) bottles. *J. Water Health* **2010**, *8*, 712–719. [CrossRef]
149. Zimmermann, L.; Dierkes, G.; Ternes, T.A.; Völker, C.; Wagner, M. Benchmarking the in vitro toxicity and chemical composition of plastic consumer products. *Environ. Sci. Technol.* **2019**, *53*, 11467–11477. [CrossRef] [PubMed]
150. Westerhoff, P.; Prapaipong, P.; Shock, E.; Hillaireau, A. Antimony leaching from polyethylene terephthalate (PET) plastic used for bottled drinking water. *Water Res.* **2008**, *42*, 551–556. [CrossRef]
151. Schmid, P.; Kohler, M.; Meierhofer, R.; Luzi, S.; Wegelin, M. Does the reuse of PET bottles during solar water disinfection pose a health risk due to the migration of plasticisers and other chemicals into the water? *Water Res.* **2008**, *42*, 5054–5060. [CrossRef]
152. Yahya, E.B.; Jummaat, F.; Amirul, A.; Adnan, A.; Olaiya, N.; Abdullah, C.; Rizal, S.; Mohamad Haafiz, M.; Khalil, H. A review on revolutionary natural biopolymer-based aerogels for antibacterial delivery. *Antibiotics* **2020**, *9*, 648. [CrossRef] [PubMed]
153. Yahya, E.B.; Amirul, A.; HPS, A.K.; Olaiya, N.G.; Iqbal, M.O.; Jummaat, F.; H.P.S., A.K.; Adnan, A. Insights into the Role of Biopolymer Aerogel Scaffolds in Tissue Engineering and Regenerative Medicine. *Polymers* **2021**, *13*, 1612. [CrossRef]
154. Asyakina, L.; Dolganyuk, V.; Belova, D.; Peral, M.; Dyshlyuk, L. The study of rheological behavior and safety metrics of natural biopolymers. *Foods Raw Mater.* **2016**, *4*, 70–78. [CrossRef]
155. Haider, H.; Ali, W.; Haydar, S.; Tesfamariam, S.; Sadiq, R. Modeling exposure period for solar disinfection (SODIS) under varying turbidity and cloud cover conditions. *Clean Technol. Environ. Policy* **2014**, *16*, 861–874. [CrossRef]
156. Sommers, M. Limitations of Solar Disinfection (SODIS) Water Treatment in Low-Income Communities. Senior Thesis, Elizabethtown College, Elizabethtown, KY, USA, 2021.
157. Rosa, G.; Clasen, T. Estimating the scope of household water treatment in low-and medium-income countries. *Am. J. Trop. Med. Hyg.* **2010**, *82*, 289–300. [CrossRef] [PubMed]
158. Michaels, H.; Benesperi, I.; Freitag, M. Challenges and prospects of ambient hybrid solar cell applications. *Chem. Sci.* **2021**, *12*, 5002–5015. [CrossRef] [PubMed]

Article

Formulation of Organic Wastes as Growth Media for Cultivation of Earthworm Nutrient-Rich *Eisenia foetida*

Mashur Mashur [1,*], Muhammad Roil Bilad [2,3], Hunaepi Hunaepi [3], Nurul Huda [4,5,*] and Jumardi Roslan [4]

1 Faculty of Veterinary Medicine, Universitas Pendidikan Mandalika, Mataram 83126, Indonesia
2 Faculty of Integrated Technologies, Universiti Brunei Darussalam, Gadong BE1410, Brunei; roil.bilad@ubd.edu.bn
3 Faculty of Applied Science and Technology, Universitas Pendidikan Mandalika, Mataram 83126, Indonesia; hunaepi@undikma.ac.id
4 Faculty of Food Science and Nutrition, Universiti Malaysia Sabah, Kota Kinabalu 88400, Sabah, Malaysia; jumardi@ums.edu.my
5 Department of Food Science and Technology, Faculty of Agriculture, Universitas Sebelas Maret, Surakarta 57126, Indonesia
* Correspondence: mashur@undikma.ac.id (M.M.); drnurulhuda@ums.edu.my (N.H.)

Abstract: Inadequate management of solid organic waste can lead to the spread of diseases and negatively affects the environment. Fermentation and vermicomposting of organic waste could have dual benefits by generating earthworm biomass for a source of animal feed protein, and, at the same time, turning the organic waste into readily used compost. This study investigated the effect of an organic waste source (as a sole source or blended with others) totaling 24 media for the cultivation of the earthworm *Eisenia foetida*. Eight media sources were applied, namely cow manure, horse manure, goat manure, broiler chicken manure, market organic waste, household organic waste, rice straw, and beef rumen content. *E. foetida* was cultivated for 40 days, then the number of cocoons, earthworms, and the total biomass weight were measured at the end of the cultivation. Results demonstrated that the media source affected *E. foetida* earthworm cultivation. The most effective media were those containing horse manure that led to the production of the highest earthworms and the highest biomass. The produced cocoons and earthworms were poorly correlated with an r-value of 0.26 and *p*-value of 0.21. Meanwhile, the number and weight of the earthworms correlated well with an r-value of 0.784 and *p*-value of <0.01. However, the average numbers and weights of the produced earthworms in the media containing horse manure, cow manure, goat manure, and non-blended organic waste were insignificant. Overall results suggest that blended organic wastes can undergo composting to produce nutrient-rich earthworm biomass while turning the solid organic waste into readily used compost.

Keywords: biomass; protein source; *Eisenia foetida*; vermicomposting; organic waste; cultivation media

Citation: Mashur, M.; Bilad, M.R.; Hunaepi, H.; Huda, N.; Roslan, J. Formulation of Organic Wastes as Growth Media for Cultivation of Earthworm Nutrient-Rich *Eisenia foetida*. *Sustainability* 2021, 13, 10322. https://doi.org/10.3390/su131810322

Academic Editors: Mohd Rafatullah and Masoom Raza Siddiqui

Received: 24 July 2021
Accepted: 8 September 2021
Published: 15 September 2021

Publisher's Note: MDPI stays neutral with regard to jurisdictional claims in published maps and institutional affiliations.

Copyright: © 2021 by the authors. Licensee MDPI, Basel, Switzerland. This article is an open access article distributed under the terms and conditions of the Creative Commons Attribution (CC BY) license (https://creativecommons.org/licenses/by/4.0/).

1. Introduction

Humans, livestock, and crops produce approximately 38 billion metric tons of organic waste worldwide annually [1]. Such a vast amount of solid waste can have significant impacts on the disposal and methane emission from the anaerobic fermentation process. The management and safe disposal of these wastes has become a global priority. Moreover, the open dumping of organic waste also facilitates the breeding of disease vectors and creates environmental pollution issues. Fortunately, adequately processed organic waste can be used for agriculture and industries. Composting is a simple, sustainable option and is most economical for handling organic waste. Although composting has been adopted as a primary tool for on-site waste decomposition, it has a few shortcomings, and long retention time requires frequent aeration, etc. [2]. Organic wastes are naturally transformed into plant nutrients by a variety of soil decomposers involving bacteria, fungi, and earthworms [3].

Vermicomposting is a process for the stabilization of organic material through the joint actions of earthworms and microorganisms [4]. In this process, microorganisms (bacteria and fungi) are responsible for the biodegradation of organic matter, while earthworms are drivers of the process. Earthworms act as mechanical blenders. The biological activities of the earthworms lead to modification of growth media in terms of biological, physical, and chemical condition, reducing its C:N ratio gradually, increasing the exposed surface area to microorganisms and eventually making it more favorable for microbial activities and further decomposition [5]. Earthworms maintain aerobic conditions, ingest organic solid, partially convert organics into earthworm biomass and metabolite products, and expel the remaining partially stabilized product. The vermicomposting yields product with higher nutrient availability than the traditional composting systems. The nutrients in vermicompost are also readily taken up by the plants [6]. Apart from producing high-quality compost, the earthworm biomass can also be used to supply nutrition in livestock and aquaculture industries.

Eisenia foetida is an earthworm that has a high advantage in reproduction and overhauling organic matter into nutrient-rich biomass [7]. Gunya et al. (2016) [8] reported that the dried *E. fetida* contains about 45.8% saturated, 22.2% monounsaturated, 31% polyunsaturated, 23.5% of n−6 and 8.3% of n−3 fatty acids. Furthermore, it has tolerable crude fibre levels (10.9%) suitable for fish digestion, and as such promotes high-protein assimilation efficiency when used as a component of fish feed [9]. In other reports, the nutrient content of earthworm *E. foetida* and the *Lumbricus rubellus* mixture are dry matter 12.9–25%, crude protein 58.2–71%, crude fat 2.3–10%, crude fiber 0.73–3.3%, carbohydrates 21%, ash 5.2–10%, calcium 0.33–0.8%, phosphorus 0.7–1.0% and total energy 17 MJ/kg [10–14]. Meanwhile, the essential amino acid content (in g/100 g protein) are phenylalanine 3.5–5.1, valine 4.4–5.2, methionine 1.5–3.6, isoleucine 4.2–5.3, threonine 4.8–6.0, histidine 2.2–3.8, arginine 6.1–7.3, lysine 6.6–7.5, leucine 6.2–8.2 and tryptophan 2.1. The content of non-essential amino acids (in g/100 g protein) are cysteine 1.8–3.8, tyrosine 2.2–4.6, aspartic acid 10.5–11.0, glutamic acid 13.2–15.4, serine 4.2–5.8, glycine 4.3–4.8, alanine 5.4–6.0 and proline 5.1 [8,15]. *E. foetida* biomass is considered among the promising non-conventional protein sources for animal and fish feed ingredients thanks to its high protein levels, proper amino acid profile, high reproduction rate, low mortalities, fast growth and ease of production [16].

Earthworm cultivation for biomass production is highly attractive given the high nutritional value of its flour as a source of animal protein and amino acids for animal and fish feed. High earthworm biomass is generally obtained from the production of a high number of cocoons (earthworm eggs). The earthworm life cycle is divided into four stages, namely cocoon production, incubation, hatching, and growth [17,18]. The cocoons produced by earthworms are influenced by population density, temperature, humidity, and the energy content available in the feed/media. Some earthworm species naturally produce cocoons throughout the year when the soil is moist, feed reserves are sufficient and other environmental factors are favorable. The earthworm *E. foetida* can produce 14 cocoons in 70 days or one every five days [19]. The number of earthworms that hatched ranged from 1–7, with an average of 3.9 per cocoon. *E. foetida* reaches sexual maturity at the age of four weeks; this is marked by the formation of a clitelum. At that age, earthworms can mate and produce cocoons at the age of 35 days, so the entire time required for one life cycle is 40–60 days. The life span of the earthworm *E. foetida* is estimated at 4.5 years [20].

One component of *E. foetida* earthworm cultivation technology that can increase cocoon and biomass production is the use of earthworm cultivation media that is suitable for earthworm life. A good earthworm media comes from organic waste. Organic waste is the material left over from the activities of human, animal, and plant life that are wasted and have no economic value. Materials used as earthworm cultivation media must be able to retain moisture, porosity and contain sufficient food substances, including protein, carbohydrates, minerals and vitamins, fat, and crude fiber [4,21,22].

The potential of *E. foetida* biomass as a nutrient-rich feed source has long been recognized. Therefore, many reports are available on the cultivation performances of *E. foetida* in various media (i.e., cattle and goat manure [23], various organic wastes [24], etc.). The cultivation performances seemed to differ slightly and were affected by the nutritional content of the media. However, most of those reports focus on the vermi composting aspect of organic waste management. Recently, vermicomposting of different types of waste using *E. foetida* has been reported. The use of those waste for cultivation of E. foetida increased the organic nitrogen, organic carbon and phosphorous content significantly [7]. More recently, vermi composting using *E. foetida* was reported for conversion of vegetable solid waste amended with wheat straw, cow dung, and biogas slurry [25]. It resulted in agronomic potentials of nutrient-rich vermicompost with acceptable C:N ratio ranges ($\geq$1:20), demonstrating that *E. foetida* facilitated conversion of organic wastes into nutrient-rich biofertilizer if mixed with bulking materials in appropriate ratios. This study focused more on the exploration of organic waste directed to nutrient-rich biomass production yields. Eight main sources of the cultivation media (and 16 other combinations) were evaluated to assess their suitability as cultivation media for *E. foetida* biomass production.

This study investigated the effect of cultivation media on the produced number and the biomass weight of the *E. foetida* earthworm. Eight sources of media were applied, namely cow manure, horse manure, goat manure, broiler chicken manure, market organic waste, household organic waste, rice straw, and beef rumen content. They were also blended to form a total of 24 cultivation media. After media preparation, *E. foetida* was cultivated for 40 days. At the end of cultivation, the number of cocoons, earthworms, and the total weight were measured. The relationships between media source and composition with the production of cocoon, earthworm, and biomass were later analyzed and also linked with the nutritional composition of the media.

2. Materials and Methods

2.1. Preparation of the Cultivation Media

The compositions of all growth media evaluated in this study are summarized in Table 1. The base media for the cultivation were cow manure, horse manure, goat manure, broiler chicken manure, cow rumen contents, wet market organic waste, household organic waste, rice straw, and some combinations thereof. Few criteria were used as the basis for selecting the organic waste base media, namely: its local availability in the field (potential), not competing for its use for basic human/animal needs, and a good source of carbon (C) and nitrogen (N) for the earthworm growth building block and nutrition.

The preparation of the growth media was carried out in stages, as follows below. The market and the household organic waste were separated from glass, metal, and plastic materials, leaving only the organic fraction. Most of the organic fraction contained spoiled vegetables, fruit, and food waste. The organic waste was washed with clean water to remove the adhering dirt and to release odors. The organic material was then finely chopped or blended to sizes of 2–3 cm, then further grounded and screened with a mesh number of 18, resulting in a maximum particle size of 1 mm to ease the composting. Ten kg of each type of waste was then fermented. After collection from the farm, fresh rice straw waste was chopped or blended to sizes of 2–3 cm, then finely grounded. Later, 10 kg of the fine rice straw waste was composted. Ten kg of cow manure, horse manure, goat manure, broiler chicken manure, and the cow's rumen contents were then placed in a barrel for composting.

After composting of the organic waste media, 2.5 kg of each medium was separated for cultivation. Each homogeneous medium was used for worm cultivation without mixing with others. These were cow manure (CM), horse manure (HM), goat manure (GM), broiler chicken manure (BM), market organic waste (MW), household organic waste (HW), rice straw (RS), and beef rumen content (RC), totaling eight growth media. The blended media were formed by mixing two media with equal weight composition (wt.%). They were CM + MW (CW-MW), CM + HW (CM-HW), CM + RS (CM-RS), CM + RC (CM-RC), HM + MW (HM-MW), HM + HW (HM-HW), HM + RS (HM-RS), HM + RC (HM-RC), GM + MW (GM-MW), GM + HW (GM-HW), GM + RS (GM-RS), GM + RC (GM-RC),

BM + MW (BM-MW), BM + HW (BM-HW), BM + RS (BM-RS) and lastly BM + RC (BM-RC), totaling 16 media. Each growth medium was put into a nest box, dosed with 0.3 wt.% of lime (to maintain the pH close to normal value) followed by blending, after which it underwent aerobic composting for 21 days. During the composting, the media was stirred once a week. After media preparation, their nutrition contents were analyzed approximately according to a method developed by Henneberg and Stohmann [26] to determine the content of crude protein, fat, crude fiber, and ash as detailed elsewhere [27]. The analytical method was developed to provide a top level, very broad classification of food components [27].

Table 1. Details of the growth media evaluated in the present work.

No.	Code	Media	Source of Collection
1	CM	Cow manure	Local farmer
2	HM	Horse manure	Local farmer
3	GM	Goat manure	Local farmer
4	BM	Broiler chicken manure	Local farmer
5	MW	Market organic waste	Wet market
6	HW	Household organic waste	Household waste
7	RS	Rice straw	Local rice farm
8	RC	Beef rumen content	Local slaughterhouse
9	CM-MW	CM + MW	
10	CM-HW	CM + HW	
11	CM-RS	CM + RS	
12	CM-RC	CM + RC	
13	HM-MW	HM + MW	
14	HM-HW	HM + HW	
15	HM-RS	HM + RS	
16	HM-RC	HM + RC	The media were from from equal wt.%
17	GM-MW	GM + MW	
18	GM-HW	GM + HW	
19	GM-RS	GM + RS	
20	GM-RC	GM + RC	
21	BM-MW	BM + MW	
22	BM-HW	BM + HW	
23	BM-RS	BM + RS	
24	BM-RC	BM + RC	

2.2. Cultivation Process

The study was conducted by cultivating earthworm species of *E. foetida* collected from the Zoology Laboratory of IPB University. The cultivation was done during September–December 2019 by using cultivation media detailed in Table 1. The cultivations were carried out in a cage (nest box) made from thatched roofs and bamboo. All nest boxes were placed on a plastered floor and were assigned a pre-randomized code. They were covered tightly to avoid predators, to reduce water evaporation, and to maintain moist conditions. After completion of the media preparation each nest box was filled with 2.5 kg of growth media.

The earthworms were introduced into each nest box with a stocking density of 10 g per kg of growth media. The earthworm broodstocks were placed into the middle of the piled media through a top hole and were cultivated for 40 days. The medium was mixed on days 8, 15, 22, 29, and 36 of the cultivation. The harvesting of earthworms was carried out on day 40, whereby the data on cocoon and biomass (number and weight) of the earthworms was collected.

During the research, measurements of pH, humidity, and temperature of the media were carried out every two days at 1:00 p.m. during the media composting and over the entire duration of the earthworm cultivation. In addition, the daily temperature of the environment inside the nest box was measured three times a day: at 6:00 a.m., 1:00 p.m., and 8:00 p.m. If the temperature of the media or the environment in the cage increased, it was sprayed with water. The temperature was maintained in the range of 18–27 °C.

2.3. Experimental Design and Statistical Analysis

The formulation of the media was done based on a completely randomized design, resulting in 24 combinations (as listed in Table 1). For each media, the cultivation was done in triplicate, resulting in a total of 72 nests. The data on mortality rate, cocoon production, earthworm production, and mass of the produced earthworms were analyzed statistically. The significance of each parameter was evaluated using the one-way analysis of variance (ANOVA) at 95% confidence intervals ($p < 0.05$). Then, the Tukey HSD post hoc test was performed to identify which pairs of mean were significantly different. The Pearson coefficient of correlation was applied to identify the relation between the cocoon and earthworm number, as well as the total weight of the produced biomass.

3. Results and Discussion
3.1. Mortality Rate

Figure 1 shows the mortality rate of the inoculated *E. foetida* earthworm. Four media were found not suitable because no earthworms survived at the end of the cultivation. These media were MW, BM, BM-MW and BM-HW. They were excluded from the results presented in subsequent figures. Figure 1 also shows that media containing the broiler chicken manure led to a significantly high earthworm mortality of 81.9 ± 36.3%. No earthworm survived in BM, BM-MW, and BM-HW media; the mortality rates in BM-RS and BM-RC media were 27.4 ± 2.5% and 68.1 ± 5.0%, respectively. Media containing the horse manure showed an average mortality rate of 16.6 ± 3.9%, followed by the ones constituted of the cow manure and the goat manures with average mortality rates of 3.2 ± 3.4% and 1.3 ± 2.6%. Interestingly, despite showing 100% mortality when used as the sole component of the growth media, the market organic waste can still be used in combination with other media to lower the mortality rate, suggesting that blending of organic waste from different sources can be optimized to support the growth of the earthworm *E. foetida*.

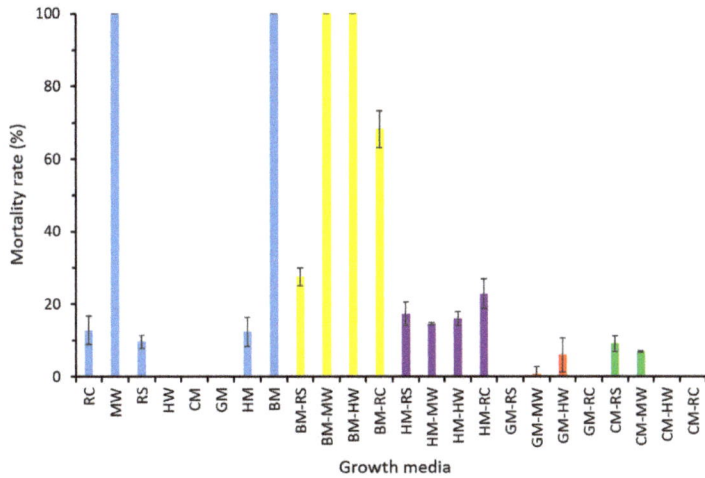

Figure 1. Earthworm mortality rate under different cultivation media.

The summary of the compositions of all media tested in this study is detailed in Tables 2 and 3. The four media with high mortality rates had relatively high phosphorous content (of 0.91–1.13%) which were much higher than the rest, with an average of 0.49%.

Exposure to a high concentrations of phosphate (i.e., iron phosphate) increased earthworm mortality, and surviving individual worms gained less mass [28].

Table 2. Composition of the cultivation media.

Cultivation Media	Water (%)	Ash (%)	Protein (%)	Fat (%)	Fiber (%)
RC	16.96	14.99	17.6	0.04	33.76
MW	21.41	17.16	18.66	2.69	33.86
RS	17.61	26.88	8.33	0.07	20.83
HW	28.38	16.71	21.12	2.73	23.72
CM	15.37	51.97	10.62	0.54	16.21
GM	19.69	21.15	17.84	0.92	32.9
HM	17.74	31.54	13.2	0.14	25.73
BM	23.87	24.35	24.93	1.25	16.53
BM-RS	18.76	33	15.49	0.15	17.27
BM-MW	21.82	26.3	22.29	2.42	9.47
BM-HW	25.94	23.15	24.29	1.56	21.17
BM-RC	18.49	21.33	14.79	0.52	25.95
HM-RS	16.07	30.28	9.39	0.62	26.24
HM-MW	15.73	30.21	9.86	0.34	27.44
HM-HW	16.89	29.06	9.27	0.55	27.4
HM-RC	15.62	26.24	10.44	0.13	31.73
GM-RS	17.16	24.34	15.9	0.46	31.58
GM-MW	18.78	19.37	19.71	0.89	35.24
GM-HW	20.51	18.45	18.71	2.05	35.65
GM-RC	18.66	20.44	17.89	0.63	36.99
CM-RS	14.85	38.1	10.86	0.35	15.9
CW-MW	15.52	41.74	11.73	0.05	21.54
CM-HW	18.67	40.3	13.73	0.17	16.2
CM-RC	16.08	33.74	11.5	0.18	18.44

Table 3. Macronutrient composition of the cultivation media.

Cultivation Media	N (%)	P (%)	K (%)	C (%)	C/N (-)	Organic (%)
RC	2.82	0.78	0.7	42.23	14.98	70.63
MW	2.99	0.4	1.66	46.02	15.39	79.16
RS	1.33	0.09	0.95	40.62	30.54	69.87
HW	3.38	0.48	1.97	46.27	13.69	79.59
CM	1.7	0.49	1.11	26.68	15.69	45.89
GM	2.85	0.41	1.39	43.81	15.37	75.35
HM	2.11	0.74	1.03	38.11	18.06	65.55
BM	3.99	1.13	1.5	42.03	10.53	72.29
BM-RS	2.48	0.91	1.46	37.22	15.01	64.02
BM-MW	3.57	0.99	1.6	40.94	11.48	70.42
BM-HW	3.89	1.01	1.54	42.69	10.97	73.43
BM-RC	2.37	1.09	1.2	43.71	18.44	75.17
HM-RS	1.5	0.43	0.86	38.76	25.84	66.67
HM-MW	1.58	0.68	1.2	38.77	24.54	66.69
HM-HW	1.48	0.74	1.2	39.41	26.63	67.79
HM-RC	1.67	0.79	0.93	40.93	24.54	70.48
GM-RS	2.54	0.26	1.26	42.03	16.55	72.3
GM-MW	3.15	0.37	1.62	44.79	14.22	77.05
GM-HW	2.99	0.37	1.68	45.31	15.15	77.93
GM-RC	2.86	0.52	1.32	44.2	15.45	76.02
CM-RS	1.74	0.3	1.1	34.59	19.88	59.15
CM-MW	1.88	0.33	1.26	32.37	17.22	55.67
CM-HW	2.2	0.53	1.32	33.13	15.06	57.05
CM-RC	1.84	0.67	1	36.81	20.21	63.31

3.2. Cocoon Production

Figure 2 shows the number of produced cocoons after 40 days of cultivation. The top three highest cocoon productions were achieved by media from blended sources. There were CM-HW, GM-RC, and SW-MW, with the average number of cocoon production of 318.3 ± 0.6, 296.0 ± 1.0, and 217.7 ± 0.6, respectively. The respective average cocoon production for cultivation in media containing the goat manure, the cow manure, the horse manure, the non-blended and the broiler chicken manure were 180.8 ± 68.4, 160.1 ± 103.7, 90.3 ± 17.0, 83.7 ± 46.1, and 7.0 ± 4.2, which was in the range of earlier report [29]. Media containing broiler chicken manure were less attractive for cultivating *E. foetida* judging by both the mortality and the cocoon production rates. The low numbers of produced cocoons can be justified by the low number of surviving earthworm broodstocks cultivated in the chicken manure containing media (see Figure 1).

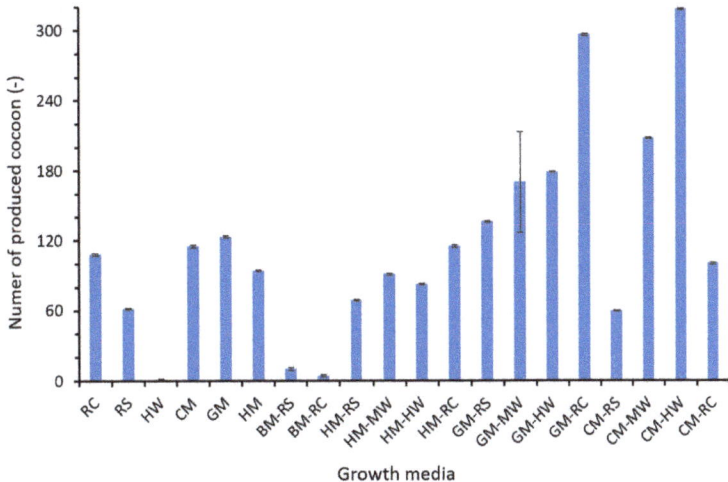

Figure 2. The rate of cocoon production under different cultivation media.

Results in Figure 2 suggest that the media source for *E. foetida* cultivation had a significant effect on the number of the produced cocoons, as also reported recently [29]. The cow and goat manures showed a higher amount of cocoon production because *E. foetida* has a natural habitat similar to the one suitable for decomposition of organic matter by soil microorganisms [21,30,31]. The presence of these microorganisms can increase the ability of earthworms to digest organic matters in their digestive tract [20].

The cocoons produced varied in terms of shape, size, weight, and color. The cocoons of *E. foetida* were generally oval to round, with an average length of 4.103 mm and a width of 2.6 mm. The weight of the cocoons also varied depending on size, with an average weight of 1.3 mg/unit, which is significantly lower than the one reported earlier (12–23 mg/unit) [32]. The color of the cocoons varied from beige (light yellow) to dark yellow or light brown, and some were even dark brown. The color of the cocoons generally depends on the age of cocoons. Newly produced cocoons were cream or light yellow and even very clear (close to white). However, with increasing age the cocoons, the color turns yellow or brown, and even before hatching the color of the cocoons approaches dark brown.

In terms of nutrition, based on the results of the proximate analysis in Tables 2 and 3, the crude protein content of the CM-HW medium was 13.73%, fat 0.17%, crude fiber 16.20%, N 2.20%, P 0.53%, K 1.32%, C 33.13%, C/N ratio of 15.06 and organic matter of 57.05%. The GM-RC contained nutrients of 17.89%, crude protein, 0.63% fat, 36.99% crude fiber, 2.86% N, 0.52% P, 1.32% K, 44.20% C, C/N ratio of 15.45 and organic matter of 76.02%. Meanwhile, the nutrient content of SW-MW was 11.73% of crude protein, 0.05% of fat, 21.54% of crude fiber, 1.88% of N, 0.33% of P, 1.26% of K, 32.37% of organic C; C/N ratio of 17.22% and organic matter of 55.67%. When compared with HW, which produced the lowest cocoons, it contained 21.12% of crude protein, 2.73% of fat, 23.72% of crude fiber, 3.38% of N, 0.48% of P, 1.97% of K, 46.27% organic C, C/N ratio of 13.69 and organic matter of 79.59%.

Analysis of the nutritional data for the mixture of *E. foetida* earthworm cultivation media listed in Tables 2 and 3 showed that to produce the most cocoons, *E. foetida* required crude protein, fat, crude fiber, N, P, K, organic C, C/N ratio and organic matter of 11.73–17.89%, 0.05–0.63%, 16.20–36.99%, 1.88–2.20%, 0.33–0.52%, 1.26–1.32, 32.37–44.20%, 15.06–17.22 and 55.67–76.02%, respectively. The application of a mixture of media containing too much protein (>18%) would not increase cocoon production. This is in accordance with an earlier report [14] that found that the best feed for earthworms contained 9–15% protein under a neutral pH. Furthermore, either excess protein interfered with the digestive system of the *E. foetida* earthworm, or protein poisoning occurred in the form of swelling of the cache, thus affecting the health of the earthworms and ultimately affecting their productivity, and even causing death [33,34]. The cultivation media of MW contained 21.12% protein, which made it unsuitable for *E. foetida* earthworm cultivation, as shown in Figures 1 and 2.

The rate of cocoon production is also affected by the nutrient content of the cultivation media. Among these nutrients, phosphorus (P) is positively correlated with the production of cocoons (eggs). This is in line with earlier reports [11,35] that the minerals that play a major role in the process of egg formation are calcium and phosphorus. Phosphorus is important in energy metabolism, carbohydrates, amino acids and fats, fatty acid transport, and coenzyme parts. Therefore, in the selection of organic waste as a medium or feed for earthworms, it is necessary to pay attention to the content of these two minerals, especially when aiming to produce a high number of cocoons. However, phosphorous in the form of iron phosphate increased earthworm mortality, and surviving individuals gained less mass [28]. As shown in Section 3.1, phosphorous content of less than 1% is recommended to avoid the poor survival rate of the earthworm broodstock.

In addition to media nutrients content, cocoon production during cultivation of *E. foetida* was also influenced by environmental factors such as temperature, pH, and humidity [3,30,36]. For the CM-HW that resulted in the highest cocoon production, the average temperature of the media was 28.92 °C, pH 6.46, and with a relative humidity of 56.55%. In addition, earthworm communities are generally very sensitive to physicochemical properties of the media, which directly or indirectly influence the earthworm's survival. The difference in physicochemical properties of media at different sites contributed to the formation of population patches for the earthworm species [37].

3.3. Cocoon Hatching

Data in Figure 3 show that the use of various types of organic waste as media material for the cultivation of *E. foetida* had a significant effect on biomass production, both in number and weight. When viewed as the media content, the average earthworm production from the highest to lowest were 1168.7 ± 383.3, 620.9 ± 489.4, 578.7 ± 328.6, 508.0 ± 291.1, and 47.5 ± 13.9 for the horse manure, the non-blended, the goat manure, the cow manure, and the broiler chicken manure, respectively. Analysis using the Tukey HSD test confirmed that the numbers of produced earthworms for the first four media were insignificant, with a Q-critical of 4.11 higher than the Q-statistic range of 0.55–3.65 for all possible pairs.

Figure 3 shows the ratio of the produced cocoon to the produced earthworm under different cultivation media. It shows a very large variability with an average value of 11.0 ± 18.7 earthworm/cocoon. The large variability was due to the extremely high earthworm/cocoon ratio for the HW medium. The produced cocoons were 0.7 ± 0.6, and the produced earthworms were 58.0 ± 1.0. When it was excluded, the average value of the earthworm/cocoon became 7.0 ± 5.7. It is much higher than the one reported elsewhere [19], in which the average number of earthworms/cocoons was 3.9.

Results of the Pearson coefficient of correlation suggest a positive but poor correlation between the number of the cocoons and the number of the earthworms, with an r-value of 0.26 and *p*-value of 0.21. Apart from showing a clear difference in the number of produced earthworms, the size and weight of the produced earthworms are also more important in determining the yield of biomass production during cultivation. The results of each type of organic waste as a medium for 24 types of organic waste indicate that the earthworm *E. foetida* has different abilities in producing biomass depending on the type of organic waste used as cultivation media. The findings are in line with the variability shown for the mortality rate and the cocoon production.

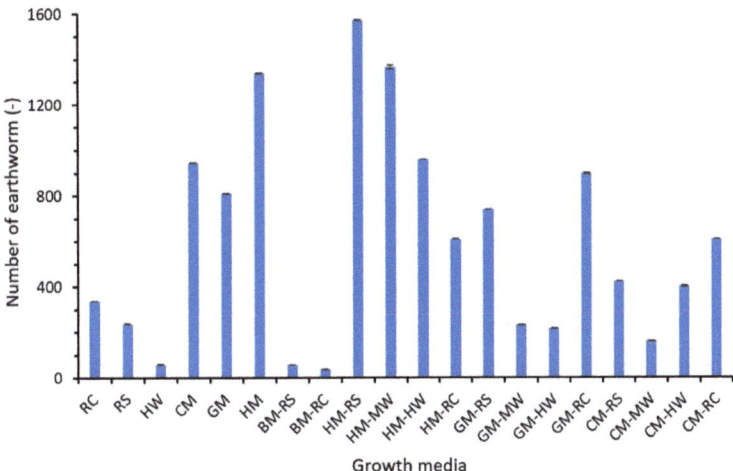

Figure 3. The transformation of the cocoon into an earthworm under different cultivation media.

3.4. Biomass Production

Figure 4 shows that the total mass of biomass produced is affected by both the number and the weight of the earthworms. Analysis using the Pearson coefficient of correlation between the number of earthworms and the total weight resulted in an r-value of 0.78 and *p*-value of <0.01. Indeed, more earthworms produced would lead to higher total biomass. The trend on the total weight of produced biomass is also generally in line with the number of the produced earthworms, as shown from the highest to lowest group of 1062 ± 327, 368 ± 256, 335 ± 197, and 111 ± 110 earthworms. In general, a higher number of earthworms lead to a higher yield of biomass. However, a few exceptions can be observed. For example, CM-RC, HM-RS, and HM-HW showed the three highest total biomass weights, with an insignificant difference based on the Tukay HSD test. However, there was substantial difference in the number of earthworms of 896, 1571, and 958 with an average biomass weight of 60, 34 and 54 mg/earthworm, respectively. HW media had among the lowest number of earthworms at 58.0, but still produced a relatively high biomass amount of 39.7 ± 6.4 g, corresponding to the specific weight of earthworms of 684 ± 109 mg/earthworm.

Figure 4 shows the total weight and the number of produced biomass. Based on the total weight of the earthworms produced, they can be classified into four groups that had significant differences from the Tukey HSD analysis. The first group constituted of eight media, namely GM-RC, HM-RS, HM-HW, CM, HM, CM-RC, HM-MW, and GM with an average total biomass weight of 50.8 ± 2.3 g/nest box. The second group consisted of five media of GM-RS, CM-HW, CM-RS, HW, and GM-HW with an average total biomass weight of 41.8 ± 2.3 g/nest box. The third group constituted four media of GM-MW, RC, HM-RC, and SW-MW with an average total biomass weight of 36.3 ± 0.5 g/nest box. Lastly, the media with the significantly lowest average total biomass weight of 16.6 ± 2.6 g/nest box constituted RS, BM-RS, and BM-RC.

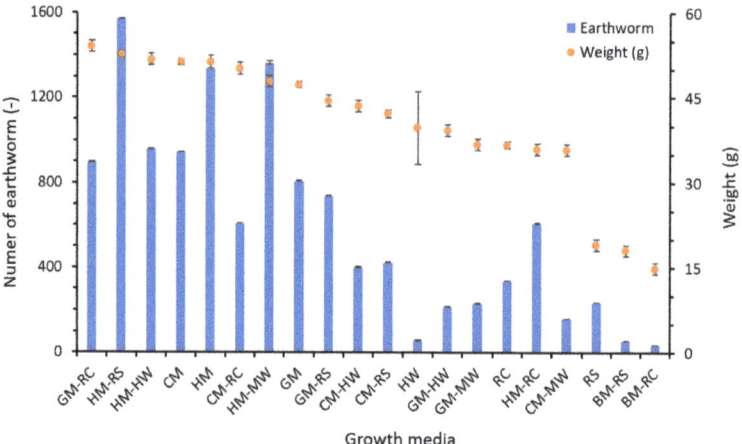

Figure 4. Number of the grown earthworm and their total weight for cultivation in different cultivation media.

In terms of the content of the media, the highest biomass production was obtained for cultivation in media containing horse manure, goat manure, cow manure, non-blended, and, lastly, chicken manure, with an average total biomass weight of 47 ± 8, 44 ± 8, 43 ± 6, 41 ± 12 and 16 ± 2 g/nest. Statistical analysis on the effect of the media source revealed that the biomass production rates of all media were insignificant except for the cultivation in media containing broiler chicken manure. The media contained horse manure, cow manure, goat manure, and non-blended sources yielded about similar total biomass, as ascribed from the Tukey HSD test. The Q-statistic of 4.11 was much higher than Q-critical ranged at 0.15–1.46 for all possible pairs. The similarities in the efficacy of the media can be attributed to their similarities in nutrient contents listed in Tables 2 and 3.

The yields of the biomass production were positive (greater than the weight of the inoculum) for all media in groups 1, 2, and 3. The biomass yields for the media in group 4 were negative, suggesting that the total mass of the harvested biomass was less than the one inoculated. By looking into the yield it seems that only group 1, which consisted of eight media, could be considered attractive for *E. foetida*. This resulted in a biomass yield of 103% higher than the one inoculated.

The highest amount of biomass production in group 1 was also influenced by environmental factors (pH, temperature, and humidity) and adequate media nutrition or feed so that earthworms could produce optimally. The average pH of the eight media was 6.44; the average media temperature was 27.65 °C and the media humidity was 57.5%. This is in line with an earlier report [38] that the earthworm *E. foetida* can reproduce at a temperature of 32 °C with an optimum temperature of 28 °C. The preferred humidity of the media needed for earthworms ranged from 50 to 80% [39].

The nutrient content of the eight media with the highest biomass-producing media was protein, with a content of 9.27–17.83%, a fat content of 0.14–0.63%, and crude fiber of 16.21–36.99%. Based on the results of this study, it appears that the environmental conditions (pH, temperature, and humidity) of the media are the best conditions for optimum production. Likewise, the nutrient content contained in the media or feed also meets the needs both in quantity and quality, so that earthworms can produce the most biomass. This is in line with the earlier results [14] that showed that the best feed for earthworms is one containing 9–15% protein with a neutral pH.

Among the four sources of media that resulted in the highest weight of the produced earthworms, the media containing horse manure showed the highest number and weight of the produced earthworms. These particular media had a characteristic of the highest C/N ratio of 24–54–26.63, which is significantly higher than the media containing goat and cow manure with C/N ratios of 14.22–16.55 and 15.06–20.21, respectively. *E. foetida* has advantages when compared to other species, especially its high ability to reproduce and convert organic matter as food. Earthworms can break down organic matter up to twice their body weight per day [40]. With the high ability of earthworms to break down organic matter and reduce pungent odours, earthworms can also be used as an alternative to prevent environmental pollution, especially that caused by livestock waste, market, and household wastes [41].

The overall results show that different sources of organic waste can be used as an effective medium to cultivate the earthworm *E. foetida*. The highest biomass production group consisted of unblended media (CM, HM, and GM) and blended media (CM-RC, HM-MW, GM-RC, HM-RS, HM-HW). The management of the unblended media can be done proximate to the farm in order to minimize the transportation costs in a composting and cultivation zone. The produced earthworm biomass can be used directly or sold for protein source in animal feed. For the blended media, additional transportation and man-power is required to blend the media according to the desired composition before they can be fermented and used for earthworm cultivation. The additional income from the earthworm biomass sales can enhance the economic sustainability of cow, horse and goat farming.

Data from Figure 4 also show that apart from yielding lower biomass, several media (CTRT50, CM-RS, HW, GM-HW, GM-MW and RC) show a higher specific weight (g/unit earthworm). It means that cultivation in those media lead to production of less but heavier earthworms. To maximize the biomass production, future research can be focused on a two stage cultivation consisting of breeding followed by growth. Apart from that, nutritional analysis of the produced earthworms are required in order to project their potential application. It would also be interesting to explore whether delicate control of cultivation conditions (i.e., temperature, moisture content, mixing, etc.) would lead to an increase in biomass yield and alter the process to be economically attractive. Finally, technoeconomic analysis on the production of compost from domestic waste and/or an animal farm is required. It can be treated as a by product of the composting site or of an animal farm that enhances economic competitiveness, or it can also be treated as a by product of vermicompost, as reported earlier [42].

4. Conclusions

This study demonstrated that most organic waste can be used as a medium for *E. foetida* cultivation. Some organic wastes tested in this study, namely MW, BM, BM-MW, and BM-HW lead to full mortality of the broodstocs. The other media resulted in variable cocoon production, earthworm and earthworm biomass. The findings suggest that common organic waste contains sufficient nutrients for *E. foetida* earthworm growth. However, none of the tested media yielded distinctively higher final biomass production. Yet, cultivation in certain media lead to substantially high numbers of cocoons and earthworms. This finding opens the possibility to engineer the media (i.e., cultivation in stages in different media) to maximize the final biomass yield. Moreover, technoeconomical analysis needs to be conducted to assess the feasibility of biomass production from common organic compost and the potential of co-production of the biomass from animal farms.

Author Contributions: M.M. contributed to the original idea of the study, performed all the experiments, and prepared the manuscript draft; H.H. and M.R.B. performed data and statistical analysis; N.H. and J.R. revised the manuscript. All authors have read and agreed to the published version of the manuscript.

Funding: This research received no external funding. The article processing charge was supported by the Universiti Malaysia Sabah.

Institutional Review Board Statement: Not applicable.

Informed Consent Statement: Not applicable.

Data Availability Statement: Not applicable.

Conflicts of Interest: The authors declare no conflict of interest.

References

1. Ghosh, C. Integrated Vermi-Pisciculture—An Alternative Option for Recycling of Solid Municipal Waste in Rural India. *Bioresour. Technol.* **2004**, *93*, 71–75. [CrossRef] [PubMed]
2. Nair, J.; Sekiozoic, V.; Anda, M. Effect of Pre-Composting on Vermicomposting of Kitchen Waste. *Bioresour. Technol.* **2006**, *97*, 2091–2095. [CrossRef] [PubMed]
3. Oyedele, D.J.; Schjønning, P.; Amusan, A.A. Physicochemical Properties of Earthworm Casts and Uningested Parent Soil from Selected Sites in Southwestern Nigeria. *Ecol. Eng.* **2006**, *28*, 106–113. [CrossRef]
4. Adhikary, S. Vermicompost, the Story of Organic Gold: A Review. *Agric. Sci.* **2012**, *3*, 905–917. [CrossRef]
5. Edwards, C.A. (Ed.) *Earthworm Ecology*, 2nd ed.; CRC Press: Boca Raton, FL, USA, 2004; ISBN 978-0-8493-1819-1.
6. Lim, P.N.; Wu, T.Y.; Shyang Sim, E.Y.; Lim, S.L. The Potential Reuse of Soybean Husk as Feedstock of Eudrilus Eugeniae in Vermicomposting. *J. Sci. Food Agric.* **2011**, *91*, 2637–2642. [CrossRef]
7. Garg, P.; Gupta, A.; Satya, S. Vermicomposting of Different Types of Waste Using *Eisenia foetida*: A Comparative Study. *Bioresour. Technol.* **2006**, *97*, 391–395. [CrossRef]
8. Gunya, B.; Masika, P.J.; Hugo, A.; Muchenje, V. Nutrient Composition and Fatty Acid Profiles of Oven-Dried and Freeze-Dried Earthworm *Eisenia foetida*. *J. Food Nutr. Res.* **2016**, *4*, 343–348. [CrossRef]
9. Chaves, R.C.; de Paula, R.Q.; Gücker, B.; Marriel, I.E.; Teixeira, A.O.; Boëchat, I.G. An Alternative Fish Feed Based on Earthworm and Fruit Meals for Tilapia and Carp Postlarvae. *Rev. Bras. Biociênc.* **2015**, *13*, 15–24.
10. Loh, T.C.; Fong, L.Y.; Foo, H.L.; Thanh, N.T.; Sheikh-Omar, A.R. Utilisation of Earthworm Meal in Partial Replacement of Soybean and Fish Meals in Diets of Broilers. *J. Appl. Anim. Res.* **2009**, *36*, 29–32. [CrossRef]
11. Edwards, C.A.; Blaxter, K.L.; Fowden, L. Production of Feed Protein from Animal Waste by Earthworms. *Philos. Trans. R. Soc. Lond. B Biol. Sci.* **1985**, *310*, 299–307. [CrossRef]
12. Parolini, M.; Ganzaroli, A.; Bacenetti, J. Earthworm as an Alternative Protein Source in Poultry and Fish Farming: Current Applications and Future Perspectives. *Sci. Total Environ.* **2020**, *734*, 139460. [CrossRef]
13. Stafford, E.A.; Tacon, A.G.J. The Nutritional Evaluation of Dried Earthworm Meal (*Eisenia foetida*, Savigny, 1826) Included at Low Levels in Production Diets for Rainbow Trout, Salmo Gairdneri Richardson. *Aquac. Res.* **1985**, *16*, 213–222. [CrossRef]
14. Fosgate, O.T.; Babb, M.R. Biodegradation of Animal Waste by *Lumbricus terrestris*. *J. Dairy Sci.* **1972**, *55*, 870–872. [CrossRef]
15. Hayati, S.N.; Herdian, H.; Damayanti, E.; Istiqomah, L.; Julendra, H. Profil Asam Amino Ekstrak Cacing Tanah (Lumbricus Rubellus) Terenkapsulasi Dengan Metode Spray Drying. *J. Teknol. Indones.* **2011**, *34*, 1–7.
16. Musyoka, S.N.; Liti, D.M.; Ogello, E.; Waidbacher, H. Utilization of the Earthworm, *Eisenia fetida* (Savigny, 1826) as an Alternative Protein Source in Fish Feeds Processing: A Review. *Aquac. Res.* **2019**, *50*, 2301–2315. [CrossRef]
17. Nahmani, J.; Hodson, M.E.; Black, S. Effects of Metals on Life Cycle Parameters of the Earthworm *Eisenia fetida* Exposed to Field-Contaminated, Metal-Polluted Soils. *Environ. Pollut.* **2007**, *149*, 44–58. [CrossRef]
18. Žaltauskaitė, J.; Sodienė, I. Effects of Cadmium and Lead on the Life-Cycle Parameters of Juvenile Earthworm *Eisenia fetida*. *Ecotoxicol. Environ. Saf.* **2014**, *103*, 9–16. [CrossRef] [PubMed]
19. Hatanaka, K.; Ishioka, Y.; Furuichi, E. Cultivation of *Eisenia fetida* using dairy waste sludge cake. In *Earthworm Ecology*; Satchell, J.E., Ed.; Springer: Dordrecht, The Netherlands, 1983; pp. 323–329, ISBN 978-94-009-5967-5.
20. Minnich, J. *The Earthworm Book: How to Raise and Use Earthworms for Your Farm and Garden*; Rodale Press: Emmaus, PA, USA, 1977; ISBN 0-87857-193-2.
21. Huang, K.; Li, F.; Wei, Y.; Chen, X.; Fu, X. Changes of Bacterial and Fungal Community Compositions during Vermicomposting of Vegetable Wastes by *Eisenia foetida*. *Bioresour. Technol.* **2013**, *150*, 235–241. [CrossRef] [PubMed]
22. Vodounnou, D.S.J.V.; Kpogue, D.N.S.; Tossavi, C.E.; Mennsah, G.A.; Fiogbe, E.D. Effect of Animal Waste and Vegetable Compost on Production and Growth of Earthworm (*Eisenia fetida*) during Vermiculture. *Int. J. Recycl. Org. Waste Agric.* **2016**, *5*, 87–92. [CrossRef]
23. Loh, T. Vermicomposting of Cattle and Goat Manures by *Eisenia foetida* and Their Growth and Reproduction Performance. *Bioresour. Technol.* **2005**, *96*, 111–114. [CrossRef]
24. Domínguez, J.; Velando, A.; Ferreiro, A. Are *Eisenia fetida* (Savigny, 1826) and *Eisenia andrei* Bouché (1972) (Oligochaeta, Lumbricidae) Different Biological Species? *Pedobiologia* **2005**, *49*, 81–87. [CrossRef]
25. Suthar, S. Vermicomposting of Vegetable-Market Solid Waste Using *Eisenia fetida*: Impact of Bulking Material on Earthworm Growth and Decomposition Rate. *Ecol. Eng.* **2009**, *35*, 914–920. [CrossRef]
26. Henneberg, W.; Stohmann, F. Über Das Erhaltungsfutter Volljährigen Rindviehs. *J. Landwirtsch* **1859**, *3*, 485–551.
27. Greenfield, H.; Southgate, D.A.T. *Food Composition Data: Production, Management, and Use*; FAO: Rome, Italy, 2003; ISBN 978-92-5-104949-5.
28. Langan, A.M.; Shaw, E.M. Responses of the Earthworm *Lumbricus terrestris* (L.) to Iron Phosphate and Metaldehyde Slug Pellet Formulations. *Appl. Soil Ecol.* **2006**, *34*, 184–189. [CrossRef]
29. Esmaeili, A.; Khoram, M.R.; Gholami, M.; Eslami, H. Pistachio Waste Management Using Combined Composting-Vermicomposting Technique: Physico-Chemical Changes and Worm Growth Analysis. *J. Clean. Prod.* **2020**, *242*, 118523. [CrossRef]

30. Abbott, I.; Parker, C.A. Interactions between Earthworms and Their Soil Environment. *Soil Biol. Biochem.* **1981**, *13*, 191–197. [CrossRef]
31. Edwards, C.A.; Fletcher, K.E. Interactions between Earthworms and Microorganisms in Organic-Matter Breakdown. *Agric. Ecosyst. Environ.* **1988**, *24*, 235–247. [CrossRef]
32. Monroy, F.; Aira, M.; Domínguez, J.; Velando, A. Seasonal Population Dynamics of *Eisenia fetida* (Savigny, 1826) (Oligochaeta, Lumbricidae) in the Field. *Comptes Rendus Biol.* **2006**, *329*, 912–915. [CrossRef]
33. Luo, Y.; Zang, Y.; Zhong, Y.; Kong, Z. Toxicological Study of Two Novel Pesticides on Earthworm *Eisenia foetida*. *Chemosphere* **1999**, *39*, 2347–2356. [CrossRef]
34. Rao, J.V.; Kavitha, P. Toxicity of Azodrin on the Morphology and Acetylcholinesterase Activity of the Earthworm *Eisenia foetida*. *Environ. Res.* **2004**, *96*, 323–327. [CrossRef]
35. Zhu, X.; Lian, B.; Yang, X.; Liu, C.; Zhu, L. Biotransformation of Earthworm Activity on Potassium-Bearing Mineral Powder. *J. Earth Sci.* **2013**, *24*, 65–74. [CrossRef]
36. Singh, S.; Singh, J.; Vig, A.P. Earthworm as Ecological Engineers to Change the Physico-Chemical Properties of Soil: Soil vs Vermicast. *Ecol. Eng.* **2016**, *90*, 1–5. [CrossRef]
37. Singh, S.; Sharma, A.; Khajuria, K.; Singh, J.; Vig, A.P. Soil Properties Changes Earthworm Diversity Indices in Different Agro-Ecosystem. *BMC Ecol.* **2020**, *20*, 27. [CrossRef] [PubMed]
38. Gates, G.E. Burmese Earthworms: An Introduction to the Systematics and Biology of Megadrile Oligochaetes with Special Reference to Southeast Asia. *Trans. Am. Philos. Soc.* **1972**, *62*, 1–326. [CrossRef]
39. Yadav, K.D.; Tare, V.; Ahammed, M.M. Vermicomposting of Source-Separated Human Faeces for Nutrient Recycling. *Waste Manag.* **2010**, *30*, 50–56. [CrossRef] [PubMed]
40. Haukka, J.K. Growth and Survival of *Eisenia fetida* (Sav.) (Oligochaeta: Lumbricidae) in Relation to Temperature, Moisture and Presence of Enchytraeus Albidus (Henle) (Enchytraeidae). *Biol. Fertil. Soils* **1987**, *3*, 99–102. [CrossRef]
41. Mashur, M. Produksi kokon dan biomassa cacing tanah *Eisenia foetida* pada berbagai media budidaya limbah peternakan. *Biosci. J. Ilmiah. Biol.* **2020**, *8*, 48. [CrossRef]
42. Garg, V.K.; Yadav, Y.K.; Sheoran, A.; Chand, S.; Kaushik, P. Livestock Excreta Management through Vermicomposting Using an Epigeic Earthworm *Eisenia foetida*. *Environmentalist* **2006**, *26*, 269–276. [CrossRef]

Article

Image Processing of UAV Imagery for River Feature Recognition of Kerian River, Malaysia

Emaad Ansari [1], Mohammad Nishat Akhtar [1], Mohamad Nazir Abdullah [2], Wan Amir Fuad Wajdi Othman [2,*], Elmi Abu Bakar [1,*], Ahmad Faizul Hawary [1] and Syed Sahal Nazli Alhady [2]

1 School of Aerospace Engineering, Universiti Sains Malaysia, Engineering Campus, Nibong Tebal 14300, Malaysia; zemaadansari@gmail.com (E.A.); nishat@usm.my (M.N.A.); aefaizul@usm.my (A.F.H.)
2 School of Electrical and Electronic Engineering, Universiti Sains Malaysia, Nibong Tebal 14300, Malaysia; eemnazir@usm.my (M.N.A.); sahal@usm.my (S.S.N.A.)
* Correspondence: wafw_othman@usm.my (W.A.F.W.O.); meelmi@usm.my (E.A.B.); Tel.: +60-196691441 (W.A.F.W.O.); +60-16-4939687 (E.A.B.)

Abstract: The impact of floods is the most severe among the natural calamities occurring in Malaysia. The knock of floods is consistent and annually forces thousands of Malaysians to relocate. The lack of information from the Ministry of Environment and Water, Malaysia is the foremost obstacle in upgrading the flood mapping. With the expeditious evolution of computer techniques, processing of satellite and unmanned aerial vehicle (UAV) images for river hydromorphological feature detection and flood management have gathered pace in the last two decades. Different image processing algorithms—structure from motion (SfM), multi-view stereo (MVS), gradient vector flow (GVF) snake algorithm, etc.—and artificial neural networks are implemented for the monitoring and classification of river features. This paper presents the application of the k-means algorithm along with image thresholding to quantify variation in river surface flow areas and vegetation growth along Kerian River, Malaysia. The river characteristic recognition directly or indirectly assists in studying river behavior and flood monitoring. Dice similarity coefficient and Jaccard index are numerated between thresholded images that are clustered using the k-means algorithm and manually segmented images. Based on quantitative evaluation, a dice similarity coefficient and Jaccard index of up to 97.86% and 94.36% were yielded for flow area and vegetation calculation. Thus, the present technique is functional in evaluating river characteristics with reduced errors. With minimum errors, the present technique can be utilized for quantifying agricultural areas and urban areas around the river basin.

Keywords: image processing; unmanned aerial vehicle; feature recognition; image segmentation; color space; floods; sediment

1. Introduction

Floods have been the most damaging disasters in Malaysia for a long time. The rise in sea level due to climate change, population growth, and abrupt urbanization have given rise to frequent flood disasters in many countries around the globe. Those living near the seashore and rivers are the most affected due to floods, which often force them to rehabilitate and disturbs their life. One of the factors that leads to degradation of the environment and represents a principal cause of floods in Malaysia is the fact that human activities, including the rapid development of densely populated flood plains, destruction of trees, and encroachment on flood-prone areas for development are viewed only as a positive development, turning a blind eye to their negative aspects [1]. Floods constitute 49% of the total disasters faced by Malaysia post-independence [2]. The consequences of floods in Malaysia include loss of human and animal life, destruction of infrastructures, and crumbling health conditions. In Malaysia, the National Security Council is responsible for flood management by implementing new techniques for flood control, forecasting, warning, and eviction, thus handling pre-disaster, disaster, and post-disaster activities [3]. The Malaysian National Security Council still requires additional overview of floods by researchers and government agencies to assist in decision making [4]. Researchers are finally coming up with community-based flood mitigation policies for inhabitants of towns located in the vicinity of different rivers that are prone to floods [5].

The backwash of floods disturbing the lives of Malaysians has led to more researchers and government agencies escalating the aim of developing enhanced and upgraded flood tracking systems. Quickly predicting the temporal level rise of river water and spatial mapping of flood consequences forms the core of flood monitoring systems. Early detection of flood after-effects and timely measures being taken by government agencies and disaster relief teams in advance can save many lives and protect resources from extreme destruction. All these factors are directly dependent on the availability of suitable data for developing a well-grounded flood monitoring system.

Various researchers have applied different techniques to analyze remotely sensed data for flood mapping that are captured by satellite platforms or UAV high-resolution imagery. Annis et al. [6] compared the digital elevation model (DEM) produced by data obtained using UAVs and light detection and ranging (LiDAR) satellites. Their results, which were determined via quantitative analyses, proved that DEMs acquired using UAVs have very high resolution along with a low surveying cost and require less time compared to DEMs procured by field visits or LiDAR satellites. Tamminga et al. [7] used small drones to investigate 3D dynamic changes due to floods on the Elbow River, Canada. Due to floods, a sediment flux mobility of a complex nature was confirmed in the study. UAVs and satellite images are also useful for examining riparian vegetation along the river basin [8]. Another advantage of UAVs is that they fill the gap between space-borne and field data by quickly providing high spatial resolution data [9]. This edge of UAVs over data acquired through other means helps in measuring river water levels with a better overall accuracy using a canny method [10] or DEM generation [11]. For better validation and calibration of flood models, georeferenced information regarding river shorelines, river geometry, and vegetation along the river basin attained using UAVs provides high temporal resolution, which is limited in the case of satellite images [12]. Post-flood surveys of high-water marks and river cross-sections using small, unmanned aircrafts have proven to be more accurate and effective compared to ground surveys [13]. Ephemeral rivers flowing in ungauged regions are essential for ecosystem balance and runoff. The techniques for measuring the peak discharge of these transient rivers are limited. UAV technology merged with the incipient motion of stones is useful for calculating the peak discharges of such short-lived rivers [14]. Flash floods occurring in ephemeral streams have a direct relationship with human activities, and the magnitude of destruction caused by flash floods is better calibrated by combining field survey data with UAV observations [15]. From the existing literature, it is evident that UAV imagery has a lead over satellite images and field survey data due to its ability to provide high temporal resolution information in less time with a low overall cost. Apart from this, high-resolution satellite images are costly and can sometimes be limited due to weather conditions, whereas field surveys take a long time. Therefore, the present work includes river images captured using UAVs for river feature recognition. Satellite data are also used in the present work for tertiary information, such as weather forecasting, crucial site locating, and river mapping. An amalgamation of UAV technology and satellite images is employed to collect reliable results.

Implementation of image processing algorithms for flood mapping has improved the reliability of results. Although there is no common solution for accurate flood monitoring, image processing has been proven to be a most reliable and widely used technique for analyzing digital images of river ecosystems [16]. Supervised image analysis consisting of classical image processing techniques such as multispectral classification along with pixel aggregation and shadow treatment is useful for analyzing flooded riverbeds [17]. Digital image processing is also functional as regards historical archives of river imagery to capture features of interest [18] as well as for LiDAR images [19]. Image processing of webcam images utilizing Raspberry-Pi and optical sensors is handy for flood detection [20]. Image processing algorithms are convenient for flood mapping as they can analyze terrestrial features and satellite images [21,22]. Real-time flood monitoring through the investigation of river hydromorphological features has become convenient due to video processing [23] and image processing [24] of data captured using UAVs. Cyber surveillance of an object in a flooded region for automatic monitoring and alert feedback are other applications of image processing [25]. The applications of image processing are not limited to early flood detection and flood monitoring, however. The segmentation and classification of UAV imagery are convenient for detecting the hydromorphological effects of floods in riverbeds and floodplains [26]. Plotting flood hydrographs has become possible using object-based image analysis (OBIA) with the help of image processing technology [27]. Jafari et al. [28] used live cameras and image processing techniques for real-time stream water level monitoring in this study. Apart from image processing techniques, convolutional neural networks (CNNs) [29], deep learning (DL) [30], artificial neural networks (ANN) [31], etc., for flood mapping using UAVs are other useful techniques that can be integrated, but image processing techniques have an edge over them due to the availability of high-resolution time saving algorithms [32]. Thus, from the existing literature, a range of applications of image processing for flood detection, flood monitoring, and analyzing post-flood effects is observed.

The hydromorphological feature data of the Kerian River are very limited. Therefore, this research presents the application of image processing for feature extraction in the Kerian River. The k-means algorithm for clustering along with thresholding for image segmentation are used in the present study to evaluate temporal variation in flow width and vegetation along river banks during a period of 3 months.

2. Materials and Methods

2.1. Study Area

We carried out the study in the Kerian River (Figure 1) that flows through the northern states of Peninsular Malaysia. The Kerian River is a meandering river of approximately 90 km length, originating in the Bintang Range in the northern state of Perak. This river serves as a border between the state of Kedah and Perak for approximately 73 km until it enters the Malaysian state of Penang for its remaining 17 km, where it finally discharges into the Strait of Melaka. The study region lies downstream of the Kerian River at the location 5°7'31" N, 100°29'50" E in Bandar Baharu, Kedah, where the width of the river measures 34.51 m. The width of the mouth of the Kerian River is approximately 166.71 m.

2.2. Flight Mission

The UAV used in the study is a DJI Mavic Pro fitted with a 1/2.3" complementary metal-oxide-semiconductor (CMOS) 12MP 4K camera. We tabulate the characteristics of the camera in Table 1. We carry out the flight operations twice within three months. The flight mission comprised reconnaissance of the mapped site and pre-flight fieldwork. The H83 mm × W83 mm × L198 mm copter has an upright take-off weight of 734 g. Its fully charged lithium polymer (LiPo) 3830 mAh batteries provide a maximum flight time of 27 min. The DJI Mavic Pro can withstand wind speed up to 7 m/s, which was never exceeded during the flight mission. The flight altitude while capturing the images was 70 m, which the UAV climbed with an ascent speed of 4 m/s. The resulting ground sample distance was (GSD) 2.4 cm. To observe variation in flow width and vegetation growth, we execute a flight mission to capture UAV images on 31 July 2019 at 09:04 am and 6 November 2019 at 09:20 am at the location 5°7'40" N, 100°29'48" E in Bandar Baharu, Kedah. We name the sites S1 and S2 on 31 July 2019 and 6 November 2019, respectively. The wind speed recorded was 1.67 m/s with the partly cloudy conditions at S1 and 1.11 m/s with broken clouds at S2. The temperature recorded on these two days was 29 °C and 31 °C, respectively. The level of tides was 1.69 m at S1 and 1.73 m at S2.

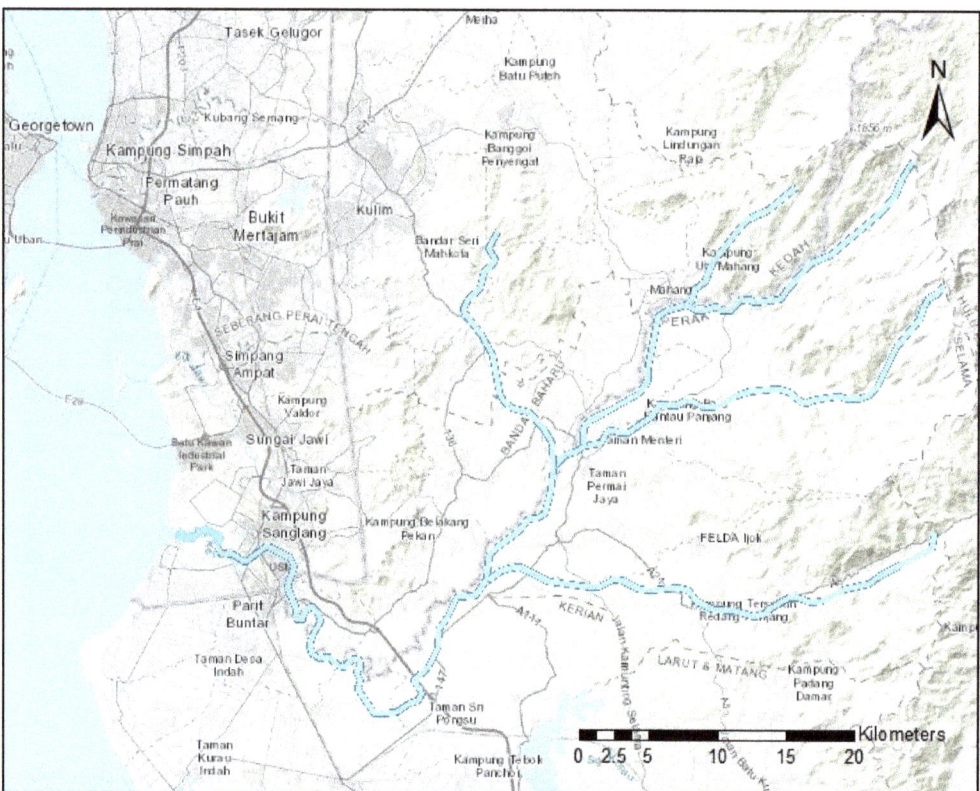

Figure 1. Localization of Kerian River and its tributaries in Malaysia.

Table 1. Characteristics of the DJI Mavic Pro camera.

Characteristics	DJI Mavic Pro Camera
Sensor Type	1/2.3″ CMOS Sensor
Million Effective Pixels	12.34
Image Size	4000 × 3000
Lens	35 mm
ISO range	100–1600

2.3. Image Processing of UAV Images

2.3.1. Implementation of Bilateral Filter

The first step in image processing was the enhancement of all UAV images, using a bilateral filter. Bilateral filter, a type of time-domain non-linear filter, reduces the noise with edge preservation by restoring the intensity of every pixel with weighted intensity values of surrounding pixels [33]. A bilateral filter uses the Gaussian distribution for calculating the weighted intensity values with an additional component, which is a function of pixel intensity difference [34]. This additional component ensures it used only weighted intensity values in computing blurred intensity values. The bilateral filter is defined as:

$$I_{filtered} = \frac{1}{W_P} \sum_{x_i \in \Omega} I(x_i) f_r(\|I(x_i) - I(x)\|) g_s(\|x_i - x\|) \qquad (1)$$

W_P is the normalization term defined by:

$$W_P = \sum_{x_i \in \Omega} f_r(\|I(x_i) - I(x)\|) g_s(\|x_i - x\|) \qquad (2)$$

In Equations (1) and (2), $I_{filtered}$ is the filtered image, I is the input image, x are the coordinates of the current pixel, Ω is the window centered in x, $x_i \in \Omega$ is another pixel, f_r is the range kernel and g_s is the spatial kernel. Allocation of the weight W_P uses f_r and g_s [34].

If a pixel located at (i, j) is to be denoised in image using surrounding pixels and we consider one of its neighbors at (k, l), then the weight assigned to a pixel at (k, l) to denoise a pixel at (i, j) is given by:

$$w(i, j, k, l) = \exp\left(-\frac{(i-k)^2 + (j-l)^2}{2\sigma_d^2} - \frac{\|I(i,j) - I(k,l)\|^2}{2\sigma_r^2}\right) \qquad (3)$$

In Equation (3), σ_d and σ_r are smoothing parameters, whereas $I(i, j)$ and $I(k, l)$ are pixel intensities at (i, j) and (k, l), respectively [34]. We calculate the denoised intensity of pixel, I_D at (i, j) after evaluating the weights and normalizing them using the equation:

$$I_D(i,j) = \frac{\sum_{k,l} I(k,l) w(i,j,k,l)}{\sum_{k,l} W(i,j,k,l)} \qquad (4)$$

Bilateral blur can fragment the image into different scales [35]. Along with edge preservation, bilateral filter finds application in denoising, tone mapping, tone management, data fusion, retinex, and texture and illumination separation [36]. The present work required denoising, which the bilateral blur serves, and hence we select it for this study. We did not detect the staircase effect and gradient reversal in the filtered image in this work, which are the limitations of the bilateral blur.

2.3.2. Clustering of Pixels Using k-Means Clustering Method

We then clustered, filtered, and enhanced images using the k-means clustering algorithm. The k-means clustering method is a clustering algorithm. In image analysis, it can cluster pixels of an image in unsupervised classification. Implementation of the k-means clustering in Python is uncomplicated, and therefore widely used by novice programmers and data scientists [37]. We implement the k-means for image pixel clustering in Python using its multiprocessing feature. The multiprocessing feature enables the parallel implementation of the k-means algorithm, reducing the execution time. The principal component of k-means clustering involves two steps, expectation and maximization. We assign each pixel to its nearest centroid in expectation, while the maximization step involves computing the mean of all pixels for each cluster and sets the new centroid. The flowchart in Figure 2 depicts the steps involved in implementing the k-means clustering algorithm.

2.3.3. Thresholding of Clustered Images

We use HSV thresholding for the analysis of clustered images post-k-means clustering. In vision and image processing, separating image intensity from color information is very important for many applications that are possible only in HSV color space. For this reason, we convert the clustered images from RGB to HSV color space.

Image segmentation is significant to gain clear perceptibility of the region of interest (ROI) [38]. We compute limits of the segmented pairs of hue (H_{lower}, H_{upper}), saturation (S_{lower}, S_{upper}), and value (V_{lower}, V_{upper}) to convert HSV color space to binary form using the equation:

$$C(x,y) = \begin{cases} 1, H_{lower} < H_{input}(x,y) < H_{upper}, \\ \{S_{lower} < S_{input}(x,y) < S_{upper},\} \\ \{V_{lower} \leq V_{input}(x,y) \leq V_{upper},\} \\ \{0, Otherwise\}. \end{cases} \qquad (5)$$

In Equation (5), $C(x, y)$ is the thresholded part. Additionally, Equation (5) shows that if the HSV values for the pixels of the input image lie between the range of lower and upper bound values, then its associated output pixel belongs to class object 1, otherwise it is null (0). The segmented image is further analyzed to compute the area of white pixels. We calibrate the percentage of white pixels to calculate the ROI in desired units.

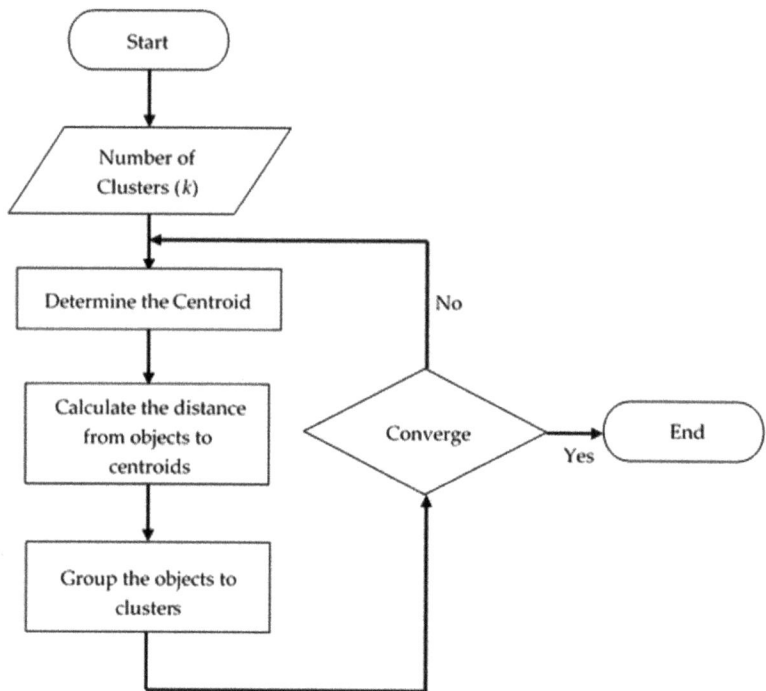

Figure 2. Flow chart of k-means clustering algorithm steps.

2.4. Evaluation of Image Processing Algorithm Effectiveness

The human evaluator or quantitative evaluation methods can calculate the efficiency and effectiveness of image processing algorithms [17]. Some of the metrics approached and extensively used by researchers for evaluating the accuracy of image processing algorithms include dice similarity coefficient (*DSC*), Jaccard index, precision, and recall. In the present work, we use *DSC* and Jaccard index to assess the effectiveness of image thresholding with the implementation of the k-means clustering. We used *DSC* and Jaccard index for evaluating variation in flow area and vegetation growth. We calculate the efficiency of clustered segmented images to manually segmented images.

The *DSC* works on binary numbers, and its value ranges from 0% to 100% for no overlap to complete overlap between two data sets of image thresholding. The *DSC* is defined as:

$$DSC = \frac{2|A \cap B|}{|A| + |B|} \quad (6)$$

where $|A|$ and $|B|$ are the number of elements of set A and set B, respectively.

The Jaccard index (*J*) is the ratio of the intersection of two data sets to the union of the two sets. It also ranges from 0 to 100%, describing the same overlap conditions as *DSC*. The Jaccard index (*J*) is given by:

$$J(A, B) = \frac{|A \cap B|}{|A \cup B|} \quad (7)$$

3. Results

We show the oblique UAV images captured in the present study and visualize the study reach for the present study along with the in-channel structure in Figure 3. Apart from this, we also capture the floodplains, water areas, low vegetation, shrub vegetation, and tree vegetation. We also captured the urban areas around the floodplain on the terrace. The present work only focuses on water area and vegetation growth in order to enumerate their variation. We cropped the study region as per the ROI, maintaining the pixel ratio. Further, in order to remove the noise, we apply a bilateral filter. On application of the bilateral filter, we obscured considerable noise and smoothened the site images. The edge preservation benefit of the bilateral filter paves the way for better clustering. Sunlight and water reflectance that are denoised to an extent by the application of the filter influences the various forms of vegetation along the river basin and flow area.

The next step post-implementation of the bilateral filter is the clustering of pixels using the k-means clustering algorithm. The necessity of clustering is that it segregates the largely distributed values of color space into desired numbers of color variation equivalent to the number of clusters (k). Figure 4 shows the color inspection in RGB color space and HSV color space. The color inspector feature of ImageJ software was used for the aforementioned purpose. Prior to clustering of pixels, it was difficult to extract the RGB or HSV limits, which lead to a decrease in the effectiveness of image segmentation. Extracting the RGB or HSV limits from the color inspector models for the water area and vegetation cover region is extremely tedious work. The researcher needs to analyze every pixel of the image for extracting the limits.

Figure 5 shows the images filtered and clustered using a bilateral filter and the k-means clustering algorithm. As per the flowchart in Figure 2, for all the site images, we selected the value of k randomly from $k = 2$ onwards until the clusters appeared unchanged. We increased the number of clusters by 2 and visualized them by human vision. Further, we increased the number of clusters in order to classify the image into an optimum number of color clusters. Finally, we clustered Figure 5a,b with $k = 16$ as the images converged at this number of clusters. Further increase in k produced unchanged cluster centroids as observed by human evaluation. Human visualization incorporated with interpreting elbow graphs produces better results in terms of accuracy in predicting an optimum number of clusters [39]. We identified the optimum number of clusters using an elbow graph in the current work.

Post clustering of images, the color inspection of HSV components becomes a simple and effective task. We converted the color space of all the images from RGB to HSV. HSV color space proves useful for robustness, removing shadows, lighting changes, etc., because of the separation of color components from the intensity. Thus, increasing the effectiveness of the image segmentation algorithm. Figure 6 shows the color inspection model in HSV color space for clustered image ($k = 16$). Compared to Figure 5, the HSV color inspector post-application of the k-means clustering algorithm (Figure 6) produces a better visual for extraction of HSV limits, which saves the energy and time of the human evaluator. We tabulated the thresholding HSV limits of water area and vegetation for sites S1 and S2 in Table 2 and used these values for image thresholding (Figure 7). Then, we counted the white pixels depicting ROI in thresholded images using a pixel counting algorithm in Python. Finally, we performed the RGB to HSV color space conversion and HSV limit evaluation using ImageJ (Version 1.53c). ImageJ allows the human evaluator to move the pointer over the image, which reflects the HSV value on the color inspector. An image with an optimum number of clusters classifies the features of the image into a clear perspective. In the present work, we clustered the water area and vegetation spread into six color components each. We got the maximum and minimum values of hue, saturation, and value component of the flow area and vegetation from Figure 6 (Table 2).

(a) Visualization of the Kerian River on 31 July 2019 (S1).

(b) Visualization of the Kerian River on 6 November 2019 (S2).

Figure 3. The Orthophoto of the Kerian River Basin and meandering structure on (**a**) 31 July 2019, and (**b**) 6 November 2019.

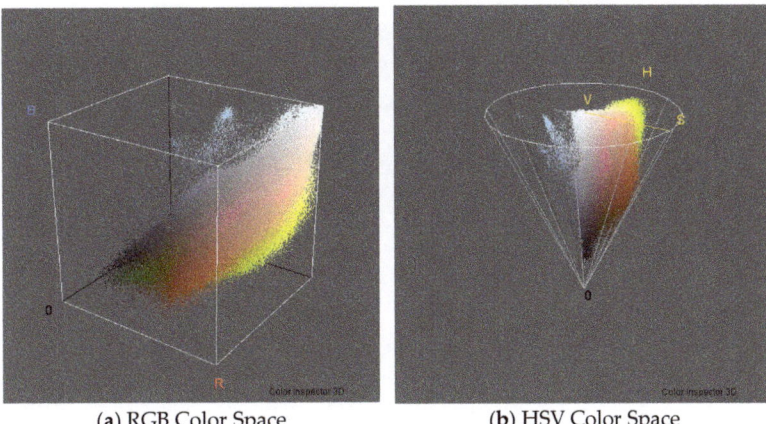

(a) RGB Color Space   (b) HSV Color Space

Figure 4. Pre-clustered color inspection model of Figure 3.

(a) Flow with emerging vegetation, shrub vegetation, flood plain and bare surface on 31 July 2019 grouped into 16 clusters.

(b) Flow area visualization with good vegetation, submerged vegetation and withered plants on 6 November 2019 converged into 16 clusters.

Figure 5. The outputs portray the implementation of bilateral filter and k-means clustering algorithm.

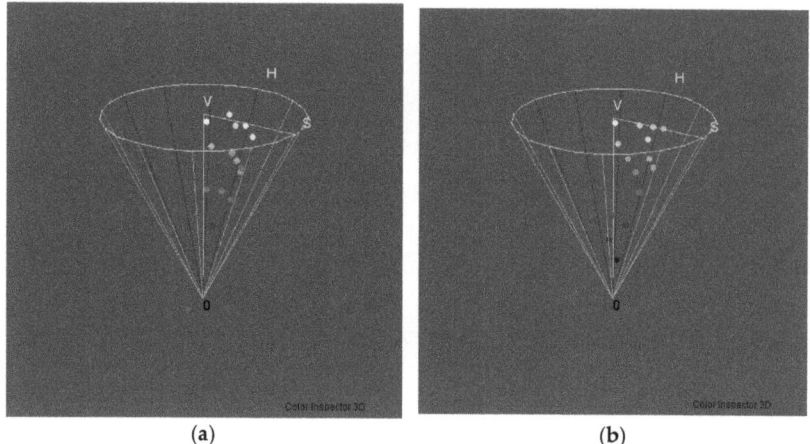

Figure 6. HSV color inspector for component limit evaluation of Figure 5 post application of k-means algorithm representing water area, vegetation and other features in (**a**) S1 and (**b**) S2.

Table 2. HSV limits of flow region and vegetation for Figure 6a,b.

Site	Feature	Minimum Limits			Maximum Limits		
		H (°)	S (%)	V (%)	H (°)	S (%)	V (%)
S1	Flow Area	34	24	100	38	31	94
	Vegetation	64	10	40	83	62	89
S2	Flow Area	25	26	95	28	49	97
	Vegetation	35	6	21	78	52	85

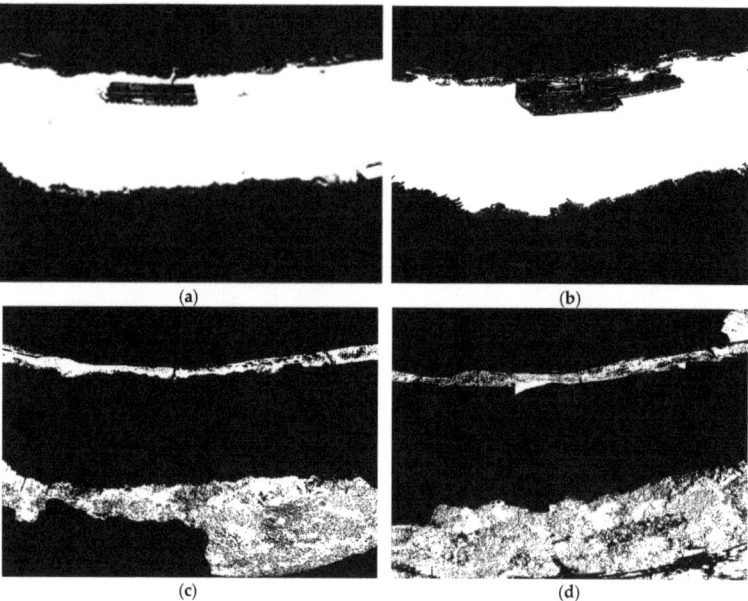

Figure 7. Thresholded Images representing the ROI and variation in Water Flow Area on (**a**) 31 July 2019 (**b**) 6 November 2019 and Vegetation cover on (**c**) 31 July 2019 (**d**) 6 November 2019 in white pixels along the river basin.

4. Discussion

4.1. Variation in Water Flow Area and Vegetation Cover

We calibrated the number of white pixels as a percentage of the total number of pixels and in terms of actual area (Table 3). The flow area varied slightly from 39.441% (4437.1 m^2) to 39.531% (4447.2 m^2). This slight variation is because of the tide difference, while all the other conditions were similar. An increase of 8.45% of the flow area will lead to overflow conditions. The vegetation growth over the period changed from 19.513% (2195.2 m^2) to 25.314% (2847.8 m^2). Malaysia having a tropical climate with year-round rainfall is one of the main reasons for this good amount of vegetation growth [40]. Vegetation significantly influenced the linkages between terrestrial ecosystems and atmospheric processes. It is workable to measure the advantages of ecosystem restoration by identifying the mechanisms through which vegetation dynamics and sensitivities to water resource availability respond at various temporal and geographical scales.

Table 3. Image segmentation results.

Site	Feature	White Pixels% (ROI)	Area (m^2)
S1	Flow Area	39.441	4437.1
	Vegetation	19.513	2195.2
S2	Flow Area	39.531	4447.2
	Vegetation	25.314	2847.8

4.2. Quantitative Evaluation

The effectiveness of thresholding post-implementation of the k-means clustering is clear from the DSC values ranging from 92.51% to 97.86% and the Jaccard index ranging from 91.39% to 94.36%. Compared to thresholding of non-clustered images where the DSC is 77.99%, the Jaccard index is 63.92% for normal flow conditions [17], the present method scores much better effectiveness for normal flow conditions. The metric values prove better overlap between the manually segmented images and the algorithm segmented images. We expect we can score a much better DSC and Jaccard index for overflow conditions since in overflow conditions the ROI becomes smoother. We encountered the commission and omission errors in the image classification technique in the present work to a small extent. During image classification, at some point, no actual change is classified as a change in the model, which leads to commission errors. While in other instances, we left some actual changes unsegmented in image thresholding, leading to omission errors [41]. We observed the combined effect of these errors in the quantitative evaluation results (Table 4). For the suggested research, the quantitative method can aid in an analysis where qualitative data is mainly in terms of analytic yield. The technique expands qualitative analyses horizons by providing researchers with a systematic way of guiding subgroup analysis within a qualitative data set.

Table 4. Quantitative evaluation results.

Site	Feature	DSC	Jaccard Index
S1	Flow Area	97.86	94.36
	Vegetation	94.91	92.20
S2	Flow Area	96.22	93.58
	Vegetation	92.51	91.39

4.3. Advantages and Limitations of UAV in River Feature Recognition

The advancement in UAV has paved the way for vertical dimension formation density and position referencing [42]. UAVs are also reliable for fluvial geomorphic study and reduce the error in observation arising in field study [43]. UAVs ensure less time consumption in data acquisition. They also have an upper hand in capturing real-time series data that we use in capturing flood events and landslides. Also, UAV technology compiles different data models produced in a single flight. Apart from being less expensive, UAV mapping provides equivalent spatial information and better operational features than airborne LIDAR [44]. However, the optical data acquisition of UAVs cannot pierce through vegetation cover and terrain features [45]. Another limitation of UAV technology is that the region with extensively low vegetation is unclassified in removal from the surface [46], but point cloud acquisition from the ground under the vegetation is possible using UAV LIDAR [47]. Another disadvantage of using small UAV platforms is that it limits the payload, while implementing drone-based imaging to capture temporal images of the same cross-sections is a tedious job.

4.4. UAV Imagery in Landscape of Kerian River and other Malaysian Rivers

The tropical climate of Malaysia and the year-round rainfall makes it difficult to analyze the river characteristics using low-cost satellite images because of heavy cloud covers [48]. UAV technology has a lead over satellite imagery because of its low operating cost. The Kerian River, for most of its upstream length, flows in an isolated region. The Kerian and other river basins of Malaysian rivers are the habitats of crocodiles, which makes it threatening for field visits. UAV imagery proves to be safe considering the climate and threats. The present study focuses only on flow area and vegetation growth using UAVs. We will get more features of the Kerian River using UAVs as inputs to the Ministry of Environment and Water, Malaysia in the future for river management. The sediment concentration by collecting sediments using samplers [49] and its validation using image processing are some gaps in this research topic. Apparently, for the Kerian River, there is limited river surveillance camera and flood does not occur frequently in this region. Therefore, direct use of a convolutional neural network (CNN) may not be a viable option for the river state classification. However, for future consideration, if enough surveillance cameras are installed at various cross-sections of the Kerian River, then we can develop a CNN model, as there will be a sufficient dataset to train the network. We may plan further work to use spectral cameras in UAVs to find the sediment concentration and composition using reflectance.

5. Conclusions

This work presented the application of the k-means clustering algorithm on UAV captured images with image thresholding for pressing out the flow area and vegetation growth. We calculated the binary white pixels to numerate the water area and vegetation cover. The results of image processing were proficient in withdrawing the river features. The dice similarity coefficient and Jaccard index for flow area and vegetation segmentation techniques fetched a value of over 90%, better than the existing techniques. Apart from this, the UAV technology is a promising tool to identify the river characteristics with high temporal information, low cost, and less time. A better understanding of river complexity may help the river management authorities in the proper planning of activities in the river basin. Our results will open a new corridor for the local officials in improving river hydro morphological conditions for sustainable river management, deploying various assessment tools.

Author Contributions: Conceptualization, E.A., M.N.A. (Mohammad Nishat Akhtar), and E.A.B.; formal analysis, E.A., M.N.A. (Mohammad Nishat Akhtar), and E.A.B.; funding acquisition, W.A.F.W.O.; methodology, E.A., M.N.A. (Mohammad Nishat Akhtar), M.N.A. (Mohamad Nazir Abdullah), E.A.B. and A.F.H.; project administration, W.A.F.W.O., E.A.B. and S.S.N.A.; supervision, M.N.A. (Mohammad Nishat Akhtar), M.N.A. (Mohamad Nazir Abdullah) and E.A.B.; writing—original draft, E.A.; writing—review and editing, E.A., W.A.F.W.O. and A.F.H. and S.S.N.A. All authors have read and agreed to the published version of the manuscript.

Funding: This study was funded by Research Creativity and Management Office, Universiti Sains Malaysia. The authors would like to acknowledge the grant RUI 1001/PAERO/8014035 and RU Top-Down 1001/PAERO/870052.

Conflicts of Interest: The authors declare no conflict of interest.

References

1. Chan, N.W. Increasing flood risk in Malaysia: Causes and solutions. *Disaster Prev. Manag. Int. J.* **1997**, *6*, 72–86. [CrossRef]
2. Shaluf, I.M.; Ahmadun, F.R. Disaster types in Malaysia: An overview. *Disaster Prev. Manag. Int. J.* **2006**, *15*, 286–298. [CrossRef]
3. Khalid, M.S.B.; Shafiai, S.B. Flood Disaster Management in Malaysia: An Evaluation of the Effectiveness Flood Delivery System. *Int. J. Soc. Sci. Humanit.* **2015**, *5*, 398–402. [CrossRef]
4. Shah, S.M.H.; Mustaffa, Z.; Yusof, K.W. Disasters Worldwide and Floods in the Malaysian Region: A Brief Review. *Indian J. Sci. Technol.* **2017**, *10*. [CrossRef]
5. Mohit, M.A.; Sellu, G.M. Mitigation of Climate Change Effects through Non-structural Flood Disaster Management in Pekan Town, Malaysia. *Procedia Soc. Behav. Sci.* **2013**, *85*, 564–573. [CrossRef]
6. Annis, A.; Nardi, F.; Petroselli, A.; Apollonio, C.; Arcangeletti, E.; Tauro, F.; Belli, C.; Bianconi, R.; Grimaldi, S. UAV-DEMs for Small-Scale Flood Hazard Mapping. *Water* **2020**, *12*, 1717. [CrossRef]
7. Tamminga, A.D.; Eaton, B.C.; Hugenholtz, C.H. UAS-based remote sensing of fluvial change following an extreme flood event. *Earth Surf. Process. Landf.* **2015**, *40*, 1464–1476. [CrossRef]
8. Morgan, B.E.; Bolger, D.T.; Chipman, J.W.; Dietrich, J.T. Lateral and longitudinal distribution of riparian vegetation along an ephemeral river in Namibia using remote sensing techniques. *J. Arid Environ.* **2020**, *181*, 104220. [CrossRef]
9. Bandini, F.; Jakobsen, J.; Olesen, D.; Reyna-Gutierrez, J.A.; Bauer-Gottwein, P. Measuring water level in rivers and lakes from lightweight Unmanned Aerial Vehicles. *J. Hydrol.* **2017**, *548*, 237–250. [CrossRef]
10. Ridolfi, E.; Manciola, P. Water level measurements from drones: A Pilot case study at a dam site. *Water* **2018**, *10*, 297. [CrossRef]
11. Hashemi-Beni, L.; Jones, J.; Thompson, G.; Johnson, C.; Gebrehiwot, A. Challenges and opportunities for UAV-based digital elevation model generation for flood-risk management: A case of princeville, north carolina. *Sensors* **2018**, *18*, 3843. [CrossRef] [PubMed]
12. Karamuz, E.; Romanowicz, R.J.; Doroszkiewicz, J. The use of unmanned aerial vehicles in flood hazard assessment. *J. Flood Risk Manag.* **2020**, *13*, 1–12. [CrossRef]
13. Forbes, B.T.; DeBenedetto, G.P.; Dickinson, J.E.; Bunch, C.E.; Fitzpatrick, F.A. Using small unmanned aircraft systems for measuring post-flood high-water marks and streambed elevations. *Remote Sens.* **2020**, *12*, 1437. [CrossRef]
14. Yang, S.; Li, C.; Lou, H.; Wang, P.; Wang, J.; Ren, X. Performance of an unmanned aerial vehicle (UAV) in calculating the flood peak discharge of ephemeral rivers combined with the incipient motion of moving stones in arid ungauged regions. *Remote Sens.* **2020**, *12*, 1610. [CrossRef]
15. Kastridis, A.; Kirkenidis, C.; Sapountzis, M. An integrated approach of flash flood analysis in ungauged Mediterranean watersheds using post-flood surveys and unmanned aerial vehicles. *Hydrol. Process.* **2020**, *34*, 4920–4939. [CrossRef]
16. Rusnák, M.; Sládek, J.; Kidová, A.; Lehotský, M. Template for high-resolution river landscape mapping using UAV technology. *Meas. J. Int. Meas. Confed.* **2018**, *115*, 139–151. [CrossRef]
17. Muhadi, N.A.; Abdullah, A.F.; Bejo, S.K.; Mahadi, M.R.; Mijic, A. Image segmentation methods for flood monitoring system. *Water* **2020**, *12*, 1825. [CrossRef]
18. Brigante, R.; Cencetti, C.; De Rosa, P.; Fredduzzi, A.; Radicioni, F.; Stoppini, A. Use of aerial multispectral images for spatial analysis of flooded riverbed-alluvial plain systems: The case study of the Paglia River (central Italy). *Geomat. Nat. Hazards Risk* **2017**, *8*, 1126–1143. [CrossRef]
19. Lane, S.N.; Widdison, P.E.; Thomas, R.E.; Ashworth, P.J.; Best, J.L.; Lunt, I.A.; Sambrook Smith, G.H.; Simpson, C.J. Quantification of braided river channel change using archival digital image analysis. *Earth Surf. Process. Landf.* **2010**, *35*, 971–985. [CrossRef]
20. Cobby, D.M.; Mason, D.C.; Davenport, I.J. Image processing of airborne scanning laser altimetry data for improved river flood modelling. *ISPRS J. Photogramm. Remote Sens.* **2001**, *56*, 121–138. [CrossRef]
21. Ariawan, A.; Pebrianti, D.; Akbar, Y.M.; Margatama, L.; Bayuaji, L. Image Processing-Based Flood Detection BT—Proceedings of the 10th National Technical Seminar on Underwater System Technology 2018; Md Zain, Z., Ahmad, H., Pebrianti, D., Mustafa, M., Abdullah, N.R.H., Samad, R., Noh, M.M., Eds.; Springer: Singapore, 2019; pp. 371–380.
22. Efremova, O.A.; Kunakov, Y.N.; Pavlov, S.V.; Sultanov, A.K. An algorithm for mapping flooded areas through analysis of satellite imagery and terrestrial relief features. *Comput. Opt.* **2018**, *42*, 695–703. [CrossRef]
23. Sultanov, A.K.; Pavlov, S.V.; Efremova, O.A. Development of a processing method of digital maps and satellite images for solving problems of emergencies. In Proceedings of the SPIE, Kazan, Russia, 6 June 2018; Volume 10774.
24. Jyh-Horng, W.; Chien-Hao, T.; Lun-Chi, C.; Shi-Wei, L.; Fang-Pang, L. Automated image identification method for flood disaster monitoring in riverine environments: A case study in Taiwan. In Proceedings of the AASRI International Conference on Industrial Electronics and Applications (IEA 2015), London, UK, 27–28 June 2015; Atlantis Press: Dordrecht, The Netharlands, 2015.
25. Zhang, Q.; Jindapetch, N.; Duangsoithong, R.; Buranapanichkit, D. Investigation of Image Processing based Real-time Flood Monitoring. In Proceedings of the 2018 IEEE 5th International Conference on Smart Instrumentation, Measurement and Application (ICSIMA), Songkla, Thailand, 28–30 November 2018; IEEE: Piscataway, NJ, USA, 2018; pp. 1–4. [CrossRef]
26. Lo, S.-W.; Wu, J.-H.; Lin, F.-P.; Hsu, C.-H. Cyber surveillance for flood disasters. *Sensors* **2015**, *15*, 2369–2387. [CrossRef] [PubMed]
27. Langhammer, J.; Vacková, T. Detection and mapping of the geomorphic effects of flooding using UAV photogrammetry. *Pure Appl. Geophys.* **2018**, *175*, 3223–3245. [CrossRef]
28. Jafari, N.H.; Li, X.; Chen, Q.; Le, C.-Y.; Betzer, L.P.; Liang, Y. Real-time water level monitoring using live cameras and computer vision techniques. *Comput. Geosci.* **2020**, *147*, 104642. [CrossRef]

29. Oga, T.; Harakawa, R.; Minewaki, S.; Umeki, Y.; Matsuda, Y.; Iwahashi, M. River state classification combining patch-based processing and CNN. *PLoS ONE* **2020**, *15*, e0243073. [CrossRef]
30. Akiyama, T.S.; Marcato Junior, J.; Gonçalves, W.N.; Bressan, P.O.; Eltner, A.; Binder, F.; Singer, T. Deep learning applied to water segmentation. *Int. Arch. Photogramm. Remote Sens. Spat. Inf. Sci. ISPRS Arch.* **2020**, *43*, 1189–1193. [CrossRef]
31. Casado, M.R.; Gonzalez, R.B.; Kriechbaumer, T.; Veal, A. Automated identification of river hydromorphological features using UAV high resolution aerial imagery. *Sensors* **2015**, *15*, 27969–27989. [CrossRef] [PubMed]
32. Cuevas, J.; Chua, A.; Sybingco, E.; Bakar, E.A. Identification of river hydromorphological features using histograms of oriented gradients cascaded to the Viola-Jones algorithm. *Int. J. Mech. Eng. Robot. Res.* **2019**, *8*, 289–292. [CrossRef]
33. Elad, M. On the origin of the bilateral filter and ways to improve it. *IEEE Trans. Image Process.* **2002**, *11*, 1141–1151. [CrossRef] [PubMed]
34. Paris, S.; Durand, F. A fast approximation of the bilateral filter using a signal processing approach. *Int. J. Comput. Vis.* **2009**, *81*, 24–52. [CrossRef]
35. Zhang, B.; Allebach, J.P. Adaptive bilateral filter for sharpness enhancement and noise removal. *Proc. Int. Conf. Image Process. ICIP* **2007**, *4*, 664–678. [CrossRef]
36. Chen, B.-H.; Tseng, Y.-S.; Yin, J.-L. Gaussian-Adaptive Bilateral Filter. *IEEE Signal. Process. Lett.* **2020**, *27*, 1670–1674. [CrossRef]
37. Akhtar, M.N.; Ahmed, W.; Kakar, M.R.; Bakar, E.A.; Othman, A.R.; Bueno, M. Implementation of Parallel K-Means Algorithm to Estimate Adhesion Failure in Warm Mix Asphalt. *Adv. Civ. Eng.* **2020**, *2020*, 8848945. [CrossRef]
38. Danish, M.; Akhtar, M.N.; Hashim, R.; Saleh, J.M.; Bakar, E.A. Analysis using image segmentation for the elemental composition of activated carbon. *MethodsX* **2020**, *7*, 1–9. [CrossRef] [PubMed]
39. Syakur, M.A.; Khotimah, B.K.; Rochman, E.M.S.; Satoto, B.D. Integration K-Means Clustering Method and Elbow Method for Identification of The Best Customer Profile Cluster. *IOP Conf. Ser. Mater. Sci. Eng.* **2018**, *336*, 12017. [CrossRef]
40. Ansari, E.; Akhtar, M.N.; Bakar, E.A.; Uchiyama, N.; Kamaruddin, N.M.; Umar, S.N.H. *Investigation of Geomorphological Features of Kerian River Using Satellite Images BT—Intelligent Manufacturing and Mechatronics*; Bahari, M.S., Harun, A., Zainal Abidin, Z., Hamidon, R., Eds.; Springer: Singapore, 2021; pp. 91–101.
41. Pierce, K.B. Accuracy optimization for high resolution object-based change detection: An example mapping regional urbanization with 1-m aerial imagery. *Remote Sens.* **2015**, *7*, 12654–12679. [CrossRef]
42. Kršák, B.; Blišťan, P.; Pauliková, A.; Puškárová, P.; Kovanič, L.; Palková, J.; Zelizňaková, V. Use of low-cost UAV photogrammetry to analyze the accuracy of a digital elevation model in a case study. *Meas. J. Int. Meas. Confed.* **2016**, *91*, 276–287. [CrossRef]
43. Mirijovský, J.; Langhammer, J. Multitemporal monitoring of the morphodynamics of a mid-mountain stream using UAS photogrammetry. *Remote Sens.* **2015**, *7*, 8586–8609. [CrossRef]
44. Niethammer, U.; James, M.R.; Rothmund, S.; Travelletti, J.; Joswig, M. UAV-based remote sensing of the Super-Sauze landslide: Evaluation and results. *Eng. Geol.* **2012**, *128*, 2–11. [CrossRef]
45. Turner, D.; Lucieer, A.; de Jong, S.M. Time series analysis of landslide dynamics using an Unmanned Aerial Vehicle (UAV). *Remote Sens.* **2015**, *7*, 1736–1757. [CrossRef]
46. Rusnák, M.; Sládek, J.; Buša, J.; Greif, V. Suitability of Digital Elevation Models Generated By Uav Photogrammetry for Slope Stability Assessment (Case Study of Landslide in Svätý Anton, Slovakia). *Acta Sci. Pol. Form. Circumiectus* **2016**, *15*, 439–449. [CrossRef]
47. Sankey, T.; Donager, J.; McVay, J.; Sankey, J.B. UAV lidar and hyperspectral fusion for forest monitoring in the southwestern USA. *Remote Sens. Environ.* **2017**, *195*, 30–43. [CrossRef]
48. Ansari, E.; Akhtar, M.N.; Abdullah, M.N.; Bakar, E.A. Design, assembly and use of a simple, economical water sampler for suspended sediment collection. *J. Phys. Conf. Ser.* **2021**, *1921*, 12096. [CrossRef]
49. Ansari, E.; Akhtar, M.N.; Bakar, E.A.; Hawary, A.F.; Alhady, S.S.N. Investigation of Suspended Sediment Samplers: A Review. *J. Phys. Conf. Ser.* **2021**, *1874*, 12018. [CrossRef]

Article

Membrane Filtration as Post-Treatment of Rotating Biological Contactor for Wastewater Treatment

Sharjeel Waqas [1,2], Muhammad Roil Bilad [3,4,*], Nurul Huda [5,6,*], Noorfidza Yub Harun [1], Nik Abdul Hadi Md Nordin [1], Norazanita Shamsuddin [4], Yusuf Wibisono [7], Asim Laeeq Khan [8] and Jumardi Roslan [5]

1. Chemical Engineering Department, Universiti Teknologi PETRONAS, Bandar Seri Iskandar 32610, Perak, Malaysia; sharjeel_17000606@utp.edu.my (S.W.); noorfidza.yub@utp.edu.my (N.Y.H.); nahadi.sapiaa@utp.edu.my (N.A.H.M.N.)
2. School of Chemical Engineering, The University of Faisalabad, Faisalabad 37610, Pakistan
3. Faculty of Applied Science and Technology, Universitas Pendidikan Mandalika, Jl. Pemuda No. 59A, Mataram 83126, Indonesia
4. Faculty of Integrated Technologies, Universiti Brunei Darussalam, Jalan Tungku Link, Gadong BE1410, Brunei; norazanita.shamsudin@ubd.edu.bn
5. Faculty of Food Science and Nutrition, Universiti Malaysia Sabah, Jalan UMS, Kota Kinabalu 88400, Sabah, Malaysia; jumardi@ums.edu.my
6. Department of Food Science and Technology, Faculty of Agriculture, Universitas Sebelas Maret, Surakarta 57126, Indonesia
7. Department of Bioprocess Engineering, Faculty of Agricultural Technology, Brawijaya University, Malang 65141, Indonesia; y_wibisono@ub.ac.id
8. Department of Chemical Engineering, COMSATS Institute of Information Technology, Lahore 54000, Pakistan; alaeeqkhan@cuilahore.edu.pk
* Correspondence: roil.bilad@ubd.edu.bn (M.R.B.); drnurulhuda@ums.edu.my (N.H.)

Citation: Waqas, S.; Bilad, M.R.; Huda, N.; Harun, N.Y.; Md Nordin, N.A.H.; Shamsuddin, N.; Wibisono, Y.; Khan, A.L.; Roslan, J. Membrane Filtration as Post-Treatment of Rotating Biological Contactor for Wastewater Treatment. Sustainability 2021, 13, 7287. https://doi.org/10.3390/su13137287

Academic Editors: Mohd Rafatullah and Masoom Raza Siddiqui

Received: 8 June 2021
Accepted: 23 June 2021
Published: 29 June 2021

Publisher's Note: MDPI stays neutral with regard to jurisdictional claims in published maps and institutional affiliations.

Copyright: © 2021 by the authors. Licensee MDPI, Basel, Switzerland. This article is an open access article distributed under the terms and conditions of the Creative Commons Attribution (CC BY) license (https://creativecommons.org/licenses/by/4.0/).

Abstract: A rotating biological contactor (RBC) offers a low energy footprint but suffers from performance instability, making it less popular for domestic wastewater treatment. This paper presents a study on an RBC integrated with membrane technology in which membrane filtration was used as a post-treatment step (RBC–ME) to achieve enhanced biological performance. The RBC and RBC–ME systems were operated under different hydraulic retention times (HRTs) of 12, 18, 24, and 48 h, and the effects of HRT on biological performance and effluent filterability were assessed. The results show that RBC–ME demonstrates superior biological performance than the standalone RBC. The RBC–ME bioreactor achieved 87.9 ± 3.2% of chemical oxygen demand (COD), 98.9 ± 1.1% ammonium, 45.2 ± 0.7% total nitrogen (TN), and 97.9 ± 0.1% turbidity removals. A comparison of the HRTs showed that COD and TN removal efficiency was the highest at 48 h, with 92.4 ± 2.4% and 48.6 ± 1.3% removal efficiencies, respectively. The longer HRTs also lead to better RBC effluent filterability. The steady-state permeability increased respectively by 2.4%, 9.5%, and 19.1% at HRTs of 18, 24, and 48 h, compared to 12 h. Our analysis of membrane fouling shows that fouling resistance decreased at higher HRTs. Overall, RBC–ME offered a promising alternative for traditional suspended growth processes with higher microbial activity and enhanced biological performance, which is in line with the requirements of sustainable development and environment-friendly treatment.

Keywords: attached growth process; biological treatment; biofilm; membrane fouling; rotating biological contactors; wastewater treatment

1. Introduction

The treatment of wastewater using biological processes is an economical, energy-efficient, and environmentally sound approach [1]. Typically, microbial aggregates are employed to biodegrade the organic compounds and nutrients in the wastewater [2]. Suspended flocs (employed in the conventional activated sludge (CAS) process) and attached biofilm (employed in trickling filters and rotating biological contactor (RBC)) are the two types of aggregates. Both the suspended flocs and the attached biofilm processes are established in full-scale wastewater treatment processes [3]. RBC—also referred to as biofilm reactors—offers a substitute to the CAS process [4].

There are two main types of RBC configurations: integral and modular [5]. The integral system, a compact decentralized unit, consists of a single unit combining the primary treatment, the RBC bioreactor, and the final clarifier. The integral units are usually contained within a package plant and have treatment capacities of ≤250 population equivalents (PE) [6]. On the other hand, the modular systems have separate independent operations for the primary treatment, the RBC bio-zone, the clarifier, and the solids treatment unit. Modular systems are centralized networks that allow for more flexible process configurations and have treatment capacities of >1000 PEs [7].

RBCs are equipped with solid media as a platform for the development of microorganisms as a biofilm [8]. Unlike the CAS process, which uses air bubbling to supply oxygen in the suspended flocs, the RBC employs mechanical rotations of the disks for contacting the biofilm adhered to them with air to provide the oxygen [9,10]. The commercial significance of membrane technology in municipal and industrial wastewater treatments is increasingly pervasive [11,12]. RBC systems have the inherent drawback of a high footprint due to space requirements for air contact and modular construction. Therefore, the application of such a system must be adjusted to meet these constraints. It is an attractive choice to facilitate the reuse of municipal and industrial wastewaters. The integration of membrane processes has been proven to be superior compared to conventional biological processes [13].

RBCs are seen as a low-cost biological process and an alternative to the more established CAS in treating wastewater [14,15]. The biofilm in RBC is, to a certain degree, resistant to toxic compounds, shock in organic loadings, and moderate changes in hydraulic loading. Biofilm also maintains a high proportion of microorganisms to ensure a high pollutant removal capacity [16,17].

Despite offering the aforementioned benefits, RBCs still suffer from operational instability and variable effluent quality, which are attributed to the biofilm dynamics and the detachment of the biofilm. To overcome these issues, various RBC configurations, including RBC/biofilm, RBC/suspended growth combination, RBC/wetland combination, and RBC membrane technology systems have been proposed [18,19]. RBCs can be installed modularly in parallel, allowing for more stable loadings. If the effluent quality is of primary interest, it can be operated in a series. Hybrid systems can also be considered where an RBC is combined with another unit to enhance the combined process stability to cater to variable loadings, boost the load capacity, or enhance the attainable effluent standard [5]. RBCs can be combined with a polishing step (tertiary treatment) or the capacity can be upgraded [20]. For instance, the RBC/wetland configuration provides a storm flow buffer and improves the discharge effluent quality [7]. In previous studies, the application of membrane technology in combination with RBCs as a post-treatment helped to enhance effluent quality and removal efficiency [21,22].

Studies with RBC systems have revealed that longer contact times improve the diffusion of the substrate into the biofilm and its consequent removal of the influent. This trend is also verified with toxic and heavy metals substrates [23]. The hydraulic retention time (HRT) is directly related to the organic and hydraulic loading of the influent wastewater. Longer HRTs facilitate the proper degradation of the substrate substances and improve removal efficiencies. The trend remains the same for toxic and heavy metal substrates. A short HRT would result in a poor treatment process due to the improper degradation of the substrate, while longer HRTs would affect the process economics. The selection of the optimum HRT is very important for obtaining the desired effluent quality at a minimal cost [24,25]. Najafpour et al. [8] studied the effect of HRTs on the performance of RBCs for the treatment of palm oil mill effluents and found that both the COD (45 to 88% removal efficiency) and total nitrogen (TN) removal efficiency increased with an increase in the HRT (10–55 h). The impact of the HRT is more for the 10–26 h than the 26–55 h interval change.

Ghalehkhondabi et al. [26] studied that chemical oxygen demand (COD) and ammonium removal rates increased with HRT, and the maximum removal rate depends on the selected HRT and the interactions between the disks' rotational speeds.

Saha et al. [13] showed that the combination of RBC with an external membrane for the treatment of oily wastewater resulted in higher treatment efficiency than a stand-alone RBC at all feed ratios and HRTs. A total petroleum hydrocarbon removal efficiency of 99% was obtained at a 24 h HRT in an RBC system combined with an external membrane bioreactor. Similarly, the micro-pollutant degradation and general treatment efficiency could be improved by altering the HRTs. According to Win et al. [27] and Zhang et al., [28], HRT changes affect the bacterial communities and thus, the system performance. Overall, it has been demonstrated that integrating an RBC with membrane technology increases the overall system performance. The attached biofilm in the RBC, with a higher biomass concentration and microbial diversity, favors the integration of membrane separation because of less contact between the biomass with the membrane, which may otherwise cause membrane fouling.

Instigated by the success of CAS incorporated into the membrane filtration, an RBC bioreactor combined with membrane filtration (RBC–ME) is proposed in this study. RBC–ME combines the conventional RBC with a membrane filtration unit placed externally on the bioreactor. This study evaluates the integration of an RBC system with membrane filtration by replacing the settling tank with a membrane filtration unit. In previous studies, the membrane filtration was placed after the settling tank in a separate vessel, which required extra space. The main functionality of an RBC–ME is the membrane placement into the settling tank immediately after the RBC bioreactor. The integration of the membrane with an RBC is expected to maintain high effluent quality and removal efficiency. The HRT is an important operation parameter and an optimized valve has a strong interaction with the effluent quality and plant economy. Limited literature is available for an RBC integrated with membrane technology; hence, the current study aims to study the effect of HRT on the COD and ammonium removal along with membrane permeability.

This study focused on the process intensification of the RBC–ME, particularly on the biological performance and the membrane fouling control. The operational parameter of the HRT was selected for this study, as it is one of the most influential parameters in the bioreactor. Alterations in HRTs could also result in variations of organic loading rates. Firstly, the biological performances, in terms of organic and nutrient compound removal of both standalone RBC and RBC–ME, were evaluated as a function of the hydraulic retention time. The membrane filtration and fouling analysis are presented at various HRTs.

2. Materials and Methods

2.1. Wastewater Preparation and Characterization

Synthetic wastewater was prepared by blending refined food leftovers (1 g/L), as suggested in previous work [29]. After mixing food leftovers with water, the mixture was left for 2 h to settle the suspended particles. The stock solution (supernatant) was then filtered through Whatman filter paper, 11 μm medium flow filter paper (Grade 1 Qualitative Filter Paper Standard Grade, GE Whatman, Kent, UK). The stock solution was then diluted to obtain the influent wastewater concentration, as summarized in Table 1. The prepared wastewater was analyzed in terms of COD, TN, ammonium, and nitrate.

Table 1. Influent characteristics for the RBC and RBC–ME bioreactors.

	Influent
COD (mg/L)	298 ± 45.6
TN (mg/L)	2.4 ± 0.2
Ammonium (mg/L)	0.92 ± 0.07
Nitrate (mg/L)	0.52 ± 0.08
Turbidity (NTU)	15.2 ± 0.6
pH	6.35 ± 0.18

RBC: rotating biological contactor, RBC–ME: rotating biological contactor–membrane external, COD: chemical oxygen demand, TN: total nitrogen.

The COD, the TN, the ammonium, and the nitrate were measured using the specific Hach digestion solution (HACH, Loveland, CO, USA) for each compound. The solution was diluted accordingly to fall into the range of the digestion vials. The values were determined through a Hach DR3900 Spectrophotometer (HACH, Loveland, CO, USA). A Hach 2100Q portable turbidimeter (HACH, Loveland, CO, USA) and a Hach HQ411D benchtop PH/MV meter (HACH, Loveland, CO, USA) were used to determine the turbidity and pH, respectively [30].

2.2. Membrane Preparation and Characterization

A polysulfone (PSF) membrane was prepared through the phase inversion technique detailed in Table 2. The PSF membrane was fabricated from a dope solution containing PSF as the polymer, (BASF-Ultrason, Mw 22 kDa), polyethylene glycol (PEG) as the additive (Sigma-Aldrich, MO, USA, Mw 10 kDa), and N,N-Dimethylacetamide (DMAc) as the solvent (Sigma-Aldrich, MO, USA) at concentrations of 12, 1, and 87 wt%, respectively. The dope solution was cast according to the phase inversion method, as described in our previous study [19]. The cast film was immersed immediately in demineralized water (acting as a non-solvent) to form a membrane sheet. A solid, thin, and porous membrane was then stored in water until further usage.

Table 2. Summary of materials for membrane fabrication.

IUPAC Name	Abbreviation	Avg. Molecular Mass	Purity *
Polysulfone	PSF	22,000 Da	100 wt%
Polyethylene glycol	PEG	9000–12,500 Da	100 wt%
N,N-Dimethylacetamide	DMAc	-	99.7 vol%
Water	H_2O	-	~100 vol%

* vol% = by volume percentage, wt% = by weight percentage.

The properties of the fabricated PSF membrane are listed in Table 3. The membrane thickness was measured using an electronic digital micrometer screw gauge (Mitutoyo 293-340-30 Digital Micrometer, Mitutoyo America Corporation, Aurora, CO, USA). The pore size was measured using a capillary flow porometer (Porolux, Nazareth, Belgium). The morphology of the membrane was analyzed, and the microstructure was acquired using scanning electron microscopy (SEM) (Zeiss, Leo 1430 VP, Carl Zeiss, Oberkochen, Germany). The static membrane surface water contact angle was determined using the sessile drop method (Mobile Surface Analyzer, KRUSS, Hamburg, Germany). All the characterization techniques were done in triplicate.

Table 3. Summary of membrane properties used in RBC–ME configuration.

Properties (Unit)	Values
Materials	Polysulfone
Thickness (mm)	0.28 ± 0.22
Mean flow pore size (μm)	0.03 μm
Surface contact angle (°)	61.8 ± 1.0
Cross-section morphology	Asymmetric
Clean water permeability (L/(m² h bar))	817 ± 35

2.2.1. Determination of Filtration Performance

The membrane filtration was done at a fixed transmembrane pressure (ΔP) of 0.1 bar with a system detailed elsewhere [31]. The low pressure for filtration not only reduces the energy cost but it is also less susceptible to membrane fouling and maintains sustainable flux, as reported elsewhere [32]. The membrane permeability (L, L/m² h bar) was calculated using Equation (1).

$$L = \frac{\Delta V}{A \, \Delta t \, \Delta P} \qquad (1)$$

where V is the volume of permeance (L), A is the membrane area (m²) and t is the filtration time (h). During the filtration test, the permeate pump was stopped temporarily and the filtration was run in a full recycle without altering the hydraulic operation parameter.

2.2.2. Membrane Fouling Analysis

The membrane fouling analysis was evaluated using the Darcy law detailed in Equations (2) and (3).

$$J = \frac{\Delta P}{\mu R_T} \qquad (2)$$

$$R_T = R_M + R_P + R_C \qquad (3)$$

where J is the permeate flux, μ is the dynamic viscosity of the permeate, R_T is the total resistance, R_M is the intrinsic membrane resistance, R_P is the pore blocking resistance, and R_C is the cake layer resistance. The R_M was calculated by the filtration of deionized water and R_{M+P} was calculated from the filtration of deionized water using the used membrane after removing the cake layer from the membrane surface.

2.3. Bioreactor Set-Up and Operation

The lab-scale RBC–ME bioreactor was fabricated in-house (Figure 1). It comprised a feed tank, a bioreactor tank, and a settling tank. The bioreactor unit was a cuboid of 25 × 25 × 30 cm, fabricated using acrylic sheets. It had a total working volume of 6.5 L, with a 40% disk submergence. The unit consisted of 5 rotating disks (D = 18 cm) separated by a 3 cm gap between adjacent disks and fixed on a stainless-steel shaft. The rotating disk had a total surface area of 2034 cm². Polyurethane sheets were cut according to the size of the disks and then glued on the disks to be used as a platform for biofilm growth. The activated sludge used to inoculate the lab-scale RBC–ME was obtained from a nearby full-scale activated sludge domestic wastewater treatment plant.

The membrane sheet was cut and fixed onto both sides of the panel onto a plate and frame filtration panel. The membrane sheet was attached to a panel consisting of a semicircle shape and resulted in an active membrane surface area of 226 cm². The membrane sheets were glued to the panel with AB epoxy glue (AB quick epoxy, HYRO, Kuala Lumpur, Malaysia). The filtration panel was ensured to be free from leakage. A spacer fabric that was placed between the two membrane sheets acted as permeate channel. The membrane permeate was evacuated through a permeate pipe that connected the permeate channel to the permeate pump.

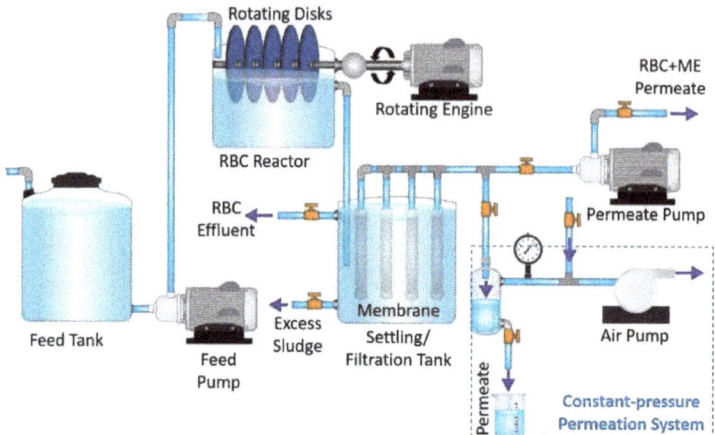

Figure 1. Schematic diagram of the laboratory-scale RBC and RBC–ME unit.

There were 2 different bioreactor configurations, one with membrane filtration and the other without membrane filtration. In the first configuration, the stand-alone RBC bioreactor followed by a settling tank was used to degrade the organics and nutrients. The treated water was flown by gravity to the settling tank and from the settling tank to a sink. The HRT was adjusted from the feed pump. In the second configuration, membrane filtration was incorporated and placed in the setting tank to allow filtration externally. This configuration is henceforth referred to as RBC–membrane external (RBC–ME). The membrane in this configuration acts as a post-treatment. The RBC–ME configuration eliminated the need for a settling tank, and thus, the addition of membrane filtration does not alter the overall plant size. To adjust the HRT in the RBC–ME, a peristaltic pump was installed to drive the permeation accordingly to meet the required HRTs.

The RBCs were run for 42 days, divided into 2 periods. During the first 15 day period, the bioreactor was operated under constant loading conditions of 17 g COD/m^2·d to grow and acclimatize the biofilm atop of the polyurethane foam surface. During this period, the biofilm was observed carefully, and the biological performance was monitored regularly. Carbonaceous bacteria responsible for COD biodegradation were expected to dominate the biofilm, as compared to nitrifying bacteria, which undergo TN removal. After the acclimatization phase, the biofilm was completely developed and was effective in degrading organics and nutrients. Any detached flocs from the rotating disks were regularly discharged.

The 2 system configurations were assessed: (1) the stand-alone RBC bioreactor and (2) the membrane placed externally on the settling tank. In the second phase of the RBC operation, the bioreactor was operated by incorporating the membrane to study the biological and filtration performance. For the RBC–ME configuration, there were no forms of membrane fouling control techniques (coarse bubble aeration, tweaking of hydrodynamics conditions) applied to annihilate membrane fouling. The feed for the RBC–ME had undergone biological treatment in the RBC bioreactor and was expected to pose low fouling potential. The effects of 2 different bioreactors (RBC and RBC–ME) on the biological and hydraulic filtration performance were then assessed.

2.3.1. Hydraulic Retention Time

The performance of both the bioreactors (RBC and RBC–ME) was assessed by varying HRTs from 12 to 48 h at a constant influent wastewater concentration. The performance of the bioreactors, in terms of biological treatment and membrane permeability, was assessed. As the HRT is directly associated with the hydraulic loading rate (HLR), this paper only discusses the effect of HRT. The bioreactor was acclimatized to a 9 h HRT, equivalent to an organic loading rate of 17 g COD/m^2.d and a 67.9 L/m^2.d of HLR. After acclimatization, the bioreactor was operated at a 12 h HRT, equivalent to a 51 L/m^2.d of HLR. The HRT was increased from 12 to 18, 24, and 48 h to determine the effects of COD, TN, turbidity, and membrane permeability.

2.3.2. Scanning Electron Microscope

After the acclimatization stage, a 1 cm^2 piece of biofilm was analyzed using SEM analysis. The biofilm sample was carefully cut from the rotating disk. The foam was then treated with formaldehyde for biofilm impregnation according to the method detailed earlier [33] to maintain the biofilm structure. The biofilm sample was then dehydrated by consecutive immersions in 20, 40, 60, 80, and 100% ethanol solution, each step for 5 min, to avoid shrinkage, followed by a drying process. The dried non-conductive sample was sprayed with conductive gold nanoparticles using an ion sputter instrument to create a conductive layer on the sample that reduces thermal damage, inhibits charging, and improves the secondary electron signal required for topographic examination in the SEM. The conductive biofilm sample was loaded onto the SEM sample stage under vacuum conditions and an electron gun shot out a beam of high-energy electrons.

3. Results
3.1. Biofilm Analysis

Figure 2 shows the biofilm developed at the surface of the disk visualized using SEM. The SEM images were obtained at 40× and 5000× magnification levels. Figure 2a shows a birds-eye view of the biofilm established on the carrier media at 40× magnification. The SEM images show the well-established biofilm of microorganisms on the media surface.

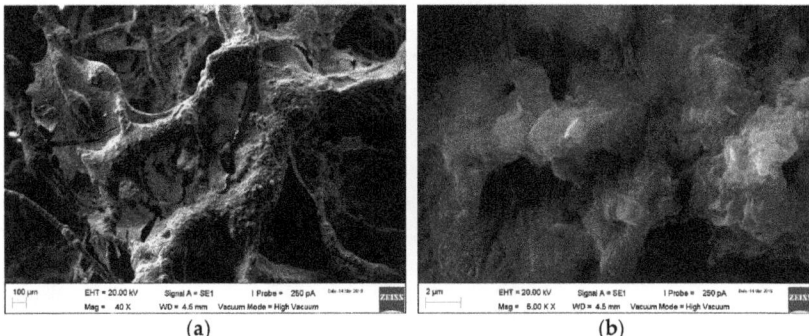

Figure 2. SEM results of biofilm developed at the surface of the rotating disk under (**a**) 40× and (**b**) 5000× magnifications.

It can be identified as a mature biofilm that occupied the sponge media surface, ascribed to its excellent biological performance in removing organics from the wastewater (detailed in Section 3.3). A mature biofilm with a characteristic mushroom formed of polysaccharides can be seen in Figure 2b. At this stage, cells start to detach and revert to planktonic cells that stick to the new surface to develop another biofilm layer.

3.2. Membrane Characterization

The properties of the applied membrane in the external filtration system are summarized in Table 3. The thickness and mean flow pore size were 0.28 ± 0.22 mm and 0.03 µm, respectively (Figure 3). The sizes of the microorganism species were much larger than the mean flow pore size of the membrane combined with the asymmetric nature of the morphology, thus ensuring complete biomass retention at the membrane surface. For an asymmetric phase-inverted membrane, the membrane pore size is dictated by the size of the pore mouth [34], which, in this context, disallows the penetration of any free biomass into the membrane structure. This advantage ensures no biomass is carried forward to the effluent, nor any suspended matter typically vulnerable in a standard settling system. However, colloidal particles and dissolved nutrients can pass through the membrane pores unless an additional layer of biofilm grows on the membrane surface, which aids in biodegradation, as often occurs in a membrane bioreactor [33]. It is worth noting that the biological performance was less affected by the membrane properties. The membrane samples were asymmetric, as revealed from their cross-section SEM image (Figure 3). The membrane surface water contact angle determines the hydrophilic/hydrophobic nature of the membrane. The membrane surface water contact angle of 61.8 ± 1.0° revealed a hydrophilic membrane. The membrane exhibited a clean water permeability of 817 ± 35 L/(m² h bar).

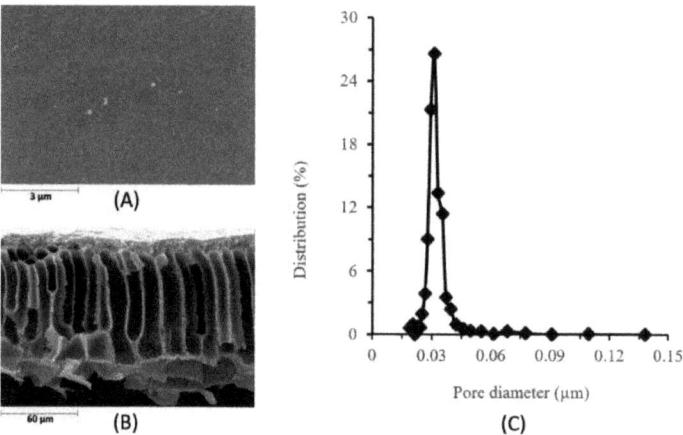

Figure 3. The surface (**A**) and cross-section SEM images (**B**), as well as its pore size distribution (**C**) of the applied membrane in the RBC–ME.

3.3. Biological Performance

Table 4 summarizes the biological performance of RBC and RBC–ME bioreactors for synthetic wastewater. Superior biological performance of the RBC incorporating the membrane showed the significance of membrane integration with the RBC. The results in Table 4 suggest that after the acclimatization period, the bioreactor stabilizes, which is depicted in the steady removal efficiencies. The bioreactors in both the RBC and RBC–ME depicted excellent removal efficiencies for COD, the ammonium, and turbidity. They also showed good performance in maintaining the pH around the neutral value.

Table 4. Effluent characteristics for the RBC and RBC–ME bioreactors employing the PSF membrane.

	RBC Effluent	RBC % Removal Efficiency	RBC–ME Effluent	RBC–ME % Removal Efficiency
COD (mg/L)	78.2 ± 7.5	72.4 ± 2.5	35 ± 8.9	87.9 ± 3.2
TN (mg/L)	1.54 ± 0.05	38.3 ± 1.9	1.41 ± 0.05	45.2 ± 0.7
Ammonium (mg/L)	0.03 ± 0.01	95.6 ± 0.8	0.01 ± 0.01	98.9 ± 1.1
Nitrate (mg/L)	1.9 ± 0.3	-	1.8 ± 0.2	-
Turbidity (NTU)	3.3 ± 0.3	78.9 ± 0.3	0.32 ± 0.03	97.9 ± 0.1
pH	6.82 ± 0.03	-	6.95 ± 0.11	-

RBC: rotating biological contactor, RBC–ME: rotating biological contactor–membrane external, COD: chemical oxygen demand, TN: total nitrogen.

The excellent biological performance in both systems can be explained as follows. Carbonaceous bacteria are responsible for the biodegradation of organic compounds, aerobically using dissolved oxygen as a terminal electron acceptor, while nitrogenous bacteria decompose the nitrogen compounds. Nitrification, an aerobic process, is a two-step process involving the oxidation of ammonium to nitrite through ammonia-oxidizing bacteria (AOB) and then the conversion of nitrite to nitrate through nitrite-oxidizing bacteria (NOB) [35]. The RBC develops abundant AOB and NOB throughout the biofilm along with carbonaceous bacteria. Nitrification occurred in the RBC without encountering any biofilm problems, while exhibiting a low biomass yield and very high sludge ages. The RBC exhibited excellent ammonium removal efficiency throughout the experimentation period. Treatment of wastewater containing a high organics concentration is typically dominated by heterotrophic bacteria that significantly diminishes nitrifier growth. Therefore, nitrogen removal occurs after organics removal during the last stage of the RBC bioreactor [36].

The microbial-rich RBC bioreactor contains a large population of microorganisms. The most abundant phyla found in the biofilm are *Proteobacteria* and *Bacteroidetes*, accounting for two-thirds of the microbial community. The oxygen-rich outer layer of the biofilm contains more *Proteobacteria*, *Bacteroidetes*, and *Nitrospira* than the inner layer [37]. The AOB and NOB are found both in the inner and outer layers of the biofilm. *Nitrosomonas* play a crucial role in oxidizing ammonia to nitrite. However, it has been found that *Nitrospira* can perform complete nitrification. Therefore, it can be argued that *Nitrosomonas* and *Nitrobacter* act as the AOB and NOB, respectively, whereas *Nitrospira* plays a role in both the AOB and NOB [38]. A high ammonium loading rate and immediate substrate accessibility result in a higher relative abundance of *Nitrosomonas* and *Nitrospira* in the outer layer.

As shown in Table 4, the RBC bioreactor exhibited good COD removal efficiency throughout the experimentation period and achieved an average removal efficiency of $72.4 \pm 2.5\%$ with a 78.2 ± 7.5 mg/L effluent value, whereas the average effluent TN concentration was 1.54 ± 0.05 mg/L with a $38.3 \pm 1.9\%$ average removal efficiency. The RBC–ME bioreactor showed a further increase in the effluent removal efficiencies thanks to the incorporation of membrane filtration. A COD removal efficiency of $87.9 \pm 3.2\%$ with a 35 ± 8.9 mg/L effluent value and an average TN removal efficiency of $45.2 \pm 0.7\%$ with a 1.41 ± 0.05 mg/L average effluent value were obtained for RBC–ME.

Despite a low influent ammonium concentration, the RBC biofilm grew nitrifying bacteria, which significantly removed $95.8 \pm 0.8\%$ ammonium, while the RBC–ME maintained a higher ammonium removal efficiency of $98.9 \pm 1.1\%$. The effluent ammonium concentration was as low as 0.03 ± 0.01 mg/L for both the RBC and RBC–ME, indicating a proficient nitrification process. An increase in the discharge nitrate value can be ascribed to the nitrifying bacteria activity [3].

Some reports have described that an RBC can undergo aerobic denitrification. A lower DO concentration at the bottom of an RBC facilitates denitrification [39]. However, a high C/N ratio and lower TN values restrict the nitrifying bacteria growth and thus, reduce the denitrification process. The system obtains a relatively lower TN removal because of lower influent quantities and strong competition between heterotrophic and autotrophic bacteria.

Higher removal efficiency for turbidity was achieved by the RBC–ME due to the membrane separation (see pore size in Table 3). The results show that the influent turbidity was 15.2 ± 0.6 NTU, which was significantly reduced to 3.3 ± 0.3 NTU and 0.32 ± 0.03 NTU in the RBC and RBC–ME, respectively, attributing to $78.9 \pm 0.3\%$ and $97.9 \pm 0.1\%$ removal efficiencies (Table 4). The effluent turbidity values for the RBC–ME are far better than the stand-alone RBC effluent. Thanks to low influent ammonium values, no substantial variations in the pH were detected during the experiments, and a neutral pH value was maintained.

Previous studies on the RBC bioreactor showed high organic and ammonium removal efficiency for both municipal and industrial wastewater [26,40]. In the self-refluxing RBC bioreactor for rural sewage treatment, a better treatment performance was obtained for a system with a 200% reflux ratio. The results show that the removal efficiency is more stable and better with reflux than without reflux. In the control with 0% reflux, the removal rates of COD, ammonium, and TN were 88.05 ± 3.17, 91.61 ± 3.26, and $41.58 \pm 5.50\%$, respectively. Under 200% reflux, the removal rates of COD, ammonium, and TN improved, especially that of TN. For the 200% reflux ratio, the removal rates of COD, ammonium, and TN were up to 93.30 ± 7.35, 97.28 ± 5.94, and 74.21 ± 9.17%, respectively [18]. A non-woven RBC evaluated for the treatment of municipal wastewater supporting both aerobic and anaerobic processes resulted in a higher TN removal efficiency. Under the optimal conditions, the removal rates of COD and TN were 83.12% and 79.13%, respectively [41]. An RBC applied for the treatment of petroleum refinery wastewater resulted in 85.76% and 99.07% COD and ammonium removal, respectively [26]. The results suggested that an RBC may be considered as a promising method for petroleum refinery wastewater treatment, especially for simultaneous COD and ammonium removal.

3.4. Effect of Hydraulic Retention Time on COD Removal

Figure 4 shows the effect of HRT (12, 18, 24, and 48 h) on COD removal efficiency for both (RBC and RBC–ME) bioreactors. The results show that higher HRTs led to better COD removal efficiency in both bioreactors. A higher retention time means that microorganisms have a longer time to biodegrade the organics present in the wastewater [42]. As depicted in Figure 4, the maximum COD removal efficiencies of 80.9 ± 2.3% and 92.4 ± 2.4% were obtained for the RBC and the RBC–ME, respectively.

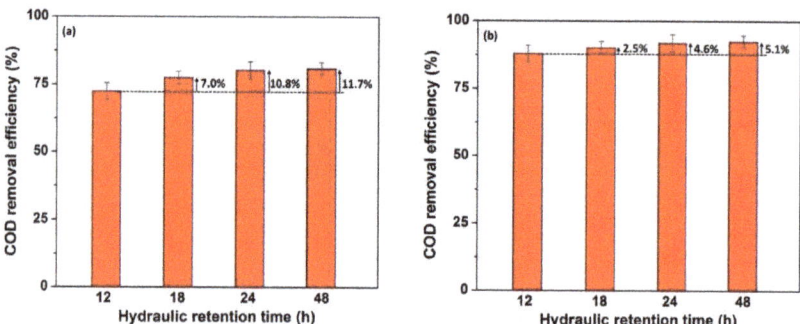

Figure 4. Effect of the HRT on the COD removal efficiency for the (**a**) RBC and (**b**) RBC–ME bioreactors.

In the RBC, COD removal efficiency increased from 72.4 ± 3.1% to 80.9 ± 2.3% as the HRT increased from 12 to 48 h. Increments of 7.0%, 10.8%, and 11.7% in COD removal efficiencies were observed for the HRTs of 18, 24, and 48 h, compared to 12 h (Figure 4a). On the other hand, RBC–ME exhibited an increase in COD removal efficiency from 87.9 ± 3.1% to 92.4 ± 2.4% as the HRT increased from 12 to 48 h. As shown in Figure 4b, an increase of 2.5%, 4.6%, and 5.1% in COD removal efficiency was observed for the HRTs of 18, 24, and 48 h, compared to 12 h. A higher COD removal efficiency confirms the effectiveness of membrane integration with an RBC. Such an advantage can be attributed to the presence of the biofilm on the membrane surface that further degrades the organics when the feed passes through it. The bioreactors perform well under higher HRTs; however, the HRT becomes limiting as it is directly related to the overall treatment capacity. Hence, careful selection of the optimum HRT becomes an essential part of the RBC design.

3.5. Effect of Hydraulic Retention Time on TN Removal

Figure 5 depicts the effect of TN removal efficiency at different HRTs (12, 18, 24, and 48 h) for both (RBC and RBC–ME) bioreactors. The results show that higher HRTs led to an increase in TN removal efficiency in both bioreactors. A higher retention time means that nitrifying microorganisms have prolonged contact time to biodegrade the nitrogen compounds present in the wastewater. Maximum TN removal efficiencies of 41.5 ± 0.8% and 48.6 ± 1.3% were obtained at a 48 h HRT for the RBC and RBC–ME, respectively.

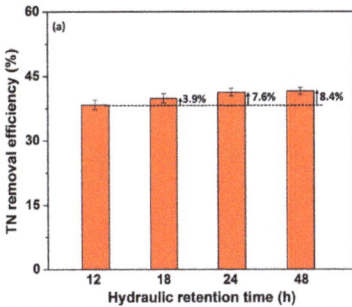 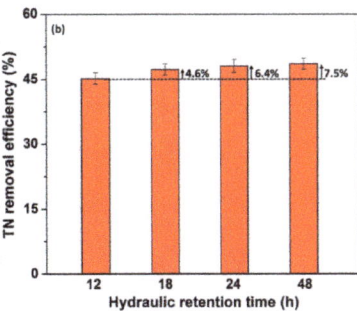

Figure 5. Effect of the HRT on the TN removal efficiency for the (**a**) RBC and (**b**) RBC–ME bioreactors.

In the RBC, TN removal efficiencies of 38.3 ± 1.2%, 39.8 ± 1.1%, 41.2 ± 0.9%, and 41.5 ± 0.8% were observed at the HRTs of 12, 18, 24 and 48 h, respectively. The RBC–ME exhibited an increase in TN removal efficiency from 45.2 ± 1.4% to 48.6 ± 1.3% as the HRT increased from 12 to 48 h (Figure 5). As shown in Figure 5b, increments of 4.6%, 6.4%, and 7.5% in TN removal efficiency were observed for the HRTs of 18, 24, and 48 h, compared to 12 h. As the HRT increases, the wastewater is retained longer in the bioreactor, allowing the microorganisms to biodegrade more nutrients. The results agree with the previous research that showed a long HRT led to a better degradation performance of municipal and industrial wastewater, recalcitrant pharmaceuticals, and micro-pollutants [43,44]. Ghalehkhondabi et al. [26] studied the performance of a four-stage RBC bioreactor for a petroleum refinery wastewater treatment. The increase in the HRT and reduction in HLR resulted in an increase in the organic and ammonium removal, and the maximum removal efficiencies of COD and ammonium obtained were 85.76% and 99.07%, respectively. In principle, a longer HRT is ideal for complete nitrification and high-strength wastewater treatment. Nevertheless, if the HRT of the reactor operation is shortened for operational or economic reasons, the influent concentration and microbial community could play a compensatory role in the biodegradation performance.

3.6. Effect of Hydraulic Retention Time on Turbidity

Figure 6 depicts the effect of turbidity removal efficiency at different HRTs (12, 18, 24, and 48 h) for both (RBC and RBC–ME) bioreactors. The results show that higher HRTs led to an increase in turbidity removal efficiency in both bioreactors. The maximum turbidity removal efficiencies of 84.2 ± 0.8% and 98.6 ± 1.3% were obtained at 48 h HRT for the RBC and RBC–ME, respectively. As the HRT increases, the wastewater is retained longer in the bioreactor, allowing the microorganisms to digest the suspended solid or allowing the solids to settle in the bioreactor. In the RBC, the turbidity removal efficiency increased from 78.9 ± 0.3% to 84.2 ± 0.8% as the HRT increased from 12 to 48 h. On the other hand, the RBC–ME exhibited a rise in turbidity removal efficiency from 97.9 ± 0.1% to 98.6 ± 1.3% as the HRT increased from 12 to 48 h. A higher turbidity removal efficiency confirms the effectiveness of membrane integration with an RBC. The membrane intercepts all the solids that otherwise contribute to the turbidity.

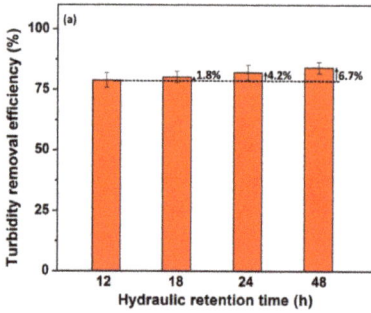 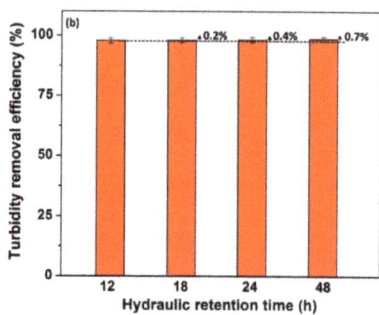

Figure 6. Effect of HRT on the turbidity removal efficiency for; (**a**) RBC and (**b**) RBC–ME bioreactors.

3.7. Membrane Permeability versus Hydraulic Retention Time in RBC–ME

Figure 7 shows the membrane permeability for short-term filtration in the RBC–ME bioreactor configuration. Membrane fouling is inevitable in almost all membrane processes [40,45]. This implies that a decline in permeability corroborates membrane fouling as a function of filtration time. As shown in Figure 7, the membrane permeability decreases sharply at the start of filtration, mainly owing to membrane pore blocking and irreversible adsorption of foulant. After that, permeability decreases steadily due to the slower rate of the deposition of foulant at the membrane surface, indicating reversible fouling. After reaching monolayer foulant adsorption, the affinity of foulant toward the membrane surface seems to be weaker as well. Toward the end of the filtration cycle, steady-state permeability is attained, which is ascribed to the development of a cake layer categorized as reversible fouling on a membrane that requires physical cleaning of the membrane [46,47]. The general trend is similar to the filtration of all parameters. A small rate of permeability decrease was still observed at the end of the filtration test, which is attributed to the buildup of foulant materials on the membrane surface. This occurred because no means of membrane fouling control was applied for the membrane filtration, an issue that can be addressed in a follow-up study. The current study was focused on the relative membrane fouling propensity of the RBC effluent under different operational HRTs.

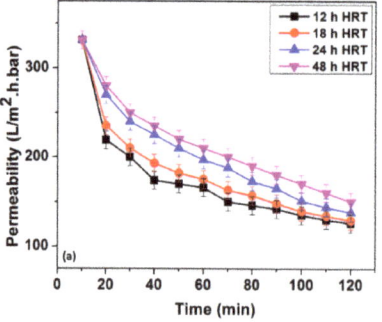 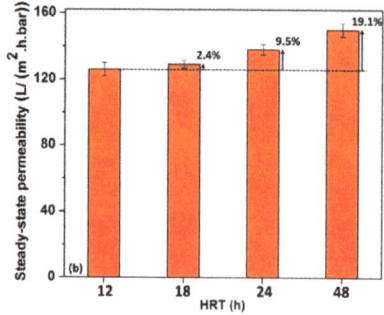

Figure 7. Evolution of the permeability as a function of filtration time at different HRTs for the RBC–ME configuration (**left**) and the summary of the steady-state permeabilities (**right**).

Figure 7 reveals the outcome of different HRTs on membrane permeability for the RBC–ME configuration. The steady-state permeabilities of 126, 129, 138, and 150 L/(m^2 h bar) were respectively attained at HRTs of 12, 18, 24, and 48 h, corresponding to filtration fluxes of 12.6, 12.9, 13.8, and 15.0 L/(m^2 h). Higher membrane permeability at an HRT of 48 h shows higher efficiency of organic compounds and suspended solids removal. This, in turn, reduces the membrane fouling and could potentially lower operating costs. The steady-state permeability increments of 2.4%, 9.5%, and 19.1% were obtained at 18, 24, and 48 h, respectively, compared to the 12 h HRT. This finding means that a higher HRT benefits from microbial degradation activity and nutrient removals coupled with better effluent filterability. Reduced membrane fouling mainly arose from the increase in the HRT and the reduction in the concentration polarization nearby the membrane surface due to a lower hydraulic loading rate in accordance with previous studies [13].

Reversible membrane fouling can be effectively controlled through optimizing parameters. In this study, the application of different HRTs has proven to be highly effective for membrane fouling control. Optimizing the parameters not only dampens the membrane fouling but also increases the effluent quality. The enhanced HRT significantly alleviated membrane fouling potential. The RBC–ME experienced severe fouling, as no fouling control technique was applied. Nevertheless, it is worth mentioning the trade-off of a high HRT and membrane fouling control that needs to be managed. A high HRT leads to a higher bioreactor volume and hence, higher investment costs. On the other hand, low membrane fouling propensity at high HRTs leads to lower costs associated with membrane fouling control as well as a lower membrane investment cost if the filtration is run at higher fluxes.

3.8. Membrane Fouling Analysis

Figure 8 shows the fouling resistance distribution in the RBC–ME bioreactor at different HRTs. The filtration resistances were measured at the end of each experiment. With an increase in the HRT from 12 to 18 h, the total fouling resistance (R_T) decreased from 3.25×10^{12} to 3.18×10^{12} m^{-1}, which further decreased to 2.97×10^{12} and 2.73×10^{12} m^{-1} at HRTs of 24 and 48 h, respectively. Since a constant membrane resistance (R_m) value of 4.55×10^{11} m^{-1} was obtained for all HRTs, both the pore blocking resistance and cake layer resistance played a crucial role in the total membrane fouling. The highest cake layer resistance (R_c) of 1.61×10^{12} m^{-1} at a 12 h HRT consisted of 49.6% of R_T, which decreased to 1.54×10^{12} m^{-1} at an HRT of 18 h, accounting for 48.6% of R_T. The R_c further decreased to 1.33×10^{12} m^{-1} and 1.09×10^{12} m^{-1} at HRTs of 24 and 48 h, respectively, consisting of 44.8% and 40% of R_T.

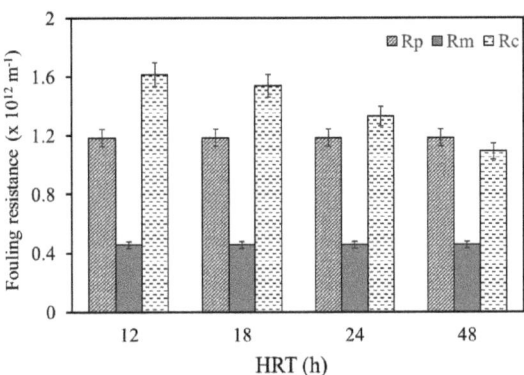

Figure 8. Fouling resistance distribution in the RBC–ME bioreactor at different HRTs.

A shorter HRT leads to insufficient degradation time for the organics and nutrients and results in the generation of more sludge, thereby increasing filtration resistance at shorter HRTs. Shorter HRTs also result in a higher secretion of extracellular polymeric substances and soluble microbial products, hence promoting membrane fouling [48,49]. The results of the membrane fouling analysis emphasized the importance of the HRT and recommended that the decline in membrane permeability at higher HRTs causes less membrane fouling and hence, increases the membrane permeability.

4. Conclusions

This study reports a high-performance membrane integrated with an RBC bioreactor as an efficient wastewater treatment process. The aim of the present research was to substitute the suspended growth system with an attached growth system. Therefore, the RBC, as an attached growth system, was coupled with an external UF membrane to treat synthetic domestic wastewater. The attached growth bioreactor creates the biofilm on the support media that provides better treatment efficiency than the suspended growth bioreactor due to the accumulation of a high microbial population over a large surface area. Therefore, better performance can be achieved by combining such a biofilm reactor as an RBC with a membrane, compared to suspended growth bioreactors. An increase in the HRT not only results in enhanced biological performance but also improves membrane permeability. The results show that the RBC exhibited $72.4 \pm 2.5\%$ COD, $38.3 \pm 1.9\%$ TN, $95.6 \pm 0.8\%$ ammonium, and $78.9 \pm 0.3\%$ turbidity removal efficiencies, while the RBC–ME showed better performance, with $87.9 \pm 3.2\%$ COD, $45.2 \pm 0.7\%$ TN, $98.9 \pm 1.1\%$ ammonium, and $97.9 \pm 0.1\%$ turbidity removal efficiencies. The highest COD and TN removal efficiencies were $92.4 \pm 2.4\%$ and $48.6 \pm 1.3\%$, respectively, for the RBC–ME at an HRT of 48 h. The HRT enhancements resulted in 19.1% higher permeability at 48 h compared to 12 h. The biological and filtration performance of the RBC–ME reveals the economic impact and opens a great opportunity for significant improvements to the current membrane technology. A membrane-integrated RBC poses an attractive alternative to treat wastewater in the decentralized and open-air canal systems where the problem of a large footprint is less important.

Author Contributions: S.W. performed all the experiments and prepared the manuscript draft; M.R.B. and N.Y.H. contributed to the original idea of the study, the conceptual design of the study, supervised the work, and revised the manuscript; N.A.H.M.N., N.H., J.R., N.S., Y.W. and A.L.K. revised the manuscript. All authors have read and agreed to the published version of the manuscript.

Funding: This work was funded by the Yayasan Universiti Teknologi PETRONAS (YUTP) with grant code YUTP-015LC0-210. The article processing charge was supported by the Universiti Malaysia Sabah.

Institutional Review Board Statement: Not applicable.

Informed Consent Statement: Not applicable.

Data Availability Statement: Not applicable.

Conflicts of Interest: The authors declare no conflict of interest.

References

1. Ziembińska-Buczyńska, A.; Ciesielski, S.; Żabczyński, S.; Cema, G. Bacterial community structure in rotating biological contactor treating coke wastewater in relation to medium composition. *Environ. Sci. Pollut. Res.* **2019**, *26*, 19171–19179. [CrossRef] [PubMed]
2. Huang, C.; Shi, Y.; El-Din, M.G.; Liu, Y. Performance of flocs and biofilms in integrated fixed-film activated sludge (IFAS) systems for the treatment of oil sands process-affected water (OSPW). *Chem. Eng. J.* **2017**, *314*, 368–377. [CrossRef]
3. Waqas, S.; Bilad, M.R.; Man, Z.; Wibisono, Y.; Jaafar, J.; Mahlia, T.M.I.; Khan, A.L.; Aslam, M. Recent progress in integrated fixed-film activated sludge process for wastewater treatment: A review. *J. Environ. Manag.* **2020**, *268*, 110718. [CrossRef] [PubMed]
4. Waqas, S.; Bilad, M.R.; Aqsha, A.; Harun, N.Y.; Ayoub, M.; Wirzal, M.D.H.; Jaafar, J.; Mulyati, S.; Elma, M. Effect of membrane properties in a membrane rotating biological contactor for wastewater treatment. *J. Environ. Chem. Eng.* **2021**, *9*, 104869. [CrossRef]
5. Hassard, F.; Biddle, J.; Cartmell, E.; Jefferson, B.; Tyrrel, S.; Stephenson, T. Rotating biological contactors for wastewater treatment—A review. *Process Saf. Environ. Prot.* **2015**, *94*, 285–306. [CrossRef]
6. Findlay, G. The selection and design of rotating biological contactors and reed beds for small sewage treatment plants. *Proc. Inst. Civ. Eng. Water Marit. Energy* **1993**, *101*, 237–246. [CrossRef]
7. Griffin, P.; Findlay, G. Process and engineering improvements to rotating biological contactor design. *Water Sci. Technol.* **2000**, *41*, 137–144. [CrossRef]

8. Najafpour, G.; Yieng, H.A.; Younesi, H.; Zinatizadeh, A. Effect of organic loading on performance of rotating biological contactors using palm oil mill effluents. *Process Biochem.* **2005**, *40*, 2879–2884. [CrossRef]
9. Patwardhan, A. Rotating biological contactors: A review. *Ind. Eng. Chem. Res.* **2003**, *42*, 2035–2051. [CrossRef]
10. Waqas, S.; Bilad, M.R.; Man, Z.B. Performance and Energy Consumption Evaluation of Rotating Biological Contactor for Domestic Wastewater Treatment. *Indones. J. Sci. Technol.* **2021**, *6*, 101–112. [CrossRef]
11. Cheng, S.-F.; Lee, Y.-C.; Kuo, C.-Y.; Wu, T.-N. A case study of antibiotic wastewater treatment by using a membrane biological reactor system. *Int. Biodeterior. Biodegr.* **2015**, *102*, 398–401. [CrossRef]
12. Waqas, S.; Bilad, M.R. A review on rotating biological contactors. *Indones. J. Sci. Technol.* **2019**, *4*, 241–256. [CrossRef]
13. Safa, M.; Alemzadeh, I.; Vossoughi, M. Biodegradability of oily wastewater using rotating biological contactor combined with an external membrane. *J. Environ. Health Sci.* **2014**, *12*, 117. [CrossRef] [PubMed]
14. Rana, S.; Gupta, N.; Rana, R. Removal of organic pollutant with the use of rotating biological contactor. *Mater. Today Proc.* **2018**, *5*, 4218–4224. [CrossRef]
15. Waqas, S.; Bilad, M.R.; Man, Z.B. Effect of organic and nitrogen loading rate in a rotating biological contactor for wastewater treatment. *J. Phys. Conf. Ser.* **2021**, *1793*, 012063. [CrossRef]
16. Cortez, S.; Teixeira, P.; Oliveira, R.; Mota, M. Rotating biological contactors: A review on main factors affecting performance. *Rev. Environ. Sci. Biotechnol.* **2008**, *7*, 155–172. [CrossRef]
17. Dutta, S.; Hoffmann, E.; Hahn, H. Study of rotating biological contactor performance in wastewater treatment using multi-culture biofilm model. *Water Sci. Technol.* **2007**, *55*, 345–353. [CrossRef]
18. Han, Y.; Ma, J.; Xiao, B.; Huo, X.; Guo, X. New integrated self-refluxing rotating biological contactor for rural sewage treatment. *J. Clean. Prod.* **2019**, *217*, 324–334. [CrossRef]
19. Waqas, S.; Bilad, M.R.; Man, Z.B.; Klaysom, C.; Jaafar, J.; Khan, A.L. An integrated rotating biological contactor and membrane separation process for domestic wastewater treatment. *Alex. Eng. J.* **2020**, *59*, 4257–4265. [CrossRef]
20. Mohammadi, M.; Mohammadi, P.; Karami, N.; Barzegar, A.; Annuar, M.S.M. Efficient hydrogen gas production from molasses in hybrid anaerobic-activated sludge-rotating biological contactor. *Int. J. Hydrogen Energy* **2019**, *44*, 2592–2602. [CrossRef]
21. Barambu, N.U.; Bilad, M.R.; Huda, N.; Nordin, N.A.H.M.; Bustam, M.A.; Doyan, A.; Roslan, J. Effect of Membrane Materials and Operational Parameters on Performance and Energy Consumption of Oil/Water Emulsion Filtration. *Membranes* **2021**, *11*, 370. [CrossRef]
22. Barambu, N.U.; Peter, D.; Yusoff, M.H.M.; Bilad, M.R.; Shamsuddin, N.; Marbelia, L.; Nordin, N.A.H.; Jaafar, J. Detergent and Water Recovery from Laundry Wastewater Using Tilted Panel Membrane Filtration System. *Membranes* **2020**, *10*, 260. [CrossRef]
23. Hanhan, O.; Orhon, D.; Krauth, K.; Günder, B. Evaluation of denitrification potential of rotating biological contactors for treatment of municipal wastewater. *Water Sci. Technol.* **2005**, *51*, 131–139. [CrossRef] [PubMed]
24. Najafpour, G.; Zinatizadeh, A.; Lee, L. Performance of a three-stage aerobic RBC reactor in food canning wastewater treatment. *Biochem. Eng. J.* **2006**, *30*, 297–302. [CrossRef]
25. Costley, S.; Wallis, F. Effect of flow rate on heavy metal accumulation by rotating biological contactor (RBC) biofilms. *J. Ind. Microbiol. Biotechnol.* **2000**, *24*, 244–250. [CrossRef]
26. Ghalehkhondabi, V.; Fazlali, A.; Fallah, B. Performance analysis of four-stage rotating biological contactor in nitrification and COD removal from petroleum refinery wastewater. *Chem. Eng. Process. Process Intensif.* **2021**, *159*, 108214. [CrossRef]
27. Win, T.T.; Kim, H.; Cho, K.; Song, K.G.; Park, J. Monitoring the microbial community shift throughout the shock changes of hydraulic retention time in an anaerobic moving bed membrane bioreactor. *Bioresour. Technol.* **2016**, *202*, 125–132. [CrossRef]
28. Zhang, M.; Peng, Y.; Wang, C.; Wang, C.; Zhao, W.; Zeng, W. Optimization denitrifying phosphorus removal at different hydraulic retention times in a novel anaerobic anoxic oxic-biological contact oxidation process. *Biochem. Eng. J.* **2016**, *106*, 26–36. [CrossRef]
29. Kharraz, J.A.; Bilad, M.; Arafat, H.A. Simple and effective corrugation of PVDF membranes for enhanced MBR performance. *J. Membr. Sci.* **2015**, *475*, 91–100. [CrossRef]
30. APHA. *Standard Methods for the Examination of Water and Wastewater*, 9th ed.; American Public Health Association: Washington, DC, USA, 1997.
31. Eliseus, A.; Bilad, M.; Nordin, N.; Khan, A.L.; Putra, Z.; Wirzal, M.; Aslam, M.; Aqsha, A.; Jaafar, J. Two-way switch: Maximizing productivity of tilted panel in membrane bioreactor. *J. Environ. Manag.* **2018**, *228*, 529–537. [CrossRef]
32. Bilad, M.R.; Nawi, N.I.M.; Subramaniam, D.D.; Shamsuddin, N.; Khan, A.L.; Jaafar, J.; Nandiyanto, A.B.D. Low-pressure submerged membrane filtration for potential reuse of detergent and water from laundry wastewater. *J. Water Process. Eng.* **2020**, *36*, 101264. [CrossRef]
33. Bilad, M.R.; Declerck, P.; Piasecka, A.; Vanysacker, L.; Yan, X.; Vankelecom, I.F. Treatment of molasses wastewater in a membrane bioreactor: Influence of membrane pore size. *Sep. Purif. Technol.* **2011**, *78*, 105–112. [CrossRef]
34. AlMarzooqi, F.A.; Bilad, M.; Mansoor, B.; Arafat, H.A. A comparative study of image analysis and porometry techniques for characterization of porous membranes. *J. Mater. Sci.* **2016**, *51*, 2017–2032. [CrossRef]
35. Vázquez-Padín, J.R.; Mosquera-Corral, A.; Campos, J.L.; Méndez, R.; Carrera, J.; Pérez, J. Modelling aerobic granular SBR at variable COD/N ratios including accurate description of total solids concentration. *Biochem. Eng. J.* **2010**, *49*, 173–184. [CrossRef]
36. Brazil, B.L. Performance and operation of a rotating biological contactor in a tilapia recirculating aquaculture system. *Aquac. Eng.* **2006**, *34*, 261–274. [CrossRef]

37. Hewawasam, C.; Matsuura, N.; Maharjan, N.; Hatamoto, M.; Yamaguchi, T. Oxygen transfer dynamics and nitrification in a novel rotational sponge reactor. *Biochem. Eng. J.* **2017**, *128*, 162–167. [CrossRef]
38. Van Kessel, M.A.; Speth, D.R.; Albertsen, M.; Nielsen, P.H.; den Camp, H.J.O.; Kartal, B.; Jetten, M.S.; Lücker, S. Complete nitrification by a single microorganism. *Nature* **2015**, *528*, 555–559. [CrossRef]
39. Zha, X.; Ma, J.; Lu, X. Performance of a coupling device combined energy-efficient rotating biological contactors with anoxic filter for low-strength rural wastewater treatment. *J. Clean. Prod.* **2018**, *196*, 1106–1115. [CrossRef]
40. Waqas, S.; Bilad, M.R.; Man, Z.B.; Suleman, H.; Nordin, N.A.H.; Jaafar, J.; Othman, M.H.D.; Elma, M. An Energy-Efficient Membrane Rotating Biological Contactor for Wastewater Treatment. *J. Clean. Prod.* **2020**, *282*, 124544. [CrossRef]
41. Wang, D.; Wang, G.; Yang, F.; Liu, C.; Kong, L.; Liu, Y. Treatment of municipal sewage with low carbon-to-nitrogen ratio via simultaneous partial nitrification, anaerobic ammonia oxidation, and denitrification (SNAD) in a non-woven rotating biological contactor. *Chemosphere* **2018**, *208*, 854–861. [CrossRef] [PubMed]
42. Xu, S.; Wu, D.; Hu, Z. Impact of hydraulic retention time on organic and nutrient removal in a membrane coupled sequencing batch reactor. *Water Res.* **2014**, *55*, 12–20. [CrossRef]
43. Tadkaew, N.; Hai, F.I.; McDonald, J.A.; Khan, S.J.; Nghiem, L.D. Removal of trace organics by MBR treatment: The role of molecular properties. *Water Res.* **2011**, *45*, 2439–2451. [CrossRef] [PubMed]
44. Boonnorat, J.; Techkarnjanaruk, S.; Honda, R.; Prachanurak, P. Effects of hydraulic retention time and carbon to nitrogen ratio on micro-pollutant biodegradation in membrane bioreactor for leachate treatment. *Bioresour. Technol.* **2016**, *219*, 53–63. [CrossRef]
45. Nawi, M.; Izati, N.; Bilad, M.R.; Zolkhiflee, N.; Nordin, N.A.H.; Lau, W.J.; Narkkun, T.; Faungnawakij, K.; Arahman, N.; Mahlia, T.M.I. Development of a novel corrugated polyvinylidene difluoride membrane via improved imprinting technique for membrane distillation. *Polymers* **2019**, *11*, 865. [CrossRef] [PubMed]
46. Marbelia, L.; Bilad, M.R.; Bertels, N.; Laine, C.; Vankelecom, I.F. Ribbed PVC–silica mixed matrix membranes for membrane bioreactors. *J. Membr. Sci.* **2016**, *498*, 315–323. [CrossRef]
47. Liao, Y.; Bokhary, A.; Maleki, E.; Liao, B. A review of membrane fouling and its control in algal-related membrane processes. *Bioresour. Technol.* **2018**, *264*, 343–358. [CrossRef] [PubMed]
48. Deng, L.; Guo, W.; Ngo, H.H.; Du, B.; Wei, Q.; Tran, N.H.; Nguyen, N.C.; Chen, S.-S.; Li, J. Effects of hydraulic retention time and bioflocculant addition on membrane fouling in a sponge-submerged membrane bioreactor. *Bioresour. Technol.* **2016**, *210*, 11–17. [CrossRef] [PubMed]
49. Mannina, G.; Capodici, M.; Cosenza, A.; Di Trapani, D.; Ekama, G.A. The effect of the solids and hydraulic retention time on moving bed membrane bioreactor performance. *J. Cleaner Prod.* **2018**, *170*, 1305–1315. [CrossRef]

Article

Adsorption/Desorption Capability of Potassium-Type Zeolite Prepared from Coal Fly Ash for Removing of Hg^{2+}

Yuhei Kobayashi [1], Fumihiko Ogata [1], Chalermpong Saenjum [2,3], Takehiro Nakamura [1] and Naohito Kawasaki [1,4,*]

[1] Faculty of Pharmacy, Kindai University, 3-4-1 Kowakae, Higashi-Osaka, Osaka 577-8502, Japan; 1944420001t@kindai.ac.jp (Y.K.); ogata@phar.kindai.ac.jp (F.O.); nakamura@phar.kindai.ac.jp (T.N.)

[2] Faculty of Pharmacy, Chiang Mai University, Suthep Road, Muang District, Chiang Mai 50200, Thailand; chalermpong.saenjum@gmail.com

[3] Cluster of Excellence on Biodiversity-based Economics and Society (B.BES-CMU), Chiang Mai University, Suthep Road, Muang District, Chiang Mai 50200, Thailand

[4] Antiaging Center, Kindai University, 3-4-1 Kowakae, Higashi-Osaka, Osaka 577-8502, Japan

* Correspondence: kawasaki@phar.kindai.ac.jp; Tel.: +81-6-4307-4012

Citation: Kobayashi, Y.; Ogata, F.; Saenjum, C.; Nakamura, T.; Kawasaki, N. Adsorption/Desorption Capability of Potassium-Type Zeolite Prepared from Coal Fly Ash for Removing of Hg^{2+}. *Sustainability* **2021**, *13*, 4269. https://doi.org/10.3390/su13084269

Academic Editors: Mohd Rafatullah and Masoom Raza Siddiqui

Received: 10 March 2021
Accepted: 9 April 2021
Published: 12 April 2021

Publisher's Note: MDPI stays neutral with regard to jurisdictional claims in published maps and institutional affiliations.

Copyright: © 2021 by the authors. Licensee MDPI, Basel, Switzerland. This article is an open access article distributed under the terms and conditions of the Creative Commons Attribution (CC BY) license (https://creativecommons.org/licenses/by/4.0/).

Abstract: The feasibility of using potassium-type zeolite (K-type zeolite) prepared from coal fly ash (CFA) for the removal of Hg^{2+} from aqueous media and the adsorption/desorption capabilities of various potassium-type zeolites were assessed in this study. Potassium-type zeolite samples were synthesized by hydrothermal treatment of CFA at different intervals (designated CFA, FA1, FA3, FA6, FA12, FA24, and FA48, based on the hours of treatment) using potassium hydroxide solution, and their physicochemical characteristics were evaluated. Additionally, the quantity of Hg^{2+} adsorbed was in the order CFA, FA1 < FA3 < FA6 < FA12 < FA24 < FA48, in the current experimental design. Therefore, the hydrothermal treatment time is important to enhance the adsorption capability of K-type zeolite. Moreover, the effects of pH, temperature, contact time, and coexistence on the adsorption of Hg^{2+} were elucidated. In addition, Hg^{2+} adsorption mechanism using FA48 was demonstrated. Our results indicated that Hg^{2+} was exchanged with K^+ in the interlayer of FA48 (correlation coefficient = 0.946). Finally, adsorbed Hg^{2+} onto FA48 could be desorbed using a sodium hydroxide solution (desorption percentage was approximately 70%). Our results revealed that FA48 could be a potential adsorbent for the removal of Hg^{2+} from aqueous media.

Keywords: hydrothermal activation treatment; recycling technology; heavy metal; ion exchange

1. Introduction

The 2030 agenda for sustainable development, such as clean water and sanitation (Goal 6) and life below water (Goal 14), were adopted by all member states of the United Nations in 2015 [1], to establish a sustainable society, which is a matter of global concern. In particular, heavy metal pollution has become a severe global environmental issue, including in the developing countries. Among them, mercury (Hg^{2+}), lead (Pb^{2+}), and cadmium (Cd^{2+}) are referred to as the "big three" heavy metals with the greatest potential risk to human health and water environment [2–4]. They are highly toxic to organisms [5]. Mercury (Hg) and its compounds can cause serious threats to organisms, including humans, because of their bioaccumulative properties, damaging the bones, liver, kidney, and nervous system [6–8]. The Minamata Convention on Mercury was adopted by the Intergovernmental Negotiating Committee in 2017. The International Agency for Research on Cancer categorizes methylmercury compounds as group 2B (possibly carcinogenic to humans), and metallic Hg and inorganic Hg compounds as group 3 (unclassifiable as to carcinogenicity in humans) [9]. In addition, the maximum permissible limit of Hg in drinking water as recommended by the U.S. Environmental Protection Agency and many countries are 2 µg/L and 1 µg/L, respectively [7,10,11]. Therefore, removal of Hg^{2+} from aqueous media is crucial for human health and conservation of the water environment.

Coal is one of the most abundant energy sources worldwide [12]. A previous study reported that the global trend of increasing energy production continued in 2018 [13]. In addition, it is desirable to increase coal-fired power generation by up to 46% of the total electricity production by 2030 [14]. Additionally, the demand for coal-fired power plants has increased after the Fukushima Daiichi Nuclear Power Station disaster (2017) in Japan. Accordingly, approximately 800–900 million ton per year of coal fly ash (CFA), a by-product from the combustion of coal, is generated worldwide. Although the CFA has been recycled as supplements for cement, concrete, soil conditioners, and fertilizer materials [15–17], a major portion has been disposed of in landfills. Thus, from the perspective of a sustainable society, it is necessary to develop a recycling technology for CFA.

In this study, we focused on the preparation and production of zeolites from CFA. The CFA, characterized by aluminosilicate and silicon phases, is a superior material for zeolite synthesis [18]. Zeolite is a microporous crystalline hydrated aluminosilicate characterized by a three-dimensional network of tetrahedral (aluminum and silicon) O4 units that form a system of interconnected pores [18]. The applications of CFA derived zeolite are well-known. They have been used for heavy metal removal from aqueous media [19,20], as well as for the remediation of acid mine drainage [21,22]. Thus, this conversion of CFA to zeolite is useful for the development of a sustainable society as it decreases the waste generated from coal-fired power plants. Many conversion technologies, namely fusion-assisted hydrothermal treatment [14], multi-step treatment [14], sonication [14], conventional hydrothermal treatment [23,24], and microwave irritation [25,26] have been reported in previous studies. Zeolite synthesis involves three steps: dissolution, condensation, crystallization [27]. Among the technologies, conventional hydrothermal treatment is comparatively simple and inexpensive [23].

Previously, we reported that potassium-type zeolite (K-type zeolite) prepared from CFA had characteristic physicochemical properties and showed potential in heavy metal adsorption from aqueous media [28]. In addition, previous studies have assessed the adsorption capacity and mechanism of Hg^{2+} removal using CFA [29]. However, there are no reports on the adsorption of Hg^{2+} using potassium-type zeolites prepared from CFA using conventional hydrothermal treatment. Thus, if potassium-type zeolite could be explored for the removal of Hg^{2+} from aqueous media, this alternative would contribute considerably to the waste reduction from coal-fired power plants or water conservation.

This study aimed to investigate the possibility of Hg^{2+} removal from aqueous media using K-type zeolite prepared from CFA. The effects of pH, temperature, contact time, coexistence, and selectivity on the adsorption of Hg^{2+} were assessed.

2. Materials and Methods

2.1. Materials

The standard solution of Hg^{2+} ($HgCl_2$ in 0.1 mol/L HNO_3) was purchased from FUJIFILM Wako Pure Chemical Co., Osaka, Japan. Coal fly ash (CFA) was obtained from the Tachibana-wan Power Station (Shikoku Electric Power, Inc., Tokushima, Japan). Additionally, the standard solutions of Na^+ (NaCl in water), Mg^{2+} ($Mg(NO_3)_2$ in 0.1 mol/L HNO_3), K^+ (KCl in water), Ca^{2+} ($CaCO_3$ in 0.1 mol/L HNO_3), Ni^{2+} ($Ni(NO_3)_2$ in 0.1 mol/L HNO_3), Cu^{2+} ($Cu(NO_3)_2$ in 0.1 mol/L HNO_3), Zn^{2+} ($Zn(NO_3)_2$ in 0.1 mol/L HNO_3), Sr^{2+} ($SrCO_3$ in 0.1 mol/L HNO_3), and Cd^{2+} ($Cd(NO_3)_2$ in 0.1 mol/L HNO_3) were also obtained from FUJIFILM Wako Pure Chemical Co., Osaka, Japan. Potassium-type zeolite (K-type zeolite) was prepared by hydrothermal activation treatment using CFA in potassium hydroxide solution [28]. Three grams of CFA was mixed with 3 mol/L potassium hydroxide solution (240 mL). The mixture solution was heated at 93 °C for 1(FA1), 3(FA3), 6(FA6), 12(FA12), 24(FA24), and 48(FA48) h, followed by filtering through a 0.45 µm membrane filter (Advantec MFS, Inc., Tokyo, Japan) [30]. The residue was washed with distilled water and dried at 50 °C for 24 h. Potassium hydroxide, nitric acid, and sodium hydroxide were purchased from FUJIFILM Wako Pure Chemical Co. (Osaka, Japan). All reagents were of special grade.

We had previously reported the physicochemical characteristics of K-type zeolites [28]. X-ray diffraction (XRD) and morphology analyses were performed using MiniFlex II (Rigaku, Osaka, Japan) and SU1510 (Hitachi High-Technologies Co., Tokyo, Japan), respectively. The cation exchange capacity (CEC) and pH_{pzc} were measured using the Japanese Industrial Standard Method (JIS K 1478) and the method previously reported by Faria et al. [31]. Additionally, the specific surface area and pore volume were measured using NOVA4200e (Quantachrome Instruments Japan G.K., Tokyo, Japan). The binding energy was measured using a JXA-8530F (JEOL Ltd., Tokyo, Japan). Finally, the solution pH was measured using an F-73S digital pH meter (HORIBA, Ltd., Kyoto, Japan).

2.2. Amount of Hg^{2+} Adsorbed Using FA Series

Approximately 0.01 g of each pretreated adsorbent, namely CFA, FA1, FA3, FA6, FA12, FA24, and FA48, was mixed with 50 mL of 50 mg/L Hg^{2+} solution. Subsequently, the reaction mixture was shaken at 100 rpm and 25 °C for 24 h. The resulting sample was filtered through a 0.45 μm membrane filter. The concentration of Hg^{2+} was measured using an inductively coupled plasma optical emission spectrometer (iCAP-7600 Duo, Thermo Fisher Scientific Inc., Osaka, Japan). The quantity of Hg^{2+} adsorbed was calculated using the levels before and after adsorption in Equation (1).

$$q = \frac{(C_0 - C_e)V}{W} \quad (1)$$

where q is the quantity adsorbed (mg/g); C_0 is the initial concentration (mg/L); C_e is the equilibrium concentration (mg/L); V is the solvent volume (L); and W is the weight of the adsorbent (g).

2.3. Effect of pH, Temperature, and Contact Time on the Adsorption of Hg^{2+}

First, in order to evaluate the effect of pH, FA48 (0.01 g) was added to 50 mL of the Hg^{2+} solution at 10, 30, 50 mg/L. The pH of the solution was adjusted to 2, 5, 7, 9, 11 using either nitric acid or sodium hydroxide solutions. The suspension was shaken at 100 rpm at 25 °C for 24 h, and filtered using a 0.45 μm membrane filter. Second, in order to evaluate the temperature effect, FA48 (0.01 g) was added to a 50 mL Hg^{2+} solution at 10, 20, 30, 40, 50 mg/L, and the suspension was shaken at 100 rpm at 7, 25, 45 °C for 24 h. The 7 °C of solution was prepared as follows. The sample solution was set at 5 °C in a water bath shaker personal-11 (TAITEC Co., Nagoya, Japan) in the low-temperature room at 6 °C. Finally, to evaluate the effect of contact time, FA48 (0.01 g) was added to 50 mg/L Hg^{2+} solution (50 mL). The suspension was shaken at 100 rpm and 25 °C for 0.5, 1, 3, 6, 12, 18, 21, 24, 30, 42, 48 h. The amount of Hg^{2+} adsorbed was calculated as described in Section 2.2. In addition, to evaluate the Hg^{2+} adsorption mechanism, the concentration of potassium ions released from FA48 in the adsorption isotherm experiment was measured using an iCAP-7600 Duo (Thermo Fisher Scientific Inc., Osaka, Japan).

2.4. Effect of Coexistences on the Adsorption of Hg^{2+}

In order to evaluate the selectivity of Hg^{2+} adsorption, FA48 (0.01 g) was added to the binary solution of 50 mL. The two components were Hg^{2+} and Na^+, Mg^{2+}, K^+, Ca^{2+}, Ni^{2+}, Cu^{2+}, Zn^{2+}, Sr^{2+}, or Cd^{2+}, and Hg^{2+} or individual cation concentration was 10 mg/L in a binary solution. The sample solution was shaken at 100 rpm at 25 °C for 24 h and filtered through a 0.45 μm membrane filter. The concentration of each metal was measured using an iCAP-7600 Duo. The amount adsorbed was calculated by comparing the levels before and after adsorption.

2.5. Adsorption/Desorption of Hg^{2+} Using Sodium Hydroxide Solution

To evaluate the recycling of FA48 in Hg^{2+} adsorption/desorption, FA48 (0.15 g) was added to a 150 mL Hg^{2+} solution at 250 mg/L. The suspension was shaken at 100 rpm, 25 °C for 24 h, and filtered through a 0.45 μm membrane filter. The concentration of Hg^{2+} was measured using an iCAP-7600 Duo. The amount of Hg^{2+} adsorbed was calculated as described in Section 2.2. After adsorption, FA48 was collected, dried, and used for the desorption experiment. The collected FA48 (0.05 g) was added to 50 mL sodium hydroxide solution at 10, 100, 1000 mmol/L. The suspension was shaken at 100 rpm, 25 °C for 24 h, and filtered through a 0.45 μm membrane filter. The concentration of Hg^{2+} released from FA48 was also measured using an iCAP-7600 Duo. The amount of Hg^{2+} desorbed was calculated using the levels before and after desorption. All results in this study are expressed as mean ± standard error (n = 2–3, Sections 2.2–2.5). In addition, each Figure was prepared using Microsoft Excel.

3. Results and Discussion

3.1. Properties of Potassium-Type Zeolite

Zeolites are characterized by physicochemical properties, such as specific surface area, pore volume, and CEC (Table 1). These characteristics are related to the parameters of hydrothermal treatment, such as heat temperature, pressure, solution alkalinity, activation solution to CFA ratio, and formation process [18]. In this study, six types of potassium-type zeolites were prepared using the above-mentioned method [28]. In addition, our previous study reported the physicochemical properties of potassium-type zeolites in detail [28]. The XRD patterns indicate that CFA was mainly composed of mullite and quartz. The XRD patterns of FA1, FA3, FA6 and FA12 were similar to those of CFA under our experimental conditions. Zeolite F appeared in FA24 and FA48 structures. We observed changes in the surface of FA series with the treatment time. Aluminosilicate gels were clearly produced on FA24 and FA48 surfaces. These processes were in the following order: Al and Si dissolution, geopolymer formation, crystalline structure nucleation, finally zeolite crystal growth [18]. The CEC of FA48 (8.98–11.77 mmol/g) was the highest compared to other FA series. This value of FA48 was 26–69 times higher than that of CFA. The pH_{pzc} of FA was 9.8, and that of FA24 and FA48 were 10.4. Finally, specific surface area and pore volume ($d \leqq 20$ Å) of FA48 (47.3 m²/g and 10 Å) was 34 and 100 times higher than CFA. Additionally, the value of FA48 (potassium-type zeolite) was greater than that of sodium-type zeolite [32]. Thus, these results indicate that potassium-type zeolite (FA48) can be prepared from coal fly ash by conventional hydrothermal treatment using potassium hydroxide.

Table 1. Characteristics of the zeolite samples.

Adsorbents		CFA	FA1	FA3	FA6	FA12	FA24	FA48
CEC (mmol g^{-1})	pH 5	0.34	1.98	1.17	1.63	2.27	7.90	8.98
	pH 10	0.19	0.65	1.55	2.09	3.46	11.17	11.17
pH_{pzc}		9.8	9.3	9.3	9.5	9.7	10.4	10.4
Specific surface area (m²/g)		1.4	15.1	31.5	53.3	54.5	50.3	47.3
Pore volume (μL/g)	$d \leqq 20$ (Å)	0.1	0.9	0	0.5	0.2	10.0	10.0
	$20 < d \leqq 500$ (Å)	2.0	41.9	97.4	161.5	185.0	105.0	99.0
	Total	2.2	63.0	139.0	221.0	220.0	151.0	131.0
Mean pore diameter (Å)		57.0	167.2	176.7	165.9	161.6	120.1	110.7

3.2. Adsorption of Hg^{2+}

Figure 1 shows the quantity of Hg^{2+} adsorbed by the FA series. The adsorbed Hg^{2+} was in the order CFA, FA1 (0–0.48 mg/g) < FA3 (2.2 mg/g) < FA6 (3.5 mg/g) < FA12 (4.0 mg/g) < FA24 (7.5 mg/g) < FA48 (11.6 mg/g) in the current experiment conditions. The adsorption capability of Hg^{2+} using the FA series depended on the duration of the hydrothermal activation treatment using potassium hydroxide solution. Next, we evaluated the relationship between the adsorption capacity of Hg^{2+} and the physicochemical properties of the FA series. As a result, the positive correlation coefficient between the quantity of Hg^{2+} adsorbed and CEC, specific surface area, and pore volume ($d \leqq 20$ Å) were 0.928, 0.659, and 0.882, respectively. These results indicate that CEC and pore volume strongly affect the adsorption of Hg^{2+} from aqueous solutions. Additionally, in this study, FA48 was selected to evaluate the adsorption capability for Hg^{2+} removal from aqueous solutions.

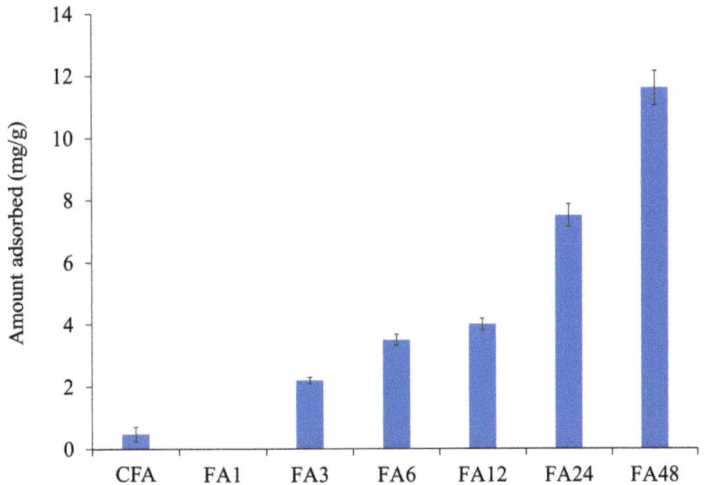

Figure 1. Quantity of Hg^{2+} adsorbed onto FA series. Initial concentration: 50 mg/L, sample volume: 50 mL, adsorbent: 0.01 g, temperature: 25 °C, contact time: 24 h, agitation speed: 100 rpm, pH: 3.0.

A comparison of the Hg^{2+} adsorption capability of FA48 with that of the other adsorbents is listed in Table 2 [24,26,32–35]. FA48 exhibited potential in Hg^{2+} adsorption from aqueous solutions compared to other reported adsorbents (except for coal gangue and multifunctional mesoporous material).

Table 2. Comparison of Hg^{2+} adsorption capacity of FA48 with other reported adsorbents.

Adsorbents	Adsorption Capability (mg/g)	pH	Temp. (°C)	Initial Concentration (mg/L)	Contact Time (h)	Adsorbent (g/L)	Ref.
Coal gangue	20.0	5.5	25	3.5	0.17	2.5	33
Microwave-assisted alkali-modified fly ash	2.7	-	25	~50	1.5	10	24
Raw coal ash zeolite sample	0.44	2.5	r.t.	10	24	1	26
Multifunctional mesoporous material	21.05	Not provided	25	Not provided	2	Not provided	34
Thiol-functionalized mesoporous silica-coated magnetite nanoparticle	9.5	6.0	22.5	8	15	8.0×10^{-5}	35
Sodium-type zeolite prepared from fly ash	7.5	3.0	25	10	24	0.2	32
FA48	11.6	3.0	25	50	24	0.2	This study

3.3. Adsorption Isotherms of Hg^{2+}

Figure 2 shows the adsorption isotherms of Hg^{2+} using FA48 at different temperatures. The quantity of Hg^{2+} adsorbed using FA48 did not significantly vary with different temperatures. Therefore, in this study, the adsorption temperature did not strongly affect the adsorption capability of FA48.

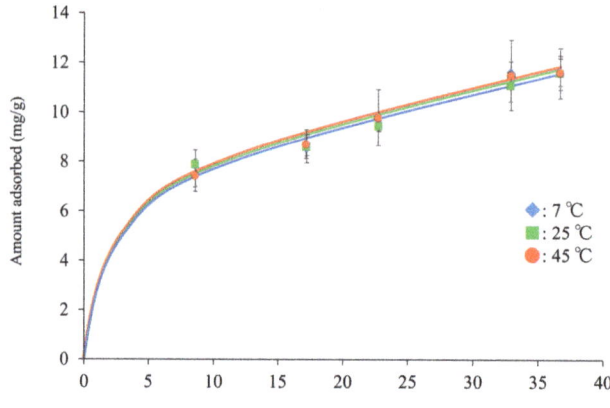

Figure 2. Adsorption isotherms of Hg^{2+} at different temperatures. Initial concentration: 10, 20, 30, 40, and 50 mg/L, sample volume: 50 mL, adsorbent: 0.01 g, temperature: 7, 25, and 45 °C, contact time: 24 h, agitation speed: 100 rpm.

Additionally, to investigate the adsorption properties and interactions, the adsorption isotherm data were evaluated using the Freundlich and Langmuir isotherm models. The Freundlich isotherm model was applied to multilayer adsorption, while the Langmuir isotherm model showed monolayer adsorption at specific homogenous sites [24].

The Freundlich isotherm model can be represented as follows [36]:

$$\log q = \frac{1}{n} \log C + \log K_F \qquad (2)$$

where q is the quantity of Hg^{2+} adsorbed (mg/g), K_F and $1/n$ are the Freundlich isotherm constants, C is the equilibrium concentration (mg/L). In general, the adsorption reaction in the aqueous phase fits this model. In the Freundlich isotherm model, the isotherm curve depends on the value of n. In particular, when the value of $1/n$ is 0.1–0.5, adsorption occurs easily, when $1/n$ is over 2, it is difficult to adsorb [37].

The Langmuir isotherm model can be represented as follows [38]:

$$\frac{1}{q} = \frac{1}{q_{max}} + \left(\frac{1}{K_L q_{max}}\right)\left(\frac{1}{C}\right) \qquad (3)$$

where K_L is the Langmuir isotherm constant (L/mg) and q_{max} is the maximum quantity adsorbed (mg/g). The Langmuir isotherm model is a theoretical model that can explain monolayer adsorption onto homogenous surfaces. In addition, this model considers adsorption sites.

Table 3 shows the Freundlich and Langmuir model constants for the adsorption of Hg^{2+} using FA48. The obtained data fitted both models (correlation coefficient of the Freundlich and Langmuir equations were $\geq$ 0.960 and $\geq$ 0.904, respectively). The maximum quantity adsorbed at 7 to 45 °C was not significantly different in this study, which is supported by the adsorption isotherm data in Figure 2. In addition, the value of $1/n$ was from 0.27 to 0.33 in this study. Therefore, the adsorption of Hg^{2+} using FA48 from aqueous solutions is more favorable.

Table 3. Freundlich model and Langmuir model constants for the adsorption of Hg^{2+}.

Sample	Temp. (°C)	Langmuir Constants			Freundlich Constants		
		q_{max} (mg/g)	K_L (L/mg)	r	$logK_F$	$1/n$	r
FA48	7	13.25	0.14	0.942	0.55	0.33	0.977
	25	12.27	0.19	0.904	0.63	0.27	0.960
	45	13.35	0.14	0.962	0.56	0.32	0.989

Finally, adsorption properties were evaluated using Sips equation (Equation (4)). The Sips model was derived from the Langmuir and Freundlich equations. This model predicts the heterogeneous adsorption system and overcoming the drawback associated with Freundlich model [39]. The Sips equation was expressed as follows:

$$\frac{1}{q_e} = \frac{1}{Q_{max}K_S}\left(\frac{1}{C_e}\right)^{1/n} + \frac{1}{Q_{max}} \quad (4)$$

where K_S is the Sips equilibrium constant (L/mg), Q_{max} is the maximum quantity adsorbed (mg/g). n is the Sips model exponent, which can be employed to describe the system's heterogeneity. If the value of n is equal to 1, this equation will become a Langmuir equation. It means a homogeneous adsorption process [40,41].

Table 4 shows the Sips model constants for the adsorption of Hg^{2+}. The value of correlation coefficient of Sips equation was from 0.841 to 0.959 under our experimental conditions. The values of Q_{max} at 7–45 °C was not significantly changed, which is similar trends to the adsorption isotherm data (Figure 2). In addition, the heterogeneous factor values (n = 0.4–1.1) indicate that heterogeneous adsorption process is related to the adsorption mechanism of Hg^{2+} using FA48.

Table 4. Sips model constants for the adsorption of Hg^{2+}.

Sample	Temp. (°C)	K_S (L/mg)	Q_{max} (mg/g)	n	r
FA48	7	0.84	11.6	1.1	0.959
	25	19.4	11.6	0.6	0.858
	45	4.1×10^2	11.6	0.4	0.841

Moreover, to evaluate the adsorption mechanism of Hg^{2+} using FA48, more detailed investigations were conducted in this study (Figure 3). First, the relationship between the quantity of Hg^{2+} adsorbed and the quantity of K^+ released from FA48 was evaluated in this study. As a result, the correlation coefficient value (r) was positive at 0.946, indicating that ion exchange with K^+ in the interlayer of FA48 was one of the mechanisms of Hg^{2+} adsorption from aqueous media. As mentioned in Section 3.2, the positive correlation coefficient between the quantity of Hg^{2+} adsorbed and the value of CEC was 0.928. These trends were similar to those reported in previous studies [28,32]. Additionally, the X-ray photoelectron spectroscopy analysis was conducted in this study. The peak intensity of Hg(5p) at 67 eV was newly detected after the adsorption of Hg^{2+}, indicating that Hg^{2+} was present on the FA48 surface after adsorption, and was not detected before adsorption. Generally, Hg(4f) peaks at 101 and 105 eV were detected after adsorption. However, Si(2p) and Hg(4f) peaks overlapped in this study. Therefore, it was difficult to elucidate and/or detect these peaks in our experiments.

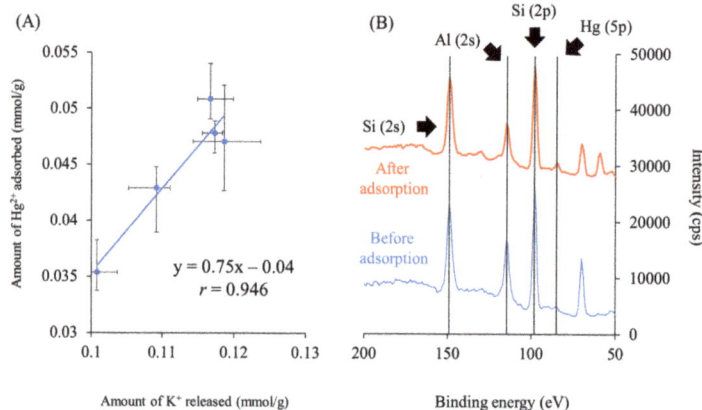

Figure 3. Relationship between the quantity of Hg^{2+} adsorbed and K^+ released (**A**) and the X-ray photoelectron spectroscopy analysis before and after adsorption of Hg^{2+} (**B**). Initial concentration: 50 mg/L, sample volume: 50 mL, adsorbent: 0.01 g, temperature: 25 °C, contact time: 24 h, agitation speed: 100 rpm.

3.4. Effect of Solution pH on the Adsorption of Hg^{2+}

In aqueous media, pH is one of the most important parameters for heavy metal removal. Thus, the solution pH strongly and directly affects the degree of metal ionization and/or metal binding on the adsorbent surface sites [42]. In this study, the quantity of Hg^{2+} adsorbed onto FA24 increased with an increase in the solution pH from 2 to 5, and decreased with further increase in pH of 5 to 7 (Figure 4). First, when the solution pH is below 3, Hg^{2+} is the dominant species; when the solution pH is over 5, $Hg(OH)_2$ is the dominant species [43,44]. In addition, $Hg(OH)^+$ exists (1–13% of the total mercury Hg^{2+}) when the solution pH is between 2 and 6. Previous studies have reported that there are several stable Hg^{2+} products related to either the equilibrium hydrolysis (such as $Hg(OH)_3^-$, $Hg(OH)_2$, $Hg(OH)^+$) or to the complexation equilibrium with chloride (such as $HgCl_4^{2-}$, $HgCl_3^-$, $HgCl_2$, $HgCl^+$). Moreover, mixed species such as $Hg(OH)Cl$ exist in aqueous media [44]. In an acidic solution (pH 2), the FA48 surface is protonated and the electrostatic repulsion between the FA48 surface (positive charge) and Hg^{2+} species (positive charge) such as $Hg(OH)^+$ and $HgCl^+$ easily occurred, resulting in low Hg adsorption. Additionally, the pH_{pzc} value of FA48 was 10.4 in this study, which supports the availability of positive charge on the FA48 surface and the low adsorption of Hg^{2+} from aqueous media. Next, similar to pH 2, at pH 5, the quantities of Hg^{2+} species such as $Hg(OH)^+$ and $HgCl^+$ decreased. Conversely, $HgCl_3^-$ species increased in aqueous media. Therefore, the quantity of Hg^{2+} adsorbed increased because of the electrostatic interaction between the FA48 surface (positive charge) and Hg^{2+} species such as $HgCl_3^-$ (negative charge). Finally, when the solution pH was over 7, the quantity of $Hg(OH)_2$ increased, and the hydroxyl ion (OH^-) also increased in the sample solution media [6,43]. Therefore, FA48 showed a low adsorption capability for Hg^{2+} under alkaline conditions.

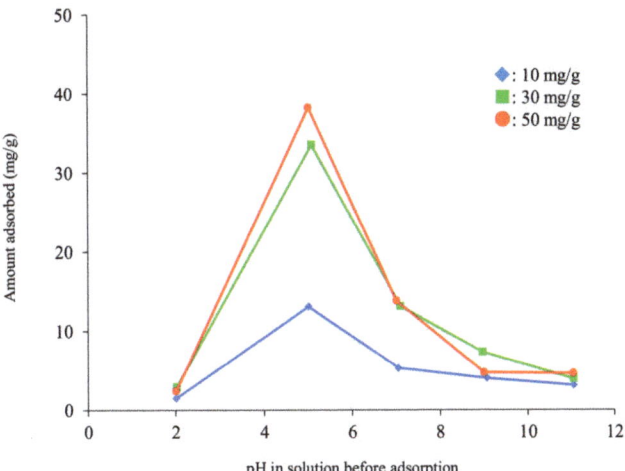

Figure 4. Effect of pH on the adsorption of Hg^{2+} onto FA48. Initial concentration: 10, 30, and 50 mg/L, sample volume: 50 mL, adsorbent: 0.01 g, temperature: 25 °C, contact time: 24 h, 100 rpm.

3.5. Effect of Contact Time on the Adsorption of Hg^{2+}

To investigate the effect of contact time on the removal of Hg^{2+} from aqueous media using FA48, the duration was varied from 0.5 to 48 h (Figure 5). Rapid adsorption was observed within 0.5 h from the start of the adsorption process, following which the rate of adsorption of Hg^{2+} fluctuated with increase in adsorption time. Finally, adsorption equilibrium was achieved at approximately 3 h under our experimental conditions. In this study, the adsorption might be mainly attributed to two factors: the interaction between Hg^{2+} and active adsorption sites, such as specific surface area and pore volume (mentioned in Section 3.2), and ion exchange with K^+ in the interlayer of FA48 (mentioned in Section 3.3).

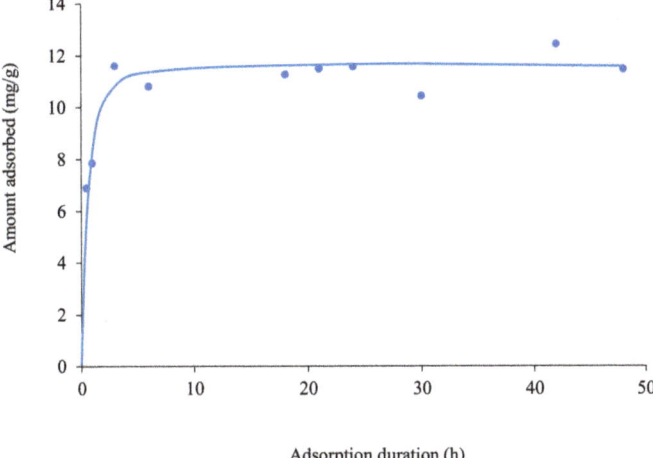

Figure 5. Effect of contact time on the adsorption of Hg^{2+} onto FA48. Initial concentration: 50 mg/L, sample volume: 50 mL, adsorbent: 0.01 g, temperature: 25 °C, contact time: 0.5, 1, 3, 6, 12, 21, 24, 30, 42, and 48 h, agitation speed: 100 rpm.

In addition, to evaluate the kinetic adsorption mechanism of Hg^{2+} using FA48, pseudo-first-order and pseudo-second-order models were selected to interpret the kinetics data using Equations (5) and (6) [36,38].

$$\ln(q_{e,exp} - q_t) = \ln q_{e,cal} - k_1 t \quad (5)$$

$$\frac{t}{q_t} = \frac{t}{q_{e,cal}^2} + \frac{1}{k_2 \times q_{e,cal}^2} \quad (6)$$

where $q_{e,exp}$ and q_t are the quantities of Hg^{2+} adsorbed at equilibrium and at time t (mg/g), respectively, $q_{e,cal}$ is the quantity of Hg^{2+} adsorbed in the calculation (mg/g), k_1 (1/h) and k_2 (g/mg/h) are the rate constants of the pseudo-first-order and pseudo-second-order models, respectively. The calculated results are shown in Table 5.

Table 5. Kinetic parameters for the adsorption of Hg^{2+} using FA48.

Adsorbents	$q_{e,exp}$	Pseudo-First-Order Model			Pseudo-Second-Order Model		
		k_1 (1/h)	$q_{e,cal}$ (mg/g)	r	k_2 (g/mg/h)	$q_{e,cal}$ (mg/g)	r
FA48	12.42	0.02	2.34	0.515	0.085	11.7	0.996

From Table 5, it is evident that the correlation coefficient (r) in the pseudo-second-order model (0.996) was significantly higher than the pseudo-first-order model (0.515), indicating that the pseudo-second-order model is more suitable for describing the adsorption kinetics of Hg^{2+} in this study. Additionally, the value of $q_{e,exp}$ was closest to the value of $q_{e,cal}$ of the pseudo-second-order model than that of the pseudo-first-order model. In addition, it is strongly suggested that the adsorption of Hg^{2+} onto FA48 is because of chemisorption, as assumed by this model [45,46].

In addition, the Elovich model (Equation (7)) was also used to describe adsorption kinetic in this study. This model describes activated adsorption, and predicts an energetically heterogeneous solid surface of adsorbent which means adsorption kinetics is not affected by interaction between the adsorbent particles [26].

$$q_t = 1/\beta \ln(\alpha\beta) + 1/\beta \ln t \quad (7)$$

where q_t is the quantity of Hg^{2+} adsorbed at time t (mg/g), α is the initial adsorption rate (mg/g/h), β is the related to the extent of surface coverage and activation energy for chemisorption (g/mg).

From the result, the value of α, β, and r (correlation coefficient) was 8.4×10^3 mg/g/h, 1.1 g/mg, and 0.888, respectively. The Elovich equation is suitable to describe adsorption behavior of Hg^{2+} using FA48 that relates to the nature of chemical sorption under our experimental conditions [47].

3.6. Selectivity for Hg^{2+} Removal from Binary Solution System

Considering the field application of FA48, the selectivity for Hg^{2+} adsorption is one of the critical parameters in this study. Therefore, the effect of coexisting ions on the adsorption capability of Hg^{2+} is shown in Table 6. In our study, Na^+, Mg^{2+}, K^+, Ca^{2+}, Ni^{2+}, Cu^{2+}, Zn^{2+}, Sr^{2+}, and Cd^{2+} were used as the components of the binary solution system, as these ions are ubiquitous in the water environment [48,49]. In this study, the removal percentage of Hg^{2+} using FA48 in a single solution system was approximately 14.0% whereas, the removal of Hg^{2+} in the binary solution system was over 11.4% (except for Na^+ and K^+), and the removal of other cations was significantly lower. A similar trend was reported in a previous study [6]. In addition, previous studies reported that the radius of the hydrated ion and/or the electronegativity of the adsorbate (Hg^{2+} in this study) strongly and directly influenced the adsorption capability in aqueous media [48,49].

Therefore, similar phenomena were observed under our experimental conditions. Finally, our results show that FA48 is useful for the selective removal of Hg^{2+} from aqueous media. Moreover, the Minamata Convention on Mercury was adopted by the Intergovernmental Negotiating Committee in 2017. Therefore, the development of removal techniques for Hg^{2+} in wastewater from anthropogenic activities such as the steel industry is very important for establishing a sustainable society. Thus, FA48 could be applied for wastewater purification including Hg^{2+} such as the steel industry.

Table 6. Adsorption capacity of Hg^{2+} in binary solution system.

Components in Binary Solution	Removal Percentage of Hg^{2+} (%)	Removal of Other Cations (%)
$Hg^{2+} + Na^+$	4.6	0
$Hg^{2+} + Mg^{2+}$	11.4	0
$Hg^{2+} + K^+$	5.1	0
$Hg^{2+} + Ca^{2+}$	11.9	0
$Hg^{2+} + Ni^+$	16.2	0.1
$Hg^{2+} + Cu^{2+}$	12.9	2.4
$Hg^{2+} + Zn^{2+}$	16.0	1.3
$Hg^{2+} + Sr^{2+}$	15.9	0
$Hg^{2+} + Cd^{2+}$	14.4	1.8

3.7. Adsorption/Desorption Capability of Hg^{2+} Using FA48

Finally, the adsorption/desorption capability of Hg^{2+} using FA48 was demonstrated in this study (Figure 6). The quantity of Hg^{2+} desorbed increased with increasing concentration of sodium hydroxide solution from 10 to 1000 mmol/L (the quantity of Hg^{2+} adsorbed was approximately 40 mg/g). The desorption percentages using 10, 100, and 1000 mmol/L sodium hydroxide solutions were 37.5%, 41.6%, and 68.3%, respectively. Therefore, adsorbed Hg^{2+} onto FA48 could be easily desorbed using a sodium hydroxide solution under our experimental conditions. Further investigations are needed to elucidate the application of FA48 in these fields.

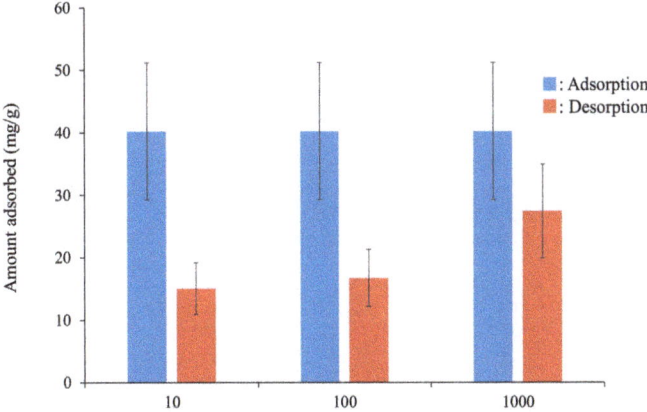

Figure 6. Adsorption/Desorption capability of Hg^{2+} using FA48. Adsorption condition; initial concentration: 250 mg/L, sample volume: 150 mL, adsorbent: 0.15 g, temperature: 25 °C, contact time: 24 h, agitation speed: 100 rpm, Desorption condition; initial concentration: 10, 100, and 1000 mmol/L, sample volume: 50 mL, adsorbent: 0.05 g, temperature: 25 °C, contact time: 24 h, agitation speed: 100 rpm.

4. Conclusions

Six types of potassium zeolites (FA1, FA3, FA6, FA12, FA24, FA48) were synthesized by hydrothermal treatment using a potassium hydroxide solution. The values of CEC, specific surface area, and pore volume ($d \leq 20$ Å) of FA48 were 26–29, 34, and 100 times higher than that of CFA, indicating that FA48 has a high potential for the removal of Hg^{2+} from aqueous media. The quantity of Hg^{2+} was in the order CFA and FA1 (0–0.48 mg/g) < FA3 (2.2 mg/g) < FA6 (3.5 mg/g) < FA12 (4.0 mg/g) < FA24 (7.5 mg/g) < FA48 (11.6 mg/g) under our experiment conditions. These adsorption behaviors were affected by the CEC and/or pore volume ($d \leq 20$ Å). In addition, the effects of pH, temperature, contact time, and coexistences on the adsorption of Hg^{2+} using FA48 were demonstrated. The optimal pH was approximately 5.0. The adsorption isotherm data or kinetics data were described by the Freundlich and Langmuir models or the pseudo-second-order model, respectively. Moreover, one of the adsorption mechanisms determined was the ion exchange with K^+ in the interlayer of FA48 (correlation coefficient = 0.946). FA48 showed selectivity for the adsorption of Hg^{2+} from a binary solution system containing Na^+, Mg^{2+}, K^+, Ca^{2+}, Ni^{2+}, Cu^{2+}, Zn^{2+}, Sr^{2+}, or Cd^{2+}. Finally, adsorbed Hg^{2+} onto FA48 was easily desorbed using a sodium hydroxide solution. It is evident that FA48 is a useful adsorbent for Hg^{2+} removal from aqueous media. These techniques may potentially aid in mitigating heavy metal pollution and thus contribute to the establishment of a sustainable society.

Author Contributions: Conceptualization, F.O. and N.K.; investigation, Y.K., C.S., and T.N.; writing—original draft preparation, Y.K. and F.O.; writing—review and editing, F.O. and N.K.; project administration, N.K. All authors have read and agreed to the published version of the manuscript.

Funding: This research received no external funding.

Institutional Review Board Statement: Not applicable.

Informed Consent Statement: Not applicable.

Data Availability Statement: Not applicable.

Conflicts of Interest: The authors declare no conflict of interest.

References

1. Sustainable Development Goals, Knowledge Platform. Available online: https://sustainabledevelopment.un.org/sdgs (accessed on 10 June 2020).
2. Kragovic, M.; Dakovic, A.; Markovic, M.; Krstic, J.; Gatta, G.D.; Rotiroti, N. Characterization of lead sorption by the natural and Fe(III)-modified zeolite. *Appl. Surf. Sci.* **2013**, *283*, 764–774. [CrossRef]
3. Hamidpour, M.; Kalbasib, M.; Afyunib, M.; Shariatmadarib, H.; Holmic, P.E.; Hansenc, H.C.B. Sorption hysteresis of Cd(II) and Pb(II) on natural zeolite and bentonite. *J. Hazard. Mater.* **2010**, *181*, 686–691. [CrossRef] [PubMed]
4. Volesky, B. *Biosorption of Heavy Metals*, 4th ed.; CRC Press: Boca Raton, FL, USA, 1990.
5. Abraham, J.; Dowling, K.; Florentine, S. Assessment of potentially toxic metal contamination in the soils of a legacy mine site in Central Victoria, Australia. *Chemosphere* **2018**, *192*, 122–132. [CrossRef] [PubMed]
6. Fu, Y.; Jiang, J.; Chen, Z.; Ying, S.; Wang, J.; Hu, J. Rapid and selective removal of Hg(II) ions and high catalytic performance of the spent adsorbent based on functionalized mesoporous silica/poly (m-aminothiophenol) nanocomposite. *J. Mol. Liq.* **2019**, *286*, 110746. [CrossRef]
7. Wang, J.; Feng, X.; Anderson, C.W.N.; Xing, Y.; Shang, L. Remediation of mercury contaminated sites—A review. *J. Hazard. Mater.* **2012**, *221*, 1–18. [CrossRef]
8. World Health Organization. *Guidelines for Drinking-Water Quality, First Addendum to Volume 1, Recommendations*; WHO Press: Geneva, Switzerland, 2006.
9. IARC Monographs on the Identification of Carcinogenic Hazards to Human. Available online: https://monographs.iarc.fr/list-of-classifications (accessed on 10 June 2020).
10. Awual, M.R.; Hasan, M.M.; Eldesoky, G.E.; Khaleque, M.A.; Rahman, M.M.; Naushad, M. Facile mercury detection and removal from aqueous media involving ligand impregnated conjugate nanomaterials. *Chem. Eng. J.* **2016**, *290*, 243–251. [CrossRef]
11. Venkateswarlu, S.; Yoon, M. Surfactant-free green synthesis of Fe_3O_4 nanoparticles capped with 3,4-dihydroxyphenethylcarbamodithioate: Stable recyclable magnetic nanoparticles for rapid and efficient removal of Hg(II) ions from water. *Dalton Trans.* **2015**, *44*, 18427–18437. [CrossRef]

12. Bukhari, S.S.; Behin, J.; Kazemian, H.; Rohani, S. Conversion of coal fly ash to zeolite utilizing microwave and ultrasound energies: A review. *Fuel* **2015**, *140*, 250–266. [CrossRef]
13. Czarna-Juszkiewicz, D.; Kunecki, P.; Panek, R.; Madej, J.; Wdowin, M. Impact of fly ash fraction on the zeolitization process. *Materials* **2020**, *13*, 1035. [CrossRef]
14. Yao, Z.T.; Ji, X.S.; Sarker, P.K.; Tang, J.H.; Ge, L.Q.; Xia, M.S.; Xi, Y.Q. A comprehensive review on the applications of coal fly ash. *Earth Sci. Rev.* **2015**, *141*, 105–121. [CrossRef]
15. Franus, W.; Wdowin, M.; Franus, M. Synthesis and characterization of zeolites prepared from industrial fly ash. *Environ. Monit. Assess.* **2014**, *186*, 5721–5729. [CrossRef]
16. Blissett, R.S.; Rowson, N.A. A review of the multi-component utilization of coal fly ash. *Fuel* **2012**, *97*, 1–23. [CrossRef]
17. Flores, C.G.; Schneider, H.; Marcillio, N.R.; Ferret, L.; Oliveira, J.C.P. Potassic zeolites from Brazilian coal ash for use as a fertilizer in agriculture. *Waste Manag.* **2017**, *70*, 263–271. [CrossRef]
18. Belviso, C. State-of-the-art applications of fly ash from coal and biomass: A focus on zeolite synthesis processes and issues. *Prog. Energy Comb. Sci.* **2018**, *65*, 109–135. [CrossRef]
19. Medina, A.; Gamero, P.; Almanza, J.M.; Vargas, A.; Montoya, A.; Vargas, G.; Izquierdo, M. Fly ash from a Mexican mineral coal. II. Source of W zeolite and its effectiveness in arsenic (V) adsorption. *J. Hazard. Mater.* **2010**, *181*, 91–104. [CrossRef]
20. Scott, J.; Guang, D.; Naeramitmarnsuk, K.; Thabuot, M.; Amal, R. Zeolite synthesis from coal fly ash for the removal of lead ions from aqueous solution. *J. Chem. Technol. Biotechnol.* **2001**, *77*, 63–69. [CrossRef]
21. Rayalu, S.; Meshram, S.U.; Hasan, M.Z. Highly crystalline faujasitic zeolites from fly ash. *J. Hazard. Mater.* **2000**, *77*, 123–131. [CrossRef]
22. Rios, C.A.; Williams, C.D.; Roberts, C.L. Removal of heavy metals from acid mine drainage (AMD) using coal fly ash, natural clinker and synthetic zeolites. *J. Hazard. Mater.* **2008**, *156*, 23–35. [CrossRef]
23. Tauanov, Z.; Tsakiridis, P.E.; Mikhalovsky, S.V.; Inglezakis, V.J. Synthetic coal fly ash-derived zeolites doped with silver nanoparticles for mercury(II) removal from water. *J. Environ. Manag.* **2018**, *224*, 164–171. [CrossRef]
24. Ma, L.; Han, L.; Chen, S.; Hu, J.; Chang, L.; Bao, W.; Wang, J. Rapid synthesis of magnetic zeolite materials from fly ash and iron-containing wastes using supercritical water for elemental mercury removal from flue gas. *Fuel Proc. Technol.* **2019**, *189*, 39–48. [CrossRef]
25. Qi, L.; Teng, F.; Deng, X.; Zhang, Y.; Zhong, X. Experimental study on adsorption of Hg(II) with microwave-assisted alkali-modified fly ash. *Powder Technol.* **2019**, *341*, 153–158. [CrossRef]
26. Attari, M.; Bukhar, S.S.; Kazemian, H.; Rohani, S. A low-cost adsorbent from coal fly ash for mercury removal from industrial wastewater. *J. Environ. Chem. Eng.* **2017**, *5*, 391–399. [CrossRef]
27. Murayama, N.; Ymamamoto, H.; Shibata, J. Mechansim of zeolite synthesis from coal fly ash by alkali hydrothermal reaction. *Int. J. Miner. Process.* **2002**, *64*, 1–17. [CrossRef]
28. Kobayashi, Y.; Ogata, F.; Saenjum, C.; Nakamura, T.; Kawasaki, N. Removal of Pb^{2+} from aqueous solutions using K-type zeolite synthesized from coal fly ash. *Water* **2020**, *12*, 2375. [CrossRef]
29. Zhou, Q.; Duan, Y.; Zhu, C.; Zhang, J.; She, M.; Wei, H.; Hong, Y. Adsorption equilibrium, kinetics and mechanism studies of mercury on coal-fired fly ash. *Korean J. Chem. Eng.* **2015**, *32*, 1405–1413. [CrossRef]
30. Okada, Y. Synthesis of zeolite using fly ash on cloased system. *Nihon Dojyou Hiryo Gakkaishi* **1991**, *62*, 1–6.
31. Faria, P.C.C.; Orfao, J.J.M.; Pereira, M.F.R. Adsorption of anionic and cationic dyes on activated carbons with different surface chemistries. *Water Res.* **2004**, *38*, 2043–2052. [CrossRef]
32. Kobayashi, Y.; Ogata, F.; Nakamura, T.; Kawasaki, N. Synthesis of novel zeolites produced from fly ash hydrothermal treatment in alkaline solution and its evaluation as an adsorbent for heavy metal. *J. Environ. Chem. Eng.* **2020**, *8*, 103687. [CrossRef]
33. Shang, Z.; Zhang, L.; Zhao, X.; Liu, S.; Li, D. Removal of Pb(II), Cd(II) and Hg(II) from aqueous solution by mercapto-modified coal gangue. *J. Environ. Manag.* **2019**, *231*, 391–396. [CrossRef]
34. Wang, C.; Tao, S.; Wei, W.; Meng, C.; Liu, F.; Han, M. Multifunctional mesoporous material for detection, adsorption and removal of Hg^{2+} in aqueous solution. *J. Mater. Chem.* **2010**, *20*, 4635–4641. [CrossRef]
35. Hakami, O.; Zhang, Y.; Banks, C. Thiol-functionalized mesoporous silica-coated magnetite nanoparticles for high efficiency removal and recovery of Hg from water. *Water Res.* **2012**, *46*, 3913–3922. [CrossRef]
36. Lagergren, S. Zur theorie der sogenannten adsorption geloster stoffie. *K. Sven. Vetensk. Handl.* **1898**, *24*, 1–39.
37. Abe, I.; Hayashi, K.; Kitagawa, M. Studies on the adsorption of surfactants on activated carbons. I. Adsorption of nonionic surfactants. *Yukagaku* **1976**, *25*, 145–150.
38. Ho, Y.S.; McKay, G. Pseudo-second order model for sorption process. *Process Biochem.* **1999**, *34*, 451–465. [CrossRef]
39. Kumar, V. Adsorption kinetics and isotherms for the removal of rhodamine B dye and Pb^{2+} ions from aqueous solutions by a hybrid ion-exchanger. *Arab. J. Chem.* **2019**, *12*, 316–329.
40. Foo, K.Y.; Hameed, B.H. Insights into the modeling of adsorption isotherms systems. *Chem. Eng. J.* **2010**, *156*, 2–10. [CrossRef]
41. Kumara, N.T.R.N.; Hamdan, N.; Peter, M.I.; Tennakoon, K.T.; Ekanayake, P. Equilibrium isotherm studies of adsorption of pigments extracted from Kuduk-kuduk (*Melastoma malabathricum* L.) pulp onto TiO_2 nanoparticles. *J. Chem.* **2014**, *2014*, 468975. [CrossRef]
42. Liu, U.; Li, Q.; Cao, X.; Wang, Y.; Jiang, X.; Li, M.; Hua, M.; Zhang, Z. Removal of uranium(VI) from aqueous solutions by CMK-3 and its polymer composite. *Chem. Eng.* **2013**, *285*, 258–266. [CrossRef]

43. Zhang, F.S.; Nriagu, J.O.; Itoh, H. Mercury removal from water using activated carbons derived from organic sewage sludge. *Water Res.* **2005**, *39*, 389–395. [CrossRef]
44. Arisa, F.E.; Beneduci, A.; Chidichimo, F.; Furia, E.; Straface, S. Study of the adsorption of mercury (II) on lignocellulosic materials under static and dynamic conditions. *Chemosphere* **2017**, *180*, 11–23.
45. Boparai, H.K.; Joseph, M.; O'Carroll, D.M. Kinetics and thermodynamics of cadmium ion removal ion removal by adsorption onto nano zerovalent iron particles. *J. Hazard. Mater.* **2011**, *186*, 458–465. [CrossRef] [PubMed]
46. Robati, D. Pseudo-second-order kinetic equations for modeling adsorption systems for removal of lead ions using multi-walled carbon nanotube. *J. Nanostructure Chem.* **2013**, *3*, 55. [CrossRef]
47. Wu, F.C.; Tseng, R.L.; Juang, R.S. Characteristics of Elovic equation used for the analysis of adsorption kinetics in dye-chitosan systems. *Chem. Eng. J.* **2009**, *150*, 366–373. [CrossRef]
48. Wang, C.; Hu, X.; Chen, M.L.; Wu, Y.H. Total concentrations and fractions of Cd, Cr, Pb, Cu, Ni and Zn in sewage sludge municipal and industrial wastewater treatment plants. *J. Hazard. Mater.* **2005**, *B119*, 245–249. [CrossRef] [PubMed]
49. Yadanaparthi, S.K.P.; Graybill, D.; Wandruszka, R.V. Adsorbents for the removal of arsenic, cadmium, and lead from contaminated waters. *J. Hazard. Mater.* **2009**, *171*, 1–15. [CrossRef] [PubMed]

MDPI
St. Alban-Anlage 66
4052 Basel
Switzerland
Tel. +41 61 683 77 34
Fax +41 61 302 89 18
www.mdpi.com

Sustainability Editorial Office
E-mail: sustainability@mdpi.com
www.mdpi.com/journal/sustainability